Physical Methods on Biological Membranes and Their Model Systems

NATO Advanced Science Institutes Series

A series of edited volumes comprising multifaceted studies of contemporary scientific issues by some of the best scientific minds in the world, assembled in cooperation with NATO Scientific Affairs Division.

This series is published by an international board of publishers in conjunction with NATO Scientific Affairs Division

A	**Life Sciences**	Plenum Publishing Corporation
B	**Physics**	New York and London
C	**Mathematical and Physical Sciences**	D. Reidel Publishing Company Dordrecht, Boston, and London
D	**Behavioral and Social Sciences**	Martinus Nijhoff Publishers The Hague, Boston, and London
E	**Applied Sciences**	
F	**Computer and Systems Sciences**	Springer Verlag Heidelberg, Berlin, and New York
G	**Ecological Sciences**	

Recent Volumes in Series A: Life Sciences

Volume 65—The Pineal Gland and Its Endocrine Role
  edited by J. Axelrod, F. Fraschini, and G. P. Velo

Volume 66—Biomagnetism: An Interdisciplinary Approach
  edited by Samuel J. Williamson, Gian-Luca Romani,
  Lloyd Kaufman, and Ivo Modena

Volume 67—Biomass Utilization
  edited by Wilfred A. Côté

Volume 68—Molecular Models of Photoresponsiveness
  edited by G. Montagnoli and B. F. Erlanger

Volume 69—Time-Resolved Fluorescence Spectroscopy in
  Biochemistry and Biology
  edited by R. B. Cundall and R. E. Dale

Volume 70—Genetic and Environmental Factors during the Growth Period
  edited by C. Susanne

Volume 71—Physical Methods on Biological Membranes and
  Their Model Systems
  edited by F. Conti, W. E. Blumberg, J. de Gier, and F. Pocchiari

Physical Methods on Biological Membranes and Their Model Systems

Edited by

F. Conti

Universitá La Sapienza
Rome, Italy

W. E. Blumberg

Bell Laboratories
Murray Hill, New Jersey

J. de Gier

The State University of Utrecht
Utrecht, The Netherlands

and

F. Pocchiari

National Institute of Health
Rome, Italy

Plenum Press
New York and London
Published in cooperation with NATO Scientific Affairs Division

Proceedings of a NATO Advanced Study Institute on
Physical Methods on Biological Membranes and Their Model Systems,
held September 2–October 2, 1982,
in Alto Villa Milicia, Italy

Library of Congress Cataloging in Publication Data

NATO Advanced Study Institute on Physical Methods on Biological Membranes
and their Model Systems (1982: Alto Villa Milicia, Italy)
    Physical methods on biological membranes and their model systems.

    (NATO advanced science institutes series. Series A, Life sciences; v. 70)
    "Published in cooperation with NATO Scientific Affairs Division."
    Bibliography: p.
    Includes index.
    1. Membranes (Biology)—Congresses. 2. Biological models—Congresses. I.
Conti, F. (Filippo) II. Title. III. Series.
QH601.N38   1982   574.87′5   83-22909
ISBN 978-1-4684-7540-1   ISBN 978-1-4684-7538-8 (eBook)
DOI 10.1007/978-1-4684-7538-8

PREFACE

     This book examines in depth the theory, instrumental innova-
tions, potential applications and limitations of the different
techniques -- NMR, EPR, DSC, spectrofluorimetry, and electron
microscopy -- that are being used to investigate the function and
structure of biological membranes and their model systems. Providing an overview of the present status of research, this book
gives evidence of the complementary nature of information generated
by different techniques, reminding one of the necessity to create
a common language and viewpoint of membranes.

                                        F. Conti
                                        W.E. Blumberg
                                        J. De Gier
                                        F. Pocchiari

CONTENTS

PHYSICAL METHODS ON BIOLOGICAL MEMBRANES

AND THEIR MODEL SYSTEMS: A PERSPECTIVE

F. Conti*, J. de Gier**, and W.E. Blumberg***

*Rome, **Utrecht and ***Murray Hill

When we try to understand living organisms, biological membranes present a problem of considerable interest. Over the years it has become clear that these lamellar structures not only form boundaries between different subcellular environments but that they also fulfill many other functions of great importance, both obvious and subtle. However, our molecular understanding of these membrane functions is still rather poor. With respect to transport phenomena, for example, all mechanistic details are missing. This is not only true when we question how newly synthesized peptide chains pass the membrane when entering the endoplasmic reticulum but also when we ask how single ions pass cell membranes. Endocytosis, exocytosis and fusion processes in general are not well understood, although interesting hypotheses are currently being presented, and some are being discussed at this Anvanced Study Institute. Other examples of membrane activities which are under discussion are the way membrane-bound enzymes are regulated and how the lipids may influence the receptor functions of membranes.

In the approach to biological membranes, the use of experimental model systems has been shown to be very useful. Membrane lipids, which are known to occur in membranes in a striking profusion of variations, can be extracted and purified. When these chemically defined lipids are dispersed in buffer solutions, the lipid molecules form a profusion of association structures which in many respects resemble a membrane organization. These systems permit one to study the properties of individual lipids and the influence of environmental conditions. It has been shown that a number of membrane proteins can be reconstituted in such a defined lipid environment. Therefore, model systems also offer the possibility of exploring protein functions and of evaluating specific lipid-protein inter-

1

actions which are important in this respect. It should be realized, however, that model systems do have limitations. For example, membrane asymmetry, which is apparent when we examine in detail the two halves of lipid bilayers and also when we consider the orientation of transmembrane proteins, is not easily reconstructed in model systems. Therefore, the tools which are being designed for membrane research not only should provide the capability of exploring membrane model systems but they should also have the ability to give information on the organization and dynamic states of the much more complicated intact biomembrane structures. Several of the physical methods being used to study biological membranes and their models are discussed in considerable detail in this volume.

Nuclear magnetic resonance (NMR) can provide information about the local environment in a very small, submolecular region of the membrane, in fact, at the level of individual atoms. NMR can be used to study both chemical composition and environmental perturbations of molecular structure. Different techniques of NMR apply to studies of static (or averaged) structural and of dynamic parameters. ^{1}H, ^{2}H, ^{13}C, and ^{31}P NMR can provide different types of information on the same membrane structure.

Electron spin resonance (ESR) also provides information on the dynamics of membranes by the effect on the freedom of movement of paramagnetic probes dispersed in the membrane or attached to proteins. The time scale for dynamic ESR spectral studies ranges from nanoseconds to milliseconds, using various instrumental techniques. This is generally a faster regime than is studied by NMR.

Low-angle and crystallographic X-ray diffraction can be used to characterize the wide variety of repeating structures that lipid-aqueous systems separate into in the immiscibility region. Cubic, hexagonal, and lamellar phases can be defined, as well as others which are, at present, less well characterized. Transformations between these phases in biological membranes as a result of changes in temperature or other environmental parameters are an extremely important determinant of both structure and function. For example, membrane permeability is strongly affected as phase transitions occur.

Electron microscopy provides intriguing pictures of membrane surfaces and fracture faces on a supramolecular scale. Two types of sample preparation are predominantly used: thin sectioning and freeze fracture. Each technique is subject to many artifacts and provides a wide scope for interpretation. Procedures for minimizing these artifacts are a major current topic of discussion.

Differential scanning calorimetry is a relatively new, sensitive microcalorimetric technique to measure latent heats of phase transitions. DSC has revealed phase transitions which up to now have been too subtle to be revealed by more standard methods.

Studies of the fluorescence of either intrinsic or extrinsic fluorescent probes in membranes or in membrane-bound proteins are very useful in the study of the spatial constitution of membrane structure as well as of membrane dynamics on a nanosecond time scale. Fluorescence depolarization and intra- and intermolecular fluorescence energy transfer can be interpreted on a physical basis in sufficiently well-defined systems.

In addition to optical absorption and fluorescence, light can be used to probe the structure of membranes by using it to cause certain chemical reactions. For example, light-initiated cross-linking can be used with photoaffinity labels to study spatial structure or associations among membrane components. Light can also be used to initiate degradative reactions, indicating the availability of conjugated bonds or other systems rendered reactive in optically excited states.

As we sit here in these lovely surroundings in the closing moments of this Advanced Study Institute, we ask ourselves, "How successful have we been in defining the applicability and limitations of these physical techniques to the study of biological membranes?" and "What new concepts have we learned about membranes by using these techniques?"

The answer to the first question is, as we might have expected, mixed. Many useful applications of the physical techniques have been discussed, both in the lectures and in "Seminar Club" contributions from the students. Various limitations and opportunities for over-interpretation have been mentioned, and we hope these will be fully developed in the manuscripts to follow. We have heard of old techniques used in new ways and of the beginnings of new techniques. We have learned of the importance of the time scale of the dynamics of membranes as determinants of their biological properties and especially of the importance of lateral diffusion, and how it is governed principally by phospholipids rather than proteins. We have seen definitive evidence that annular "ports" do not exist as static structures. We have also seen how much more forceful the concepts resulting from the research are when they can be stated in truly thermodynamic terms – and somewhat less so when only statistical concepts such as "order parameter" are employed. And so much less so when rather ill-defined concepts such as "viscosity" are used. Above all, we have been reminded of the necessity of creating a common language and viewpoint of membranes taking into account the results of a wide variety of techniques. We hope that the manuscripts will be as valuable for readers as the lectures were for those who attended the Advanced Study Institute.

HIGH RESOLUTION NMR SPECTROSCOPY IN LIQUIDS AND SOLIDS

Theodore Axenrod
Department of Chemistry
The City College of The City University of New York
New York, N.Y. 10031

INTRODUCTION

High resolution nuclear magnetic resonance (NMR) spectroscopy
has been widely used to examine molecular structures, probe intricate
conformational details and study dynamic chemical processes in solution. For samples constrained to be examined in the solid state,
either for solubility reasons or because sample properties are modified in solution, conventional liquid state NMR techniques have not
been very useful. The recent development of rare spin double-
resonance techniques together with magic angle sample spinning has
permitted high resolution spectra to be obtained for solid samples.
Of course, the advances in solid state NMR techniques are based
largely on the remarkable progress in liquid state Fourier transform
NMR that preceded it. Today NMR is among the more rapidly expanding
branches of chemical science because it is one of relatively few
techniques applicable to the study of molecules in both liquid and
solid phases. In the area of biological chemistry, comparisons of
spectral properties in the two states can provide significant new
insights into local electronic structures in important molecules
such as amino acids and proteins as well as complement crystallographic data in the interpretation of experimental results.

SCOPE

Essentially the same interactions are responsible for liquid
state and solid state spectra. The major difference is that
molecular tumbling averages these interactions in liquids while in
solids such motion is either absent or considerably reduced and
this gives rise to additional complexities in the spectrum. In

introducing the subject, it seems appropriate to review some of
the basic concepts important in pulsed Fourier transform NMR
spectroscopy. The approach adopted here will be a qualitative one
with the aim of providing a physical picture of the NMR process.
The mechanisms responsible for the line broadening in solids as well
as the versatility and limitations of the available techniques to
suppress their effects will be outlined. A few selected applications
drawn from the recent literature will illustrate the potential of
the method.

THE NMR EXPERIMENT

    The magnetic moments, μ, of nuclei having a net spin (e.g., odd
nucleons) will normally be randomly oriented, but when placed in a
static magnetic field these moments will align with the lines of
force of the field. The interaction between the applied field, B_o,
and the nuclear moment generates a torque which causes the latter
to precess about the direction of the field. For a given field
strength, as illustrated in Figure 1, each nucleus precesses at its
own characteristic frequency, ν_o, and this is determined by the
Larmor relationship $\nu_o = \gamma B_o/2\pi$ where γ, the magnetogyric ratio, is
a constant which varies from one nucleus to another.

    For nuclei of major interest such as ^1H, ^{13}C, ^{15}N, ^{19}F, and ^{31}P
only two orientations of the magnetic moments - parallel and anti-
parallel to the field - are quantum mechanically allowed. Since the
two orientations have slightly different energies a small excess of
nuclei in the lower energy state (parallel) results. The population
ratio of the two states (B - upper and A - lower) depends on the
sample's absolute temperatue, T, the strength of the magnetic field,

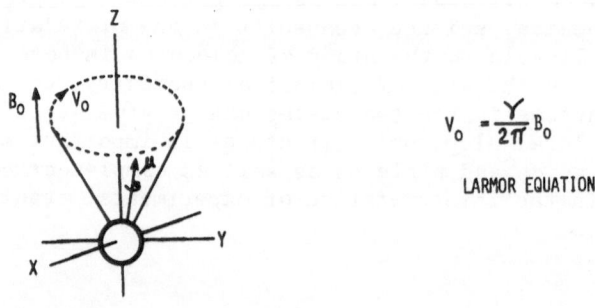

$$\nu_o = \frac{\gamma}{2\pi} B_o$$

LARMOR EQUATION

NUCLEAR PRECESSION

Figure 1. Precession of magnetic nucleus about an applied field, B_o.

B_o, and the nuclear magnetic moment, μ, according to the Boltzmann relation

$$\frac{N_B}{N_A} = e^{-\mu B_o/kT}$$

For ^{1}H, at ordinary temperatures in a field of 4.7 Tesla, the fractional excess nuclei in the lower spin state is about 1×10^{-6}. A consequence of this slight excess population in the lower spin state is the existence of a small net macroscopic equilibrium magnetization, M_o, directed along B_o (by convention the Z axis) as shown in Figure 2. This represents the resultant effect or weighted average of all the individual magnetic moments of the nuclear species in the sample.

If this net magnetization is subjected to a second field, B_1, placed at right angles to the static field, B_o, and made to rotate at a frequency ν_o, then M experiencing the resultant of the two fields will be tipped away from the original Z direction by an amount α which is dependent on the strength and duration of B_1. A common simplification is to view the spin system from a rotating frame where the entire coordinate system is taken as rotating about B_o (Z axis) at the precessional frequency of M_o. Since both the observer and the magnetic moment rotate around together, the tipping of M appears as a simple rotation of M about an apparently stationary field, B_1. In practice this displacement of M is achieved by surrounding the sample with a transmitter coil connected to a source of radio-frequency power which when pulsed creates a rotating magnetic field component in the X-Y plane as shown in Figure 3.

Immediately after M is tipped away from the Z direction and the applied pulse is discontinued the resultant magnetization will have

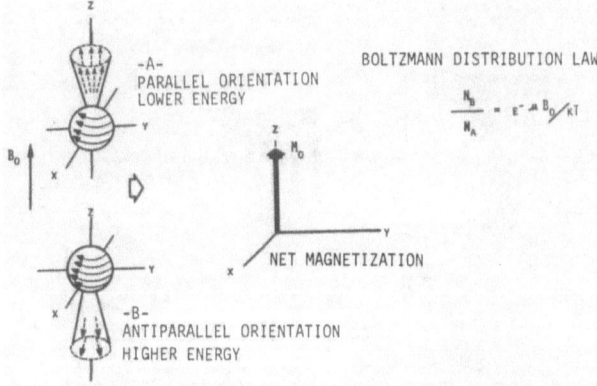

Figure 2. Net equilibrium macroscopic magnetization resulting from distribution of collection of individual nuclear moments in magnetic field, B_o.

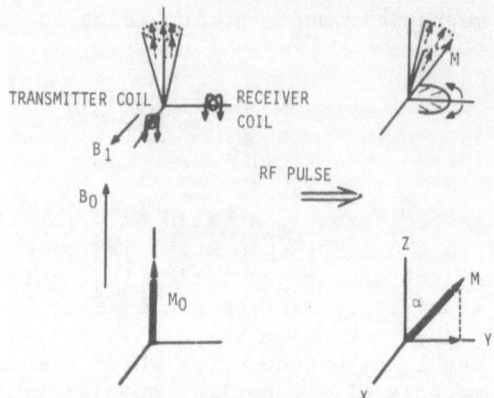

Figure 3. Effect of a radio frequency pulse on magnetization in rotating frame.

a rotating component in the X-Y plane which can be detected by the decaying sinusoidal voltage it induces in a receiver coil placed in this plane. This signal is called a free induction decay (FID). For a compound such as CH_3OH which has only one ^{13}C nucleus the FID is the simple exponentially decaying sine wave shown in Figure 4.

The tipping of the magnetization, M, corresponds to the absorption of energy as some nuclei undergo transition from the lower energy state to the higher one. In a typical NMR experiment a pulse which rotates M_0 by 90° (termed a 90-degree pulse) usually lasts between 1 and 100 μs. This pulse is of such power and encompasses a range of frequencies, for example, to simultaneously tip the magnetizations of the many dissimilar ^{13}C nuclei in a

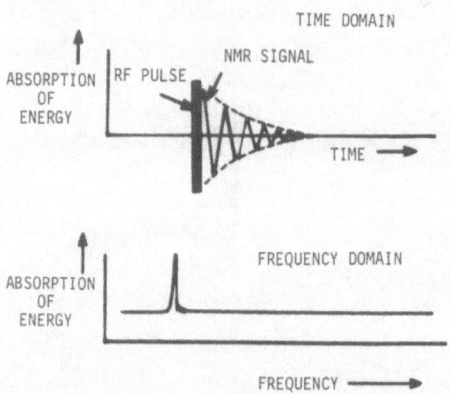

Figure 4. A simple free induction decay signal and corresponding frequency spectrum obtained by Fourier transformation.

8

molecule containing several carbons. Thus, multiple energy absorptions take place and the FID consists of super-imposed sine waves each with its characteristic frequency resulting in a complex interference pattern. To abstract the individual frequency components from the complex waveform and convert them to a frequency domain spectrum (a plot of energy absorption versus frequency) Fourier transformation, a mathematical operation, is required and this is usually conveniently performed on a minicomputer. A complex FID and its corresponding frequency domain spectrum are shown in Figure 5.

CHEMICAL SHIFTS AND SENSITIVITY

The utility of NMR as an analytical method stems directly from the fact that for a given nuclear species the resonance frequency is very sensitive to small variations in molecular structure as well as to much larger changes in going from one nuclear species to another. In complex molecules the magnetic field is subtly and non-uniformly altered in the vicinity of some nuclei as a result of shielding currents associated with neighboring electronic orbitals. These internally created secondary fields shield nuclei from the applied field to different extents giving rise to a range of chemical shifts or resonance frequencies for nuclei in different molecular environments.

Although the range of chemical shifts found for hydrogen is at least an order of magnitude smaller than that for such other biologically important nuclei as ^{13}C, ^{15}N, and ^{31}P, the opposite is true of the ease of detection or sensitivity. All nuclei have inherent sensitivities much less than hydrogen and the detection problem is frequently compounded by the low natural abundance of

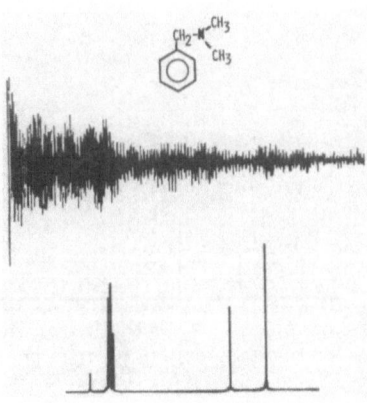

Figure 5. Proton-decoupled ^{13}C NMR spectrum of N,N-dimethyl-benzylamine and corresponding FID.

many magnetic nuclei of interest. Figure 6 compares the resonance frequencies and receptivities of some common nuclei.

We have seen that the absorption of energy (resonance) occurs when the sample is irradiated with electromagnetic energy at a frequency corresponding to the precessing macroscopic moment, M_o. In fact, both absorption and emission of energy are stimulated when a nucleus is in resonance. The probability of a particular transition, however, is proportional to the population of the energy state from which it occurs. Since the difference in energy between the two spin states at ordinary temperatures is only a few millicalories per mole, as we previously noted, this results in slightly more nuclei in the lower energy state. Moreover, from the Boltzmann distribution law we can see that both an increase in the field strength and a decrease in the temperature favor a greater population difference between the two states.

This is important because it is the net magnetic moment produced by this small population difference that determines the strength of the observable NMR signal. If the population of the two spin states becomes equal saturation results and the NMR signal disappears indicating no further net energy absorption is occurring. This can happen when, for example, high rf power B_1 and slow scanning are used in a continuous wave experiment or by pulsing too rapidly in an FT experiment. The effect of increasing the rf power on the proton signal in $CHCl_3$ is shown in Figure 7.

In part this accounts for the insensitivity of NMR relative to other spectroscopic methods and improved sensitivity is one of the advantages of higher field spectrometers. With its high natural abundance and relatively large magnetic moment it is easily understood why, historically, proton NMR developed ahead of other less receptive nuclei.

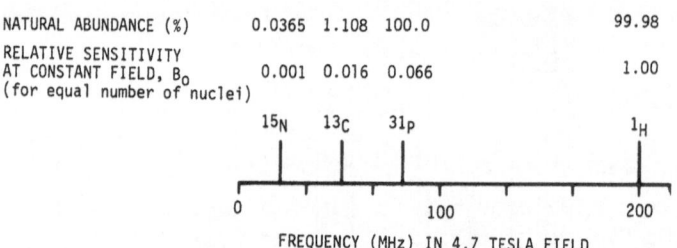

Figure 6. Nuclear properties and resonance frequencies of some common nuclei in a 4.7 Tesla field.

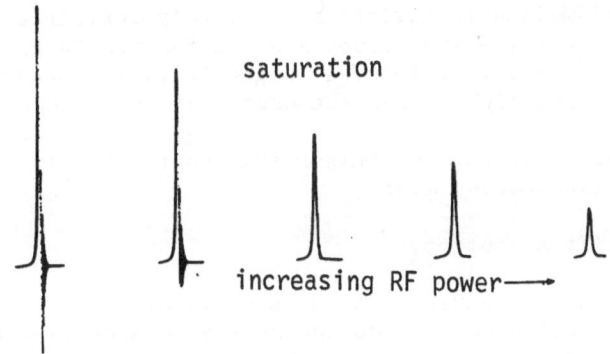

saturation

increasing RF power→

Figure 7. The effect of increasing radio frequency power on signal intensity in the ^{1}H NMR spectrum of chloroform.

To determine the spectra of low receptivity nuclei a series of repetitive pulses with signal acquisition between pulses is necessary to develop a strong FID. The successive FID's are automatically digitized and accumulated in a computer prior to Fourier transformation into the conventional spectrum.

RELAXATION PROCESSES

When the excitation pulse is discontinued the individual nuclei begin to undergo relaxation by releasing the absorbed rf energy to the surrounding lattice causing primarily increased molecular translations and rotations. In this manner the original thermal equilibrium Boltzmann population of nuclear spins is reestablished. The time, T_1, as illustrated in Figure 8, is characteristic of the time required for the magnetization to realign itself along the Z axis and depends on the mechanisms available for the transfer of energy. Although there are several mechanisms by which this transfer can occur, the common feature of all spin-lattice relaxation processes is that they depend on the existence of some molecular motion to generate a randomly varying magnetic field or electric field gradient which will have a local field component fluctuating at the Larmor frequency of the nucleus to be relaxed. It is important to note that T_1 is determined by molecular motions. Temperature and viscosity influence translational and tumbling motions whereas molecular size, shape and viscosity determine rotational motions. In solution large polymeric molecules tumble at rates comparable to the precessional frequencies providing effective relaxation mechanisms and leading to short T_1's. Small molecules undergo rotational reorientation at rates considerably faster and T_1 is found to be long, several seconds or more, because the transfer of energy is inefficient. In solids

11

where atomic and molecular motions are severely restricted T_1 can last for hours. Lattice vibrations are ineffective in providing the appropriate fluctuating internal magnetic fields because their frequencies are much higher than the usual Larmor frequencies.

Some of the important mechanisms that contribute to the spin-lattice relaxation process are:

1. Chemical Shift Anisotropy

Electrons in a molecule produce an auxiliary field the direction and magnitude of which depends on the spatial disposition of the electronic orbitals with respect to the static field. Fluctuating local fields are generated when the molecule moves in the field.

2. Internuclear Dipole-Dipole Interaction

Two neighboring magnetic nuclei will experience not only the external static field but also the local fields produced by each other the strength and direction of which depends on their respective magnetic moments, internuclear separation and orientation in the field. Constantly changing orientations of the two nuclear spins generates fluctuating local fields which stimulate relaxation of nuclei.

3. Spin Rotation Relaxation

A freely rotating molecule or molecular segment has associated with it a constant molecular magnetic moment which fluctuates when collision with other molecules occurs.

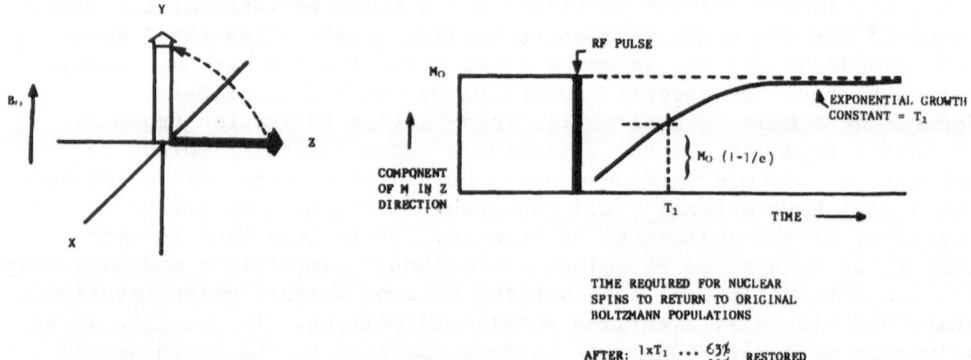

Figure 8. Spin lattice relaxation: Exponential return of magnetization to original alignment with B_o when excitation is discontinued.

Let us return to our original NMR experiment and consider a second recovery process called spin-spin relaxation, T_2, that takes place. You will recall that the NMR signal detected in the X-Y plane, that is – the FID, exponentially decreases with time. If when the excitation pulse ends all the nuclei in the sample were experiencing the identical field strength, then the individual nuclear moments would continue to precess in phase and the rotating component in the X-Y plane would disappear at a rate governed by T_1 which reflects the return alignment of M along the Z axis. In a typical sample where molecular diffusion and reorientation are occurring the precessing nuclei experience varying internuclear magnetic fields. With nuclei, thus distributed over a range of field strengths, the individual moments begin to precess at different frequencies causing dephasing, cancellation and diminishing intensity of the signal. This is shown schematically in Figure 9. It is worth noting that this spin-spin or transverse relaxation time, T_2, can be shorter than T_1, but never longer.

Moreover, since the fields created by real magnets are not uniform over the entire volume of the sample, this inhomogeneity also contributes to the dephasing of the magnetic vectors causing the FID to decay faster than it would if the field were perfectly homogeneous. The true molecular relaxation time, T_2, is inversely related to the natural line width and is distinguished from the time constant, T_2^*, which is characteristic of the rate of decay of the signal in an imperfect field. The spectral resolution as measured by the linewidth at half-height ($\nu_{\frac{1}{2}}$) is essentially governed by T_2^* according to $\nu_{\frac{1}{2}} = 1/\pi T_2^*$.

By the application of appropriate pulse sequences it is possible to identify and cancel contributions from field inhomogeneities and determine a sample's intrinsic T_2 value. For liquids T_2 can be as

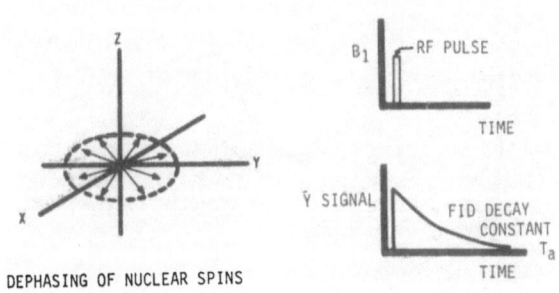

DEPHASING OF NUCLEAR SPINS

SPECTRAL RESOLUTION: $V_{\frac{1}{2}} = \dfrac{1}{\pi T_2^*}$

Figure 9. Spin-spin relaxation.

13

long as several seconds and tends to be not too different from T_1, but for solids T_2 is much smaller with values falling in the micro-second range.

NMR IN SOLIDS

Chemical Shift Anisotropy

A simple and dramatic example of the difference between the spectra of solids and liquids is found in the case of water, shown in Figure 10. In the liquid state the proton spectrum consists of a sharp resonance signal with a linewidth of about 0.1 Hz, whereas in ice the corresponding signal is some 10^6 times broader. The origin of this difference is straightforward. Characteristically, in liquids rapid and random molecular tumbling occurs and the proton frequencies are averaged to the isotropic value that is observed as a narrow line. As the rigidity of the sample lattice increases this motional averaging diminishes.

In polycrystalline samples or amorphous powders some molecules will be aligned parallel to the field, some perpendicular, and others will assume the full range of possible orientations between these limits. The interaction between the electrons surrounding a nucleus and the applied field produces an effective field at each nuclear site which depends on the orientation of the molecule in the applied field. Thus, even chemically identical nuclei in different molecules may be shielded from the external field to varying extents and the powder patterns that result reflect the three-dimensional nature of this chemical shielding anisotropy.

Figure 11 shows the theoretical absorption line shapes for poly-crystalline samples of molecules such as CO_2 having a three-fold axis

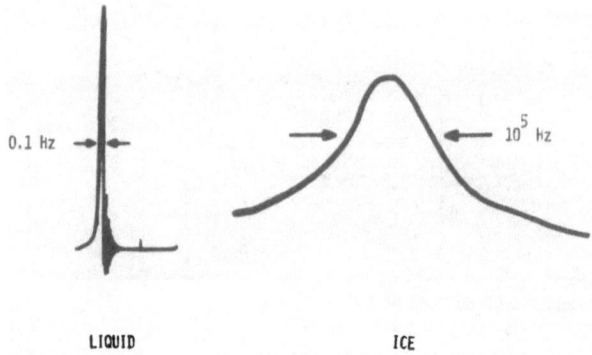

Figure 10. Comparison of ^1H NMR linewidths in liquid and solid water samples.

14

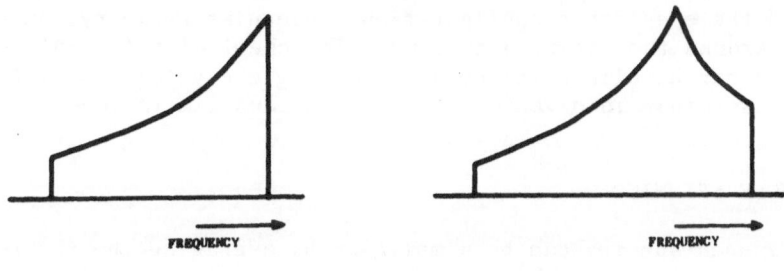

EXAMPLE: CO₂ (3-FOLD SYMMETRY AXIS)   LOWER SYMMETRY ORDER

Figure 11. Theoretical absorption line shapes for powder samples
of molecules with high and low symmetry.

of symmetry and the type of pattern exhibited when the symmetry is
lower.   In the CO_2 case the pattern consists of contributions from
all orientations with the extreme chemical shift components corresponding to those molecules whose axes are perpendicular and parallel,
respectively, to the static field.   A decidedly more complicated
and irregular shape results when the molecular symmetry is lower.

The ^{13}C spectrum of sucrose shown in Figure 12 is typical of
a solid containing molecules with many chemically inequivalent
carbon atoms and consists of a broad featureless line.   The spreading
of the resonance absorption is directly attributable to the overlap
of many patterns with different shapes and intrinsic widths due to
the individual carbon atoms.   Each pattern spans a wide range of
frequencies determined by the orientation of the molecule in the

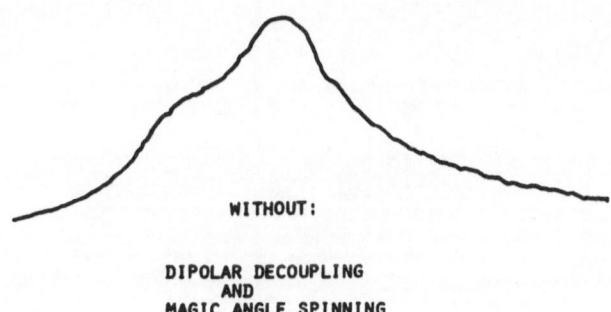

WITHOUT:

DIPOLAR DECOUPLING
AND
MAGIC ANGLE SPINNING

Figure 12. Cross polarization ^{13}C NMR spectrum of solid sucrose
without dipolar decoupling and magic angle spinning.

15

field and the electronic configuration, molecular geometry, and symmetry around the observed nucleus. The chemical shift anisotropy broadening can be eliminated by using a single crystal, but this approach is rather inconvenient and has obvious limitations.

MAGIC ANGLE SPINNING

Broadened spectra can be simplified by averaging the interactions responsible for the chemical shift anisotropy. In liquids this is done by rapid random rotational motion of the molecules. In solids this can be done by physically spinning the sample. When the rotation rate is rapid compared to the width, in Hz, of the chemical shift powder pattern narrowing occurs by an amount $\alpha \cos (3 \cos^2 \theta - 1)^{-1}$ where θ is the angle between the sample rotation axis and the static magnetic field. Clearly, maximum narrowing of the chemical shift anisotropy occurs when this angle is 54.74° (the magic angle) and the term $3 \cos^2 \theta - 1$ vanishes. The individual powder patterns then collapse to their single isotropic averages and a spectrum containing a series of sharp lines with readily discernib amplitudes and frequencies emerges.

HIGH SPEED SAMPLE ROTATION

High speed spinning is carried out with a gas-driven rotor whic may either be machined out of a solid polymer of interest or have a hollow barrel which is packed with a powdered sample. The two commo rotor designs, shown in Figure 13, are the Andrew-type which has a conical gas bearing surface and the Lowe-type which is cylindrical. Rotation is effected by circulating compressed gas between the stationary stator and the rotor whose surface is indented with flute

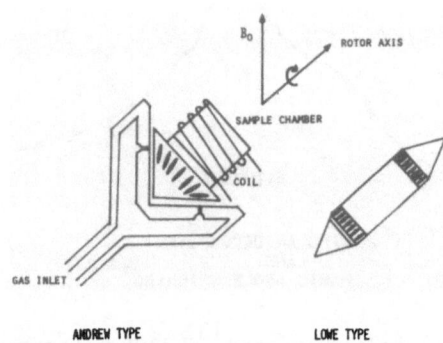

Figure 13. Magic angle rotors.

Materials that have been used to fabricate rotors include Teflon $[-(CF_2-CF_2)_x-]$, Kel-F $[-(CF_2CFCl)_x-]$, Delrin $[-(CH_2O)_x-]$ and boron nitride each of which has its own spinning characteristics as well as possible interfering signals. The monitoring of reflected light from a marked rotor surface by means of an optical fiber cable and a light sensitive diode is commonly used to measure spinning frequencies.

Sustained spinning at speeds approaching 10 kHz seems to be the present limit of technical feasibility because of rotor disintegration due to centrifugal forces, but even speeds considerably less than this are far from routine.

DIPOLAR BROADENING

Removal of the chemical shift anisotropy broadening in the spectrum of sucrose by magic angle spinning has only a negligible effect on the spectral resolution. This is because the dominant forces acting on the carbons arise from the direct nuclear dipole-nuclear dipole interaction between the ^{13}C nuclei and their nearby proton neighbors. The strength of this interaction and the local field, B_{Local}, produced by the proton at the carbon site vary inversely with the third power of the distance, R, between the two nuclei and with θ, the angle between the internuclear vector and the static magnetic field, B_o. This relationship may be expressed as

$$B_{Local} \approx \pm \frac{3\cos^2\theta-1}{R^3}$$

The effective field, B_{Local}, at the carbon is either greater or less than the static field, B_o, depending on whether the neighboring dipole is aligned with or against the field. Since the probabilities of these two alignments are almost equal the carbon resonance is split into a characteristic doublet whose width is a function of θ. For a polycrystalline sample each carbon is dipolar coupled to many neighboring protons each with somewhat different values of R and θ. The result is the overlap of the various dipolar splittings to give the broadline that we observe in the spectrum of sucrose.

In principle, the dipolar splittings can be made to collapse to a single line for each chemically distinct carbon by magic angle spinning at speeds equal to the static dipolar linewidth. In practice, however, it is generally not possible to attain the required speeds. Depending on the dipole interaction, this can vary from a few kilohertz for $^{31}P-^1H$ interactions, to as much as 100 kHz for particularly strong $^1H-^1H$ couplings, with $^{13}C-^1H$ dipolar linewidths sometimes as large as 15 kHz.

HIGH POWER DECOUPLING

Heteronuclear proton dipolar contributions to the carbon ^{13}C line-widths can be removed by an alternate approach. If while the spectrum is being recorded, the sample is simultaneously irradiated at the proton resonance frequency the result is that rapid transitions between the two proton spin states are induced effectively eliminating the dipolar field created by the protons at the carbon nuclei. Since the strength of this dipolar interaction is substantial a powerful decoupling field (∼100 W) is required. As shown in Figure 14, dipolar decoupling narrows the linewidth considerably but again no fine structure is discernible in the spectrum.

As we shall see, the combination of dipolar decoupling and magic angle spinning brings about a marked improvement in spectral resolution. The dipolar broadening is attenuated by the high power decoupling. When magic angle spinning, typically at speeds less than 4 kHz, is applied the much smaller broadening contribution from the chemical shift anisotropy is easily suppressed and the fine structure in the spectrum is uncovered. Sensitivity, however, remains a serious limitation.

SENSITIVITY ENHANCEMENT - CROSS POLARIZATION

In most chemically interesting solids containing two nuclear species it is usually the proton that is the abundant magnetic nucleus and the other is considered to be a rare nucleus. The rarity may be due to low natural abundance as with ^{13}C or it may be due to high magnetic dilution as is the case with isotopes such as ^{31}P and ^{14}N which frequently occur in a lattice rich in non-magnetic

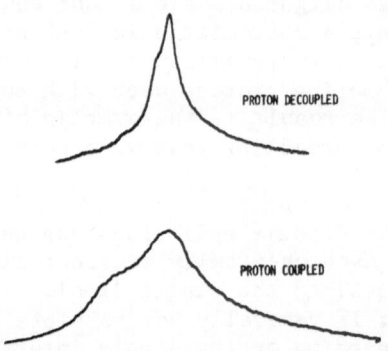

Figure 14. The effect of high power dipolar decoupling on the ^{13}C NMR cross polarization spectrum of sucrose without magic angle spinning.

nuclei (e.g., ^{16}O and/or ^{12}C). In one respect this is a fortunate
situation because it lessens the complications that might otherwise
arise from homonuclear dipolar interactions which, of course, are not
removed by proton decoupling.

Although concentration dependence is important, the more serious
difficulty hampering NMR measurements on rare spins is the exceedingly
long spin-lattice relaxation times. For ^{13}C in solids, T_1 can be of
the order of minutes imposing interpulse delays of approximately $5T_1$
in the signal accumulation and making the entire recording process
inefficient if not downright impractical. Fortunately, cross-
polarization techniques overcome the problem of low sensitivity in
the detection of the rare nucleus.

The concept of spin temperature is useful in discussing the
cross-polarization strategy. For ^{13}C nuclei in a static field the
difference in energy between the two spin alignments is $E = 2\mu B_o$.
The ratio of the populations in the two spin states is given by the
Boltzmann relation

$$\frac{N_B}{N_A} = e^{-\Delta E/kT_S}$$

where A and B refer to the lower and upper states, respectively.
Here T_S is defined as the spin temperature and this does not
necessarily correspond to the situation in which the spin system is
in thermal equilibrium with its lattice environment. Since the net
magnetization available for detection in an NMR experiment is proportional to the population difference $N_A - N_B$, a low carbon spin
temperature, implying a large polarization of the magnetic moments,
is desirable. To achieve this, advantage is taken of the fact that
the abundant protons surrounding the carbon nuclei usually undergo
spin-lattice relaxation much faster than the carbons and under
appropriate conditions magnetization can be transferred from the
protons to the carbons.

A typical ^{13}C cross-polarization experiment viewed from a
rotating frame of reference begins with the application of a 90°
pulse that tips the net proton magnetization into a plane perpendicular to the static field, B_o. Immediately following the pulse
while continuing the proton irradiation the phase of the driving
rf field, B_1, is shifted by 90° so that it is collinear with the
magnetization and the two are spin-locked to each other. Thermodynamically, this non-equilibrium condition is equivalent to a
lowering of the proton spin temperature in this rotating frame.
The large magnetization that originally corresponded to and was
locked along B_o at the beginning of the experiment is now locked
along the effective field, B_1. This field is perpendicular to the

external field, B_o, and significantly is several orders of magnitude less.

During the spin lock period the ^{13}C transmitter is turned on and thermal contact is made between the cooled protons and the warm carbon nuclei. In the spin temperature sense thermal contact means that two spins exchange energy as one undergoes a transition from a lower to higher spin state and the other makes the opposite transition. For identical nuclei in a uniform field the same quantum of energy is involved in the two transitions, but this is not the case for dissimilar nuclei. To establish the cross-polarization transfer of magnetization between ^{13}C-^{1}H pairs requires having the two nuclei simultaneously under the influence of different effective fields so that their Larmor frequencies become equal and the Hartmann-Hahn condition is satisfied

$$\gamma_H B_1 = \gamma_C B_2$$

This is accomplished by the adjustment of both the proton (B_1) and carbon (B_2) rf power levels. The oscillating field seen at one nucleus due to another will then have the correct frequency to induce mutual spin transitions.

The Hartmann-Hahn condition is maintained for a short period so that cross-polarization can occur causing a build-up of ^{13}C magnetization which is proportional to the proton magnetization previously developed. In the final step proton irradiation and magic angle spinning are maintained while the ^{13}C FID is observed in the usual manner. A schematic representation of the steps in the cross-polarization process is shown in Figure 15.

The cross-polarization technique affords two significant benefits. One is that the signal is enhanced by the abnormally large ^{13}C spin polarization that has been created and this can be as much as γ_H/γ_C or 3.98 times that obtained by a single pulse excitation. The other is that the cross-polarization process is more efficient than the normal ^{13}C spin lattice relaxation processes permitting the pulse repetition rate to be increased because the interpulse delay is now governed by the shorter proton relaxation rate.

As is evident in the ^{13}C spectrum of of crystalline sucrose shown together with its spectrum in aqueous solution in Figure 16, the combination of cross-polarization, high power decoupling and magic angle spinning results in a dramatic improvement in the resolution. Both spectra are characterized by resonances appearing at isotropic chemical shift positions for the individual carbons, although the spectra do not coincide. This probably reflects frozen molecular conformations or other magnetic inequivalences present in

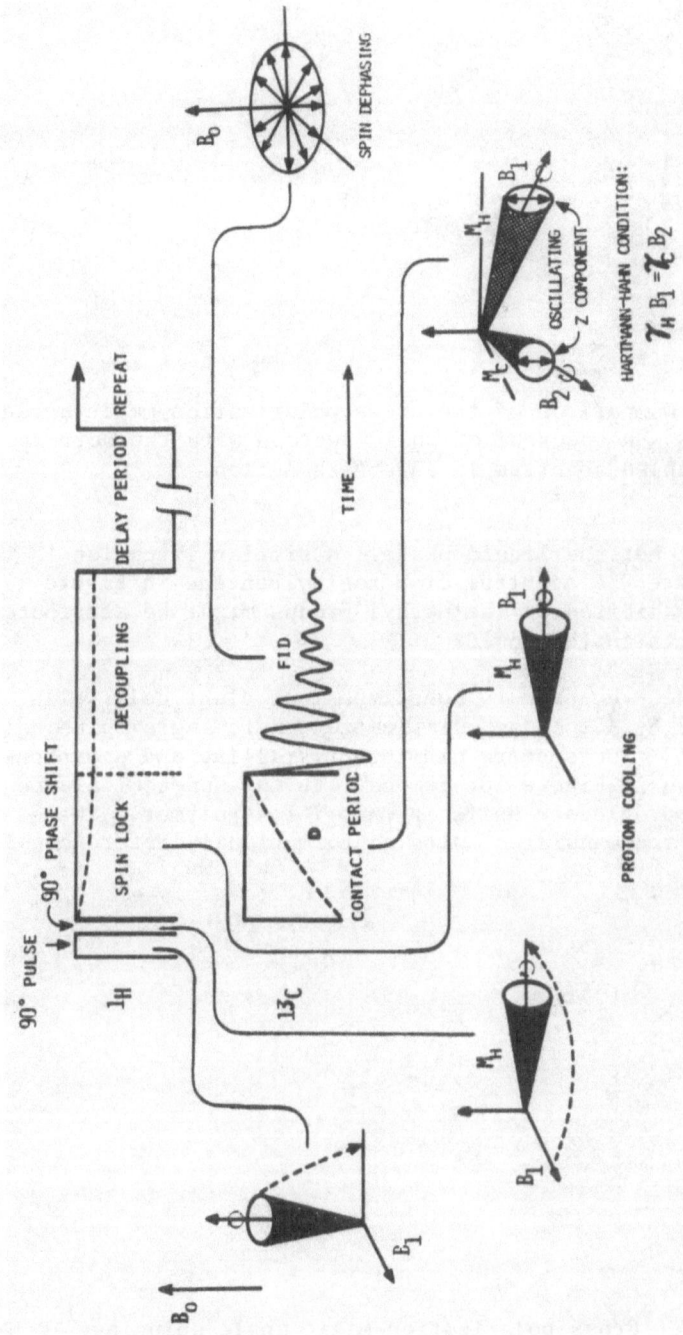

Figure 15. Cross polarization sequence.

21

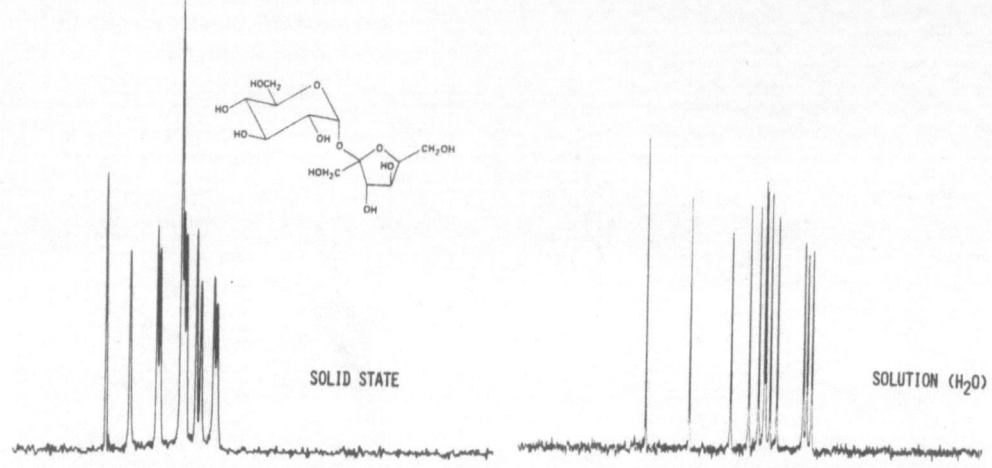

SOLID STATE

SOLUTION (H₂O)

Figure 16. Comparison of the cross polarization–magic angle spinning ^{13}C NMR spectrum of solid sucrose with the corresponding proton-decoupled spectrum in aqueous solution.

the solid but not the liquid phase. A similar situation is seen in the solid state ^{13}C spectrum of hexaethylbenzene in Figure 17 where the triplet exhibited by the methyl groups might be attributed to packing effects in the solid.

Cellulose, the primary constituent of plant cell walls, has been examined by ^{13}C cross polarization–magic angle spinning techniques.[1,2] The spectra of microcrystalline and amorphous samples, shown in Figure 18, demonstrate the non-equivalence of adjacent anhydroglucose units in this β-1,4-polymer. These observations are consistent with conformational differences in the

SOLID STATE CARBON-13 SPECTRUM

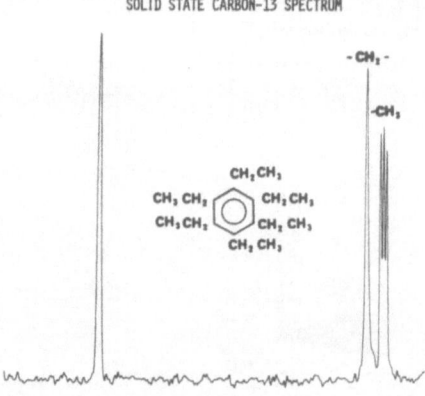

Figure 17. Cross polarization–magic angle spinning ^{13}C NMR spectrum of hexaethylbenzene.

22

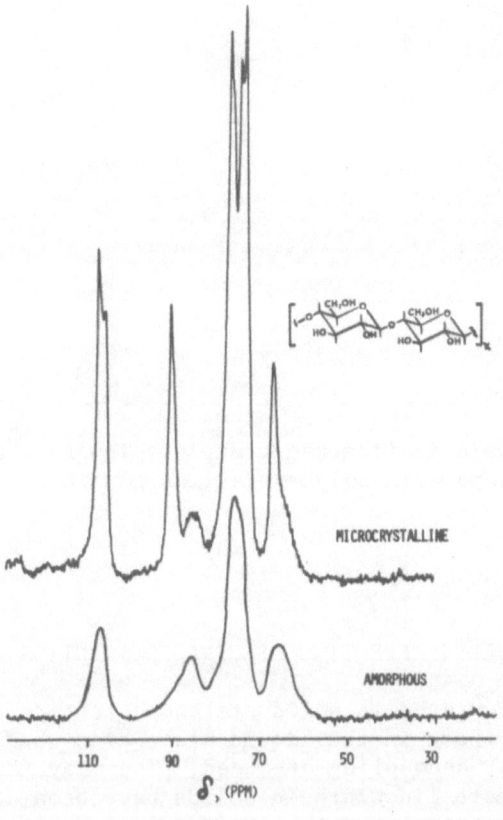

Figure 18. Cross polarization–magic angle spinning ^{13}C NMR
spectra of microcrystalline and amorphous cellulose.

chains of the important polymorphs, probably due to irregularities
in crystal packing which may distort the glucose monomer.

The ^{13}C spectra of solid amino acids have been examined by
several research groups.[3,4,5] A prominent feature of these spectra
is that the resonances of carbons bonded to nitrogen are often
split into broadened asymmetric doublets and even some resonances
of more distant carbons are affected. The ^{13}C spectra of the ^{14}N
and ^{15}N isotopomers of glycine, shown in Figure 19, clearly demonstrate the the splitting of the α-carbon resonance is due to the
influence of the ^{14}N quadrupole moment. ^{15}N has no quadrupole
moment and the ^{13}C–^{15}N dipolar couplings are averaged out by magic
angle spinning while the ^{13}C–^{14}N dipolar couplings are not.

23

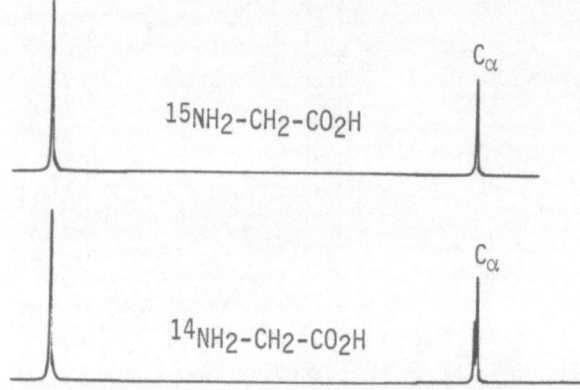

Figure 19. Cross polarization–magic angle spinning ^{13}C NMR spectra of ^{14}N and ^{15}N isotopomeric polycrystalline glycines.

SOME PRACTICAL PROBLEMS AFFECTING SPECTRAL RESOLUTION

Typically, linewidths in solid state spectra are 10 to 100 times broader than those in the liquid state even when magic angle spinning and dipolar decoupling are used. The many factors that determine the ultimate linewidth in solids have been discussed.[6] Some of these are extrinsic to the sample and in principle under the experimenter's control. Others are not.

Inaccuracy in setting the magic angle can have marked effect on the spectral resolution. Degradation of the ^{13}C linewidth in the spectrum of hexamethylbenzene caused by small deviations from the magic angle is illustrated in Figure 20.

When magic angle spinning is used to remove chemical shift anisotropy broadening and the anisotropy is larger than the spinning rate side bands arise at integral multiples of the spinning frequency. This is illustrated in the ^{13}C spectrum of hexamethylbenzene shown in Figure 21. The presence of sidebands is undesirable for two reasons. First, interpretation of the distorted spectrum is made difficult, especially when these side bands are located at or near other lines in the spectrum. Second, the sensitivity of the experiment is decreased by the sharing of signal intensity between the center bands and side bands. At higher fields where adequate spinning rates cannot be achieved this difficulty can be acute. One simple approach has been to take the spectrum at two or more spinning speeds.

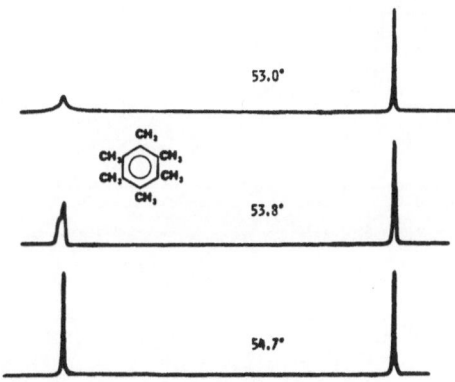

Figure 20. Effect of inaccurate setting of magic angle on ^{13}C NMR cross polarization spectrum of hexamethylbenzene.

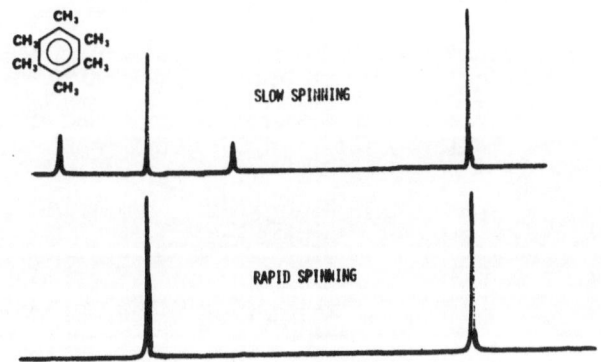

Figure 21. Effect of magic angle spinning rate on ^{13}C NMR cross polarization spectrum of solid hexamethylbenzene.

REFERENCES

1. W. L. Earl and D. L. Vander Hart, J. Amer. Chem. Soc., 102, 3251 (1980).
2. R. H. Atalla, J. C. Gat, D. W. Sindorf, V. J. Bartuska and G. E. Maciel, J. Amer. Chem. Soc, 102, 3249 (1980).
3. J. G. Hexem, M. H. Frey and S. J. Opella, J. Amer. Chem. Soc., 103, 224 (1980).
4. C. J. Groombridge, R. K. Harris, K. J. Packer, B. J. Say and S. F. Tanner, J.C.S. Chem. Comm., 174 (1980).
5. R. A. Haberkorn, R. E. Stark, H. van Willigen and R. G. Griffen, J. Amer. Chem. Soc., 103, 2534 (1981).
6. D. L. Vander Hart, W. L. Earl and A. N. Garroway, J. Mag. Reson., 44, 361 (1981).

BASIC PRINCIPLES OF DEUTERIUM AND

PHOSPHORUS-31-MAGNETIC RESONANCE

Joachim Seelig

Biocenter of the University of Basel
4056 Basel
Switzerland

ANISOTROPIC MOTIONS IN MEMBRANES

Membranes are ordered fluids and because of this have a number of properties in common with liquid crystals. Most important, the normal to the surface of the lipid bilayer constitutes an axis of motional averaging. From an optical point of view the membrane therefore behaves like an uniaxial crystal with the bilayer normal as the optical axis. This unique axis of the fluid bilayer is also called the director $\vec{n}$. For the sake of the argument let us represent the phospholipids constituting the bilayer by rigid rods. If we follow the movements of an individual rod over a long enough period of time the most probable orientation of the molecule will be parallel to the director axis. However, due to thermal energy the molecules will continuously execute angular excursions around this preferred orientation and an instantaneous picture may look as is illustrated in figure 1. As a quantitative measure for the angular excursions and fluctuations around the director $\vec{n}$ the concept of order parameters has been introduced [1]. If θ is the instantaneous angle between the rod-like molecule and the director $\vec{n}$ then the order parameter is defined as

$$S = (1/2) \ \overline{(3\cos^2\theta - 1)} \qquad (1)$$

where the bar denotes a time-average. This is actual-

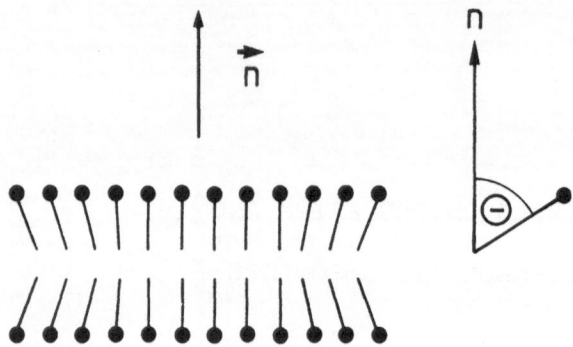

Fig. 1   Schematic representation of a lipid bilayer

ly a rather simplified representation of the order parameter formalism. It is fortunately valid for deuterium NMR where it provides a physically unambiguous
description of the average orientation of the C-D bond
vector. Due to the symmetry of the electric field gradient around the C-D bond vector <u>one</u> order parameter is
sufficient to specify the average orientation of a C-D
bond. However, there are many spectroscopic problems
where one needs more than one order parameter, in fact,
the most general case would require a 3x3 ordering
matrix (for a more detailed discussion cf. reference
[1]).

The average distribution of rod-like molecules
around the bilayer normal is described by the <u>distri-
bution function p (θ)</u>. Knowing p (θ) the order parameter S can be calculated according to:

$$S = \int_0^\pi (1/2)(3\cos^2\theta-1)p(\theta)\sin\theta d\theta / \int_0^\pi p(\theta)\sin\theta d\theta$$

Unfortunately, only S is experimentally accessible but
not the distribution function p(θ) itself. In order to
carry the molecular interpretation of S one step further, models for p(θ) must be introduced.

In the cone model it is assumed that the molecules are allowed to fluctuate around the bilayer normal in a cone of angle Θ_o but not beyond this cone:

$$p(\Theta) = \text{constant} \quad \text{for } 0 \leqslant \Theta \leqslant \Theta_o$$
$$p(\Theta) = 0 \quad \text{for } \Theta > \Theta_o$$

With this probability density the order parameter is found to be

$$S = \frac{1}{2} \cos\Theta_o (1 + \cos\Theta_o) \qquad (3)$$

This formula has been used widely for spin-label epr and more recently also for time-resolved fluorescence anisotropy data [2 - 4]. However, the shortcomings of the cone model are obvious. First, it is not understandable why there should be a sharp cut-off of allowed angular orientations, corresponding to a square-well potential within the membrane. Secondly, it is also physically unrealistic that all orientations within the allowed cone should have the same probability. A physically more useful distribution function which is experimentally supported by studies of nematic liquid crystals takes the form [5,6].

$$p(\Theta) \sim \exp\left\{ - \frac{q}{kT} \cos^2\Theta \right\} \qquad (4)$$

where kT is the thermal energy and q is an interaction potential describing the forces between the molecules. q, in turn, is related to the average order in the system ($q \sim S$).

DEUTERIUM MAGNETIC RESONANCE

Deuterium magnetic resonance (^{2}H-NMR) has proven to be particularly useful for determining the motional anisotropies in membranes. A few unique features of this method should be pointed out (for reviews see [7 - 9]). The first step is the selective deuteration of the molecules of interest. This may be achieved either by chemical synthesis or by biochemical incorporation. In the latter case, deuterated precursors are added to the growth medium of a bacterial culture and are incorporated via the metabolic pathways. One of the obvious advantages of selective deuteration is the fact that the replacement of a proton by a deuteron does not perturb the system. A second advantage is the ease of assignment of the deuterium resonances. The natural abundance of deuterium is low ($\sim$ 0.02%) and in a selectively deuterated membrane the observed signal can be assigned unambiguously. The magnetic moment of the nuclear spin of the deuteron is by a factor 6.5 smaller than that of the proton which decreases the sensitivity of the measurements. However, dipolar couplings are also reduced by the same factor. The dominant relaxation mechanism in ^{2}H-NMR is quadrupolar relaxation. For the fast correlation-time limit with $\omega_o \tau_c \ll 1$ ($\omega_o = 2\pi\nu_o$ = precession frequency) the spin-lattice relaxation time T_1 is given by the following experssion [10]:

$$\frac{1}{T_1} = \frac{3}{2} \pi^2 \left(\frac{e^2 qQ}{h}\right)^2 \tau_c \tag{5}$$

where $(e^2 qQ/h)$ is the static quadrupole coupling constant (170 kHz for a C-D bond). Using a slightly more complicated formula it is possible to evaluate T_1 over the whole range of correlation times. Two numerical examples are listed in table I.

Table I

^{2}H-NMR. Spin lattice relaxation time (T_1) and linewidth at half-height $\Delta\nu_{1/2} = (\pi T_2)^{-1}$

τ_c (ns)	T_1 (ms)	$\Delta\nu_{1/2}$ (Hz)
.01	230	1.4
1	3	124

A correlation time of .01 ns = 10^{-11} s is typical for moderately small molecules. Since the T_1 relaxation time is only 230 ms, rapid pulsing and fast data acquisition are possible. Correlation times of around 1 ns = 10^{-9} s have been observed for the glycerol backbone segments of phospholipids in bilayer membranes [11]. The spin-lattice relaxation time is very short under these circumstances (only a few ms) and close to the theoretically expected minimum value. As a second parameter table I contains the intrinsic linewidth for an isotropically tumbling molecule. Due

to the very effective quadrupole relaxation mechanism
even a moderately large molecula can give rise to rather
broad ^{2}H-NMR lines, if there is no additional intrinsic
flexibility of the deuterated segment.

^{2}H-NMR OF LIQUID CRYSTALLINE SYSTEM

The main asset of ^{2}H-NMR is the possibility to detect and to quantitate anisotropic motions in microscopically ordered systems such as bilayers or hexagonal
lipid phases. A macroscopic ordering of these systems is
not necessary instead one can work with random dispersions of lipid bilayers or whole cell suspensions.
Deuterium is a spin I = 1 nucleus and as such has
a quadrupolar moment. In a magnetic field H_0 there are
three allowed spin orientations as indicated in figure 2.
The magnetic energy levels are equally spaced and the
m = 1 → m = 0 and m = 0 → m = -1 transitions are therefore degenerate. In isotropic solutions only a single
absorption line is observed. However in anisotropic
systems there is an additional perturbation due to
quadrupolar interaction. This interaction of the electric quadrupole moment of the deuterium nucleus with

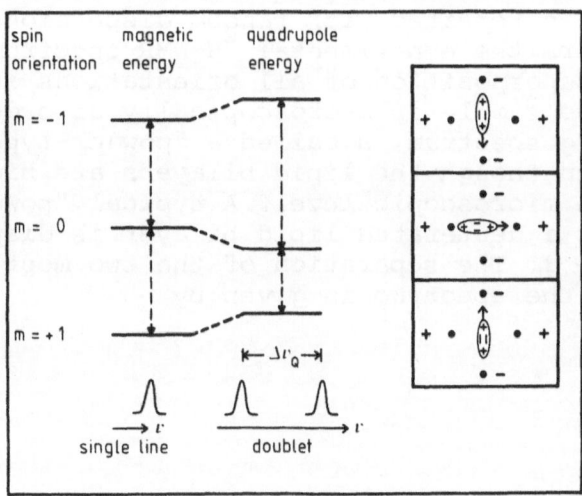

Fig. 2   Energy levels for a spin I = 1 system with
         and without quadrupole interactions
         (from [17])

the surrounding bonding electrons leads to a second order perturbation of the magnetic energy levels shifting the energy levels to different extent. The singlet is split into a doublet with the frequency separation of the two resonances being called the "quadrupole splitting". However, it should be noted that the quadrupole splitting can only be observed in solids and in microscopically ordered systems (such as bilayers or liquid crystals) but not in isotropic solutions or micelles. In the latter systems the rapid tumbling of the molecules through all angles of spaces averages out the quadrupole interactions, leading to a single absorption line.

Next let us assume that a deuterated lipid bilayer is oriented homogeneously on a supporting glass plate. The ^{2}H-NMR spectrum of this oriented bilayer would consist of two resonances with a quadrupole splitting of

$$\Delta \nu_Q(\delta) = (3/2)(e^2qQ/h) \; S_{C-D} \; (\tfrac{3}{2}\cos^2\delta - \tfrac{1}{2}) \qquad (6)$$

Clearly, the quadrupole splitting is dependent on the orientation of the bilayer normal with respect to the magnetic field as indicated by the angle δ. For a known orientation δ the order parameter S_{CD} of the deuterated C-D bond in the bilayer can be evaluated from $\Delta \nu_Q(\delta)$ as indicated by equation 6. If the same measurement is repeated with random dispersions of lipid bilayers the experimental ^{2}H-NMR spectrum consists of a superposition of all orientations in space. Since the sample is macroscopically disordered the resulting spectrum is called a "powder-type" spectrum even though the lipid bilayers are highly fluid at the microscopic level. A typical "powder-pattern" for a deuterated lipid bilayer is displayed in figure 3. The separation of the two most intense peaks in the spectrum is given by

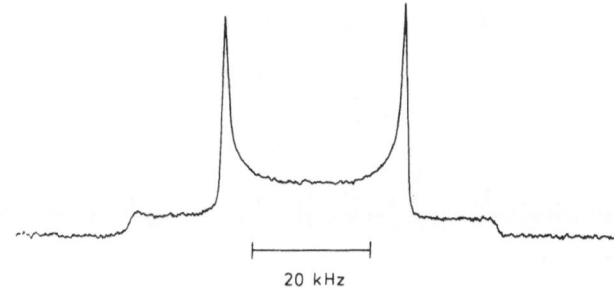

20 kHz

Fig. 3 ^{2}H-NMR spectrum of a random distribution of
         liquid crystalline bilayers (from [17]). The
         separation of the intense inner peaks
         (δ = 90°) corresponds to ca. 30 kHz

$$\Delta\nu_{powder} = (3/4)(e^2 q Q/h) \; S_{CD} \qquad\qquad (7)$$

and allows a convenient evaluation of S_{CD} even for
random dispersions of bilayers.

^{2}H-NMR OF PROTEINS

^{2}H-NMR may also be used to study motions in proteins. The type of effects which can be expected are
illustrated in figure 5 for the motion of a tyrosin
ring in a protein crystal [12]. If the ring is completely fixed in the protein crystal with no motion occuring, the experimental spectrum is rather broad and
the quadrupole splitting $\Delta\nu_{powder}$ reaches a limiting
value of about 135 kHz (figure 4A). On the other hand,
if there is a completely unhindered ring rotation
around the ring carbon-C_β bond, the spectrum is dramatically narrowed to a quadrupole splitting of about
17 kHz (4B). Finally, one could consider a two-site
flip (180° flip) of the tyrosine ring around the ring
carbon-C_β bond. In this case, the motion is no longer
axially symmetric and the shape of the powder pattern
would be changed significantly (fig 4C).

33

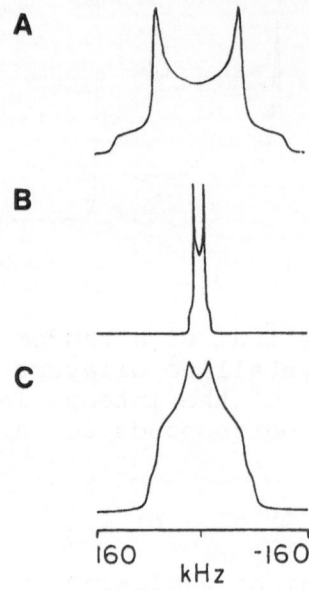

A

B

C

160   -160
kHz

Fig. 4   Theoretical lineshapes for [3,5-^{2}H$_2$]tyrosine
(from ref. [12]). A Rigid lattice. B Free
rotation around C$_\beta$ - C$_{ring}$. C Two-site jump

PHOSPHORUS-31 MAGNETIC RESONANCE

The natural abundance of ^{31}P is 100% and no synthetic labeling is necessary. ^{31}P has a nuclear spin
I = 1/2 and in an external magnetic field H$_o$ only two
spin orientations with the magnetic quantum numbers
m$_I$ = +1/2 and m$_I$ = -1/2 are allowed. The energy difference between the two spin orientations is given by

$$\omega = \gamma_{31_P} \ (1 - \sigma) \ H_o$$

The gyromagnetic ratio is γ = 10840 (rad s^{-1}G^{-1}) whereas σ is a measure of the screening of the applied
field H$_o$ via the bonding electrons. Considering the
bonding pattern around a tetrahedral PO$_4$ segment in
more detail it is obvious that the electronic screening
must be dependent on the orientation of the magnetic
field with respect to the molecular coordinate system
of the phosphate group. This can be demonstrated expe-

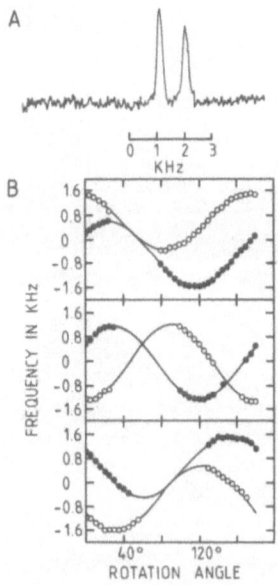

Fig. 5 ^{31}P-NMR spectra of a single crystal of phosphoethanolamine at 25°C (from [13])

rimentally with single crystals of simple phosphate
containing molecules as, for example, phosphoethanolamine [13]. The asymmetric unit cell of these crystals
contained two molecules of phosphoethanolamine with
different orientations with respect to the crystal axes
and the ^{31}P-NMR spectrum hence exhibited two resonances.
If the single crystal was rotated in the magnetic field
the positions of the resonances were shifted in a continuous fashion. This anisotropy of the chemical shielding
can only be observed in ordered systems. In solution .
the molecules tumble rapidly in space and the chemical
shielding anisotropy is averaged out completely. In
contrast, the chemical shielding anisotropy is retained in bilayer structures [14,15]. Figure 6 exhibits
the typical spectra of a fluid lipid bilayer, a hexagonal phase and an isotropic phase. The chemical shielding anisotropy is defined as the difference between
the low intensity edge ($\sigma_{\parallel}$) and the high intensity peak
($\sigma_{\perp}$). It may be noted that $\Delta\sigma = \sigma_{\parallel} - \sigma_{\perp}$ is negative for
a lipid bilayer and positive for a hexagonal phase.

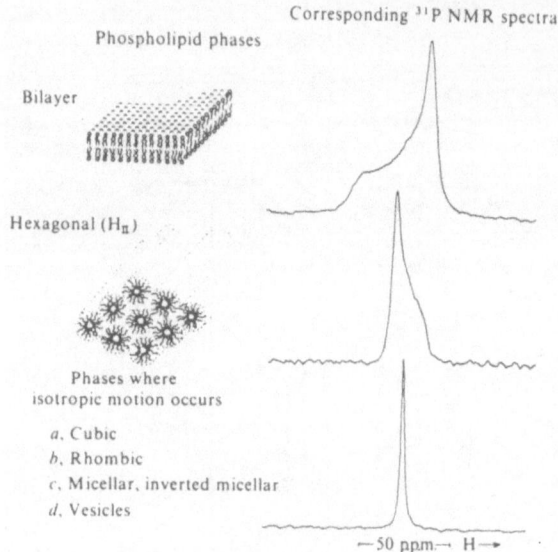

Corresponding ^{31}P NMR spectra

Phospholipid phases

Bilayer

Hexagonal (H$_{II}$)

Phases where
isotropic motion occurs

a. Cubic
b. Rhombic
c. Micellar, inverted micellar
d. Vesicles

←50 p.p.m.→ H→

Fig. 6 ^{31}P NMR spectra of liquid crystalline lipid phases (from [16])

Furthermore, the additional averaging due to rotational diffusion around the cylinder axis reduces the absolute value of the chemical shielding anisotropy by a factor of two. Thus with the molecular head group structure remaining unchanged, the transition from a bilayer to a hexagonal phase produces a dramatic change in the appearance of the ^{31}P-NMR spectrum. It is mainly for this reason that ^{31}P-NMR has become so popular for investigations of lipid polymorphism [16].

REFERENCES

[1] A. Saupe, Neuere Ergebnisse auf dem Gebiet der flüssigen Kristalle, Angew. Chem. 80:90 (1968).
[2] P. Jost, L.J. Libertini, V.C. Hebert, and O.H Griffith, Lipid spin labels in lecithin multilayers. A study of motion along fatty acid chains, J. Mol. Biol. 59:77 (1971).
[3] S. Kawato, K. Kinosita and A. Ikegami, Dynamic structure of lipid bilayers studied by nanosecond

fluorescence techniques, Biochemistry 16:2319 (1977).
[4] K. Kinosita, S. Kawato and A. Ikegami, Dynamic structure of biological membranes as probed by 1,6-diphenyl-1,3,5-hexatriene. A nanosecond fluorescence depolarization study, Biochemistry 20:4270 (1977).
[5] W. Maier and A. Saupe, Eine einfache molekulare Theorie des nematischen kristallinflüssigen Zustandes, Z. Naturforschung 13a:564 (1958).
[6] H. Schindler and J. Seelig, EPR spectra of spin labels in lipid bilayer, J. Chem. Phys. 59:1841 (1973)
[7] J. Seelig, Deuterium magnetic resonance: theory and application to lipid membranes, Quart. Rev. Biophys. 10:353 (1977).

[8] H.H. Mantsch, H. Saito and I.C.P. Smith, Deuterium magnetic resonance. Applications in chemistry, physics and biology, Progr. NMR Spectroscopy 11:211 (1977).
[9] J. Seelig and A. Seelig, Lipid conformation in model membranes and biological membranes, Quart. Rev. Biophys. 13:19 (1980).
[10] A. Abragam, "The Principles of Nuclear Magnetism", London, Oxford University Press, (1961).
[11] R. Ghosh and J. Seelig, On the interaction of cholesterol with bilayers of phosphatidylethanolamine, Biochim. Biophys. Acta 691:151 (1982).
[12] R.M. Rice, R.J. Wittebort, R.G. Griffin, E. Meirovitch, E.R. Stimson, Y.C. Meinwald, J.H. Freed and H.A. Scheraga, Rotational jumps of the tyrosine side chain in crystalline enkephalin. ^{2}H-NMR line shapes for aromatic ring motion in solids, J. Amer. Chem. Soc. 103:7707 (1981).
[13] S.J. Kohler and M.P. Klein, ^{31}P chemical shielding tensors of phosphorylethanolamine, lecithin, and related compounds: Application to head group motion in membranes, Biochemistry 15:967 (1976).
[14] W. Niederberger and J. Seelig, Phosphorus-31 chemical shift anisotropy in unsonicated phospholipid bilayers, J. Amer. Chem. Soc. 98:3704 (1976).
[15] J. Seelig, Phosphorus-31 nuclear magnetic resonance and the head group structure of phospholipids in membranes, Biochim. Biophys. Acta 505:105 (1978).
[16] P.R. Cullis and B. de Kruijff, Lipid polymorphism and the functional roles of lipids in biological membranes, Biochim. Biophys. Acta 559:399 (1979).
[17] J. Seelig, Physical properties of model membranes and biological membranes, in "Membranes and Intercellular Communications", North-Holland Publishing, Amsterdam, (1981).

37

USE OF ELECTRON SPIN RESONANCE TO

STUDY COMPLEX BIOLOGICAL MEMBRANES

Harold M. Swartz

University of Illinois College of Medicine
Urbana, Illinois

I. INTRODUCTION

This paper attempts to provide an overview of experimental considerations, possibilities, and problems in the use of spin labels to study the membranes of complex biological systems, especially living cells. There are increasing numbers of reports of such studies that approach problems of both fundamental and medical significance, but there is a lack of systematic reviews of experimental aspects of such studies of complex biological systems. It therefore seems timely and desirable to present a summary of the state of this field. There are abundant excellent reports of the basics of this technique and its application to model membranes (e.g., Berliner, 1976; Gaffney and Chen, 1977; Schreier et al., 1978) so I shall not attempt to cover that material.

This review will concentrate on the types of information that one can obtain and shall not attempt to provide a comprehensive review of the results. The main part of this paper is organized about the different experimental parameters that have been used-- there already are a surprisingly large number of such parameters. Prior to consideration of specific experimental parameters, I consider some of the general experimental concerns associated with these studies.

II. EXPERIMENTAL CONSIDERATIONS

A. Balance Between Pertinence and Accuracy/Specificity of Information

In general, one may obtain more directly interpretable

information by working with simpler, well-defined systems. The
primary motivation for using more complex systems, therefore, should
be that there is information that cannot more readily be obtained
from simpler systems. In most cases the data obtained from studies
with intact cells have more uncertainty than those obtained from
simplified preparations or model systems. The most typical
advantages that outweigh this disadvantage are the study of
functions that do not occur in units simpler than the intact cell
(e.g., changes in membrane structure associated with the cell cycle,
Lai et al., 1980) and correlative studies in which changes in
membranes are related to biological functions (e.g., changes in
membrane structure associated with phagocytosis of macrophages,
Ingraham et al., 1981). Additional studies using simpler systems
may still be required in order to obtain adequate interpretation of
the results from the more complex system. It is essential to
recognize that systems as complex as living cells have many more
possible sources of experimental variation than model systems, so if
the necessary data can be obtained with the latter it then usually
is preferable to use model systems. In general the most useful
results for investigations of properties of viable cells may be
obtained by experimental approaches that use a combination of
studies of model systems, simplified biological systems, and viable
cells.

B. Biological Effects of Spin Labels

If one accepts the assertion that studies employing whole cells
have as one of their primary rationales the need to have intact
biological functions, it follows that one must study the biological
effects of the spin labels under the experimental conditions used in
the study. Under some circumstances the spin labels have obvious
biological effects [e.g., alteration of the surface of RBC membranes
as determined by electron microscopy (Butterfield et al., 1976b) and
cell killing at high concentrations of spin labels (Newhall et al.,
1979)]. Perhaps less apparent but certainly of equal importance are
the potential effects of the physical preparations required to get
the samples into the ESR spectrometer such as: use of solvents such
as ethanol to solubilize the spin label; removal of cells from the
growth surface by scraping or proteolytic enzymes; formation of
dense concentrations of cells with consequent potential effects of
hypoxia and accumulation of metabolic products; effects of
suboptimal temperatures (usually well below optimal growth temperatures but still high enough to have the potential for some metabolic
processes to continue in an uncoordinated manner).

The principal conclusion from these considerations is that it
is essential that the investigator carefully consider whether the
experimental procedures used may effect the interpretation of the
results from the data on the spin labels. In order to answer these

40

questions it often will be necessary to carry out a number of additional experiments that study potential biological effects.

C. Location of Spin Labels

This is a potential problem in all spin label studies, including those using the simplest model systems. The possibilities of multiple and/or alternative sites of spin labels increase with the use of cells because of the many different membrane systems in cells and the capabilities of cells to transform and transport spin labels. A simple but often useful way to determine if the spin label is in the outer bilayer of the cell membrane is to use extracellular reducing agents to remove the ESR signal (Kornberg and McConnell, 1971) or oxidizing agents to restore ESR signals from reduced spin labels (Sauerheber et al., 1980). If the reducing or oxidizing agents remain in the extracellular compartment, such studies provide reliable indications of the location of the spin labels. However, the evidence for the location of the redox reagents must be considered with care (Schreier-Mucillo et al., 1976).

D. Removal of "Extraneous" Spin Label Signals

In some of the systems studied by spin labels, some of the spin label is known to go be in a second compartment that leads to an unwanted component of the ESR spectrum. When this signal is in the extracellular compartment it may be removed effectively by using a broadening agent that will remain extracellularly (Keith and Snipes, 1974; Haak et al., 1976). Usually paramagnetic metal ions are employed which, by spin-spin interaction, broaden out the ESR signal from the extracellular compartment. Nickel has been used most often, but other and perhaps less toxic ions may also be useful. When such broadening agents are used, their physiologic and toxic effects need to be considered.

E. Significance of Order Parameters and Related Measurements

It is essential to keep in mind that although the derivations of these parameters usually are based on definitive and rigorous physical principles their equivalence to molecular motions is lost when applied to the complex systems typical of biological materials (Schreier et al., 1978). Therefore these parameters should be considered strictly as relative measurements in such systems and usually should be compared only to similar measurements made by the same techniques in the same laboratory on similar samples. In practice these restrictions do not impose severe limitations on the use of the technique as long as one keeps this in mind and plans the experiments appropriately. The types of variations that one may encounter are indicated in Table 1 in which one notes that the variations between the controls obtained in three different sets of experiments exceeded the differences between control and experimental samples within any single set yet the authors still were able

Table 1 Comparison of the order parameter S for 5-NMS in normal and MMD, DMD, or CM intact erythrocytes

	S_N	S_{MMD}	S_N	S_{DMD}	S_N	
Mean	0.625	0.590	0.598	0.605	0.588	0.570
s.e.m.	0.008	0.006	0.012	0.015	0.010	0.008
n	7	9	12	10	9	9
P	<0.005		0.1<P<0.25		<0.01	

5-NMS, 5-Nitroxide Methyl Stearate; MMD, Myotonic Muscular Dystrophy; DMD, Duchenne Muscular Dystrophy; CM, Congenital Myotonia; N, Normal (from Butterfield et al., 1976a)

to conclude that there were significant differences between controls and experimentals.

III. EXPERIMENTAL PARAMETERS, INCLUDING ILLUSTRATIVE RESULTS

A. Order Parameters

The basic theory of order parameters is covered in detail in a number of publications and I will not consider it here. These parameters attempt to compare motions about the principal molecular axis that is parallel to the hydrocarbon chains in an oriented bilayer, with motion of the axes that are perpendicular to the principle axis. The ESR measurements take advantage of the anisotropy of the hyperfine splitting and g-factors of nitroxides, in combination with knowledge of the relationship of the axes of the nitroxide to those of the membrane. The x-axis of the nuclear hyperfine splitting is parallel to the N-O bond direction and the

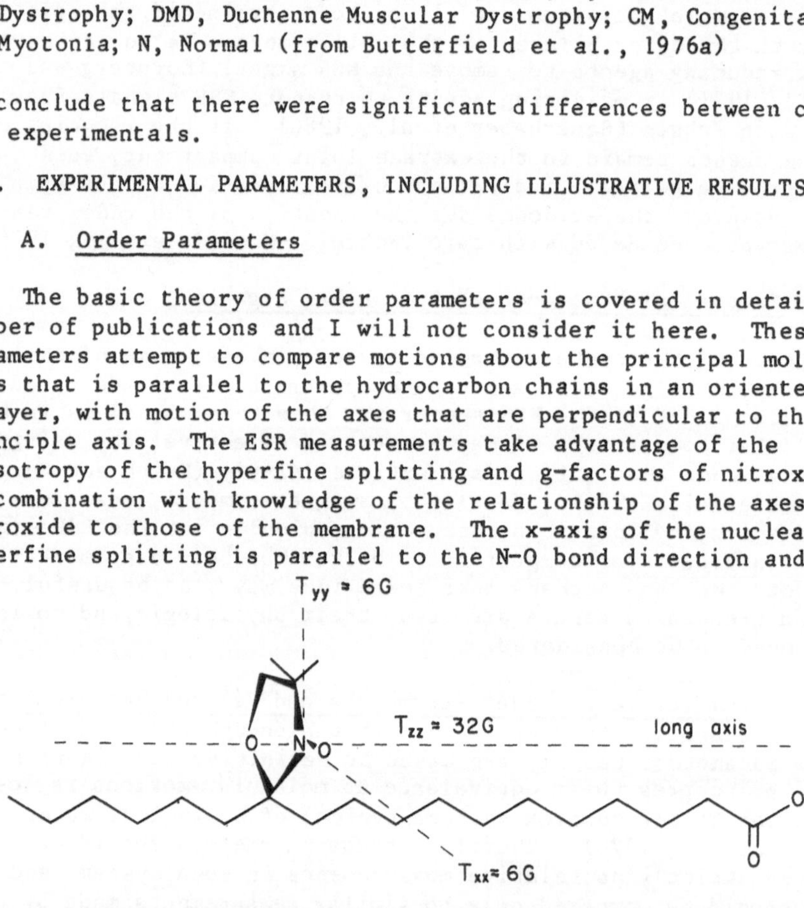

Fig. 1. Relationship of axis of fatty acid spin label to the principal axes of the unpaired electron on the nitroxide group. Note that the z-axis (hyperfine splitting T_{zz} = 32 gauss) is parallel to the long axis of the fatty acid analog and therefore is parallel to the plane of the bilayer of the membrane. The spin label in this figure is often designated as I(5,10) or 12 NS or 12 doxyl stearate or 12 DS or 12 SASL (figure from Schreier et al., 1978).

z-axis is parallel to the nitrogen π-orbital. The most common
situation is to use a spin label analog of a fatty acid which has
the z-axis of the nitroxide parallel to the hydrocarbon chains
(Fig. 1).

If the motion of the spin label is limited to motion about this axis
(the long axis of the spin labeled fatty acid) the splitting due to
Az will not be averaged out. In the ESR spectra of such spin labels
one can distinguish features (usually termed $T_{\parallel}$ and $T_{\perp}$ or $A_{\parallel}$
and $A_{\perp}$) that report on these two types of motions and then
calculate an "order" parameter, S. (Fig. 2).

In some spectra both of these features may not be distinguishable
and order parameters termed $S_{\perp}$ or $S_{\parallel}$ may be calculated using
only the features that are available. Under some conditions these

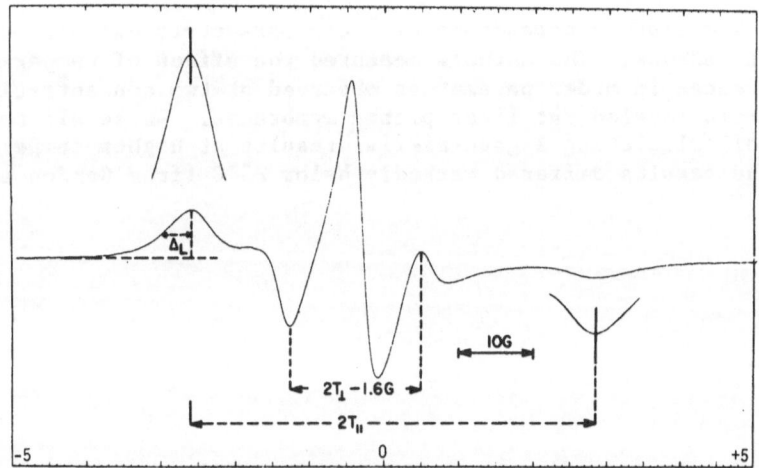

Fig. 2. Calculation of order parameters. This spectrum from
human platelets labeled with a fatty acid spin label, I(12,3)
illustrates typical experimental data for the calculation of an
order parameter. The separation in gauss of the outer peaks
(2 $T_{\parallel}$) is a close approximation of twice the hyperfine splitting
along the z-axis. The splitting marked 2 $T_{\perp}$-1.6 reflects twice
the hyperfine splitting along the x or y axes, with a 1.6 gauss
correction factor. The authors calculated S by the formula
$$S = \left(\frac{T_{\parallel} - T_{\perp}}{T_{zz} - T_{xx}}\right) \left(\frac{a_n}{a'_n}\right)$$ where T_{zz}, T_{xx}, and a_n are obtained
from experimental values for a nitroxide in a host crystal and $\frac{a_n}{a'_n}$
is a correction factor reflecting polarity differences in the
environment of the nitroxide in the membrane and the crystal (from
Sauerheber et al., 1980).

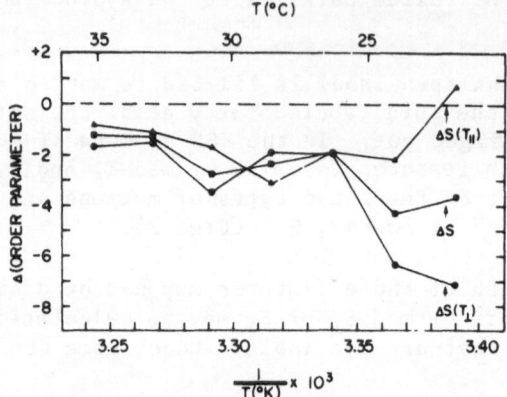

Fig. 3. Temperature dependence of order parameters calculated by
different methods. The authors measured the effect of temperature
on differences in order parameters observed at two concentrations of
I(12,3) spin labeled rat liver plasma membranes. While all three
methods of calculating S gave similar results at higher temperatures, the results differed markedly below 25°C (from Gordon et al.,
1978).

3 parameters (S, $S_{\perp}$, and $S_{\parallel}$) may give quite different results
and therefore one cannot, even within the same system, make relative
comparisons if you use different types of order parameters (Fig. 3).
This is a further extension of the point made earlier, that order
parameters generally should be considered as relative rather than
absolute parameters with respect to determination of motion. I also
remind the reader that the spectra directly report the motion of the
spin label and it requires inference to relate this to motion of
components of the membrane.

The order parameters shown in Fig. 3, are for 9 GHz. It often
is desirable to obtain spectra at 35 GHz because of the smaller
sample that is required and the clearer separation of the anisotropic features related to the three axes. With 35 GHz spectra the
A_z splitting is a convenient parameter to use to measure relative
changes in motion (Fig. 4 & 5).

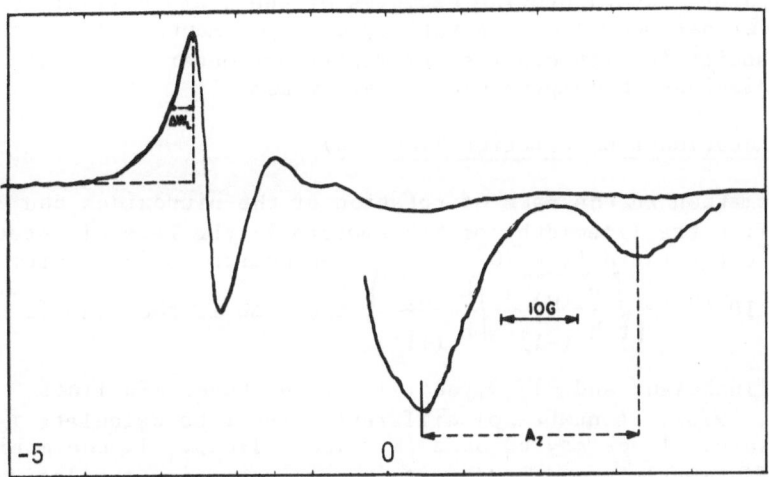

Fig. 4. 35 GHz spectra of I(12,3) spin labeled CHO cells. At this frequency one can measure A_z ($T_\parallel$) in a sample that is at least ten times smaller than that required for studies at 9 GHz. The figure also indicates how one would measure linewidth of the low field peak at 35 GHz (from Lai et al., 1980).

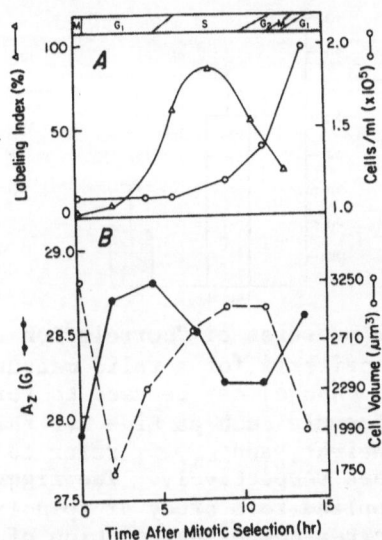

Fig. 5. The relationship between A_z and the cell cycle. It is assumed that A_z is a valid measure of relative motional freedom in the membrane. Because of the smaller volume of cells required for studies at 35 GHz, changes in motion of the I(12,3) spin label could be studied in synchronized CHO cells (from Lai et al., 1980).

Figures 3 and 5 provide examples of the usefulness of determining order parameters over a variety of experimental conditions
and in association with other experimental parameters, in order to
reach conclusions on complex biological systems.

B. <u>Rotational Correlation Time</u> (τ_c)

Information in the rate of rotation of the nitroxides can be
obtained from the linewidths of ESR spectra if the rate of rotation
is sufficiently rapid ($\tau > 10^9$ Hz) and the motion is isotropic:

$$\tau = 3.4 \times 10^{-10} \; \Delta W\left(\sqrt{\frac{h_{(0)}}{h_{(-1)}}} - \sqrt{\frac{h_{(0)}}{h_{(+1)}}}\right), \text{ where } \Delta W \text{ is the linewidth,}$$

h is the lineheight and -1, 0, and +1 are the three ESR lines
(Hemminga, 1975). A number of different methods to calculate τ_c,
using different lines may be used (Kivelson, 1960). If the motion
is not sufficiently rapid and/or it is not isotropic, a relative
parameter may be calculated and measurements made (Fig. 6).
It is particularly difficult to be certain that the motion of the
label is isotropic and therefore it may be useful to consider
virtually all τ_c calculations for membranes of cells as relative
measurements only.

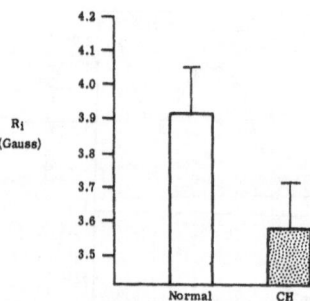

Fig. 6. Use of relative measures of "correlation times." Under
conditions in which the criteria for a valid measurement of τ_c
cannot be met, linewidth changes may be used to reflect relative
changes in motion by a formula such as Ri = W_{+1} $\sqrt{h_{+1}/h_{-1}}$ where
W = linewidth, h = lineheight, and $+_1$, $-_1$ refer to the low and
high field hyperfine lines respectively. The figure shows the
results of the method applied to a study of spin labeled
polymorphonuclear leukocytes comparing rotation of the spin label in
normal mice and mice with a syndrome (Chediak-Higashi) that involves
decreased white cell function (from Haak et al., 1979).

Table 2   2T∥ and ΔW for spin labeled N̲. gonorrhoeae strain 2686
colonial types

Colony type	2T∥ (G)	P	ΔW(G)	P	n
T1	59.7	0.480	4.4	0.11	5
T2	59.6	-----	4.2	----	5
T3	59.0	0.003	4.7	0.01	5
T4	58.8	0.004	4.6	0.02	5

n, Number of samples; ΔW, linewidth of low field line; P value for
equality with T2 (from Newhall et al., 1979).

C.   Linewidth Methods

    Mason et al., (1977) have described a method that uses the
width of the low field line as an experimental variable for
measuring changes in mobility. This appears to be a very sensitive
method, e.g., Table 2.

A potentially serious drawback of this method is the possibility
that the linewidth may also be affected by the concentration of the
spin label. In complex membranes this problem may be especially
important because of the possibility of non-uniform distribution of
the spin labels with consequent high local concentrations.

D.   Saturation Transfer Spectroscopy (STS)

    These techniques are based on the possibility of detecting
motions with correlation times of 10^{-3} to 10^{-7} seconds by observing
the effects of such motions on the microwave power saturation of
portions of the ESR spectrum (Hyde and Dalton, 1979). While the
general theoretical bases for this approach have been described in
some detail and there are some excellent theoretical treatments of
specific applications of the method, for the purposes of this
chapter it is sufficient to consider this as an empirical method of
measuring motion, and especially changes of motion, in this time
scale--a time scale that is not readily studied by other techniques.
The usual calibration for the technique is to determine the STS
spectra of spin labeled hemoglobin in solutions of known vicosities
(and hence with known rotational times of the hemoglobin molecule).
One then compares the results of studies of experimental systems
with these calibration curves. Some of the experimental parameters
can be associated with different axes of rotation and therefore
conclusions drawn in regard to the symmetry of the motion (e.g.,
Table 3).

Table 3 Diagnostic lineheight ratios and effective rotational
        correlation times (in brackets) of spin-labeled PC in gel-
        phase bilayers of phosphatidylcholines and
        phosphatidylethanolamines, T = 15°C.

	L"/L	H"/H	C'/C
DSPC	1.16 $(>10^{-3}$s.)	0.95 $(\sim 10^{-3}$s.)	0.41 $(>10^{-4}$s.)
DPPC	0.99 $(4.10^{-4}$s.)	0.73 $(3.10^{-4}$s.)	0.33 $(3.10^{-5}$s.)
DMPC	0.57 $(3.10^{-5}$s.)	0.44 $(6.10^{-5}$s.)	-0.83 $(2.10^{-7}$s.)
DPPE	0.96 $(4.10^{-4}$s.)	0.60 $(1.10^{-4}$s.)	0.24 $(2.10^{-5}$s.)
DMPE	0.89 $(2.10^{-4}$s.)	0.58 $(1.10^{-4}$s.)	0.48 $(>10^{-4}$s.)

DSPC, DPPC, DMPC: dipalmitoyl, distearoyl, and dimyristoyl
phosphatidylcholine; DPPE, DMPE: dipalmitoyl, dimyristoyl
phosphatidylethanolamine; L"/L, H"/H: parameters sensitive to
motions of the long axis of the molecule; C'/C: parameter sensitive
to rotations around the long molecular axis. Note that the motion
for DMPE appears to be considerably more isotropic than for DMPC--
this correlates with the known thermal transitions of these systems
(from Marsh and Watts, 1980).

E. Polarity

The magnitude of the splitting of the three principal lines of
the nitroxide increases with the polarity of the medium and
therefore, can be used to determine the polarity of the environment
of the spin label. If the spin label partitions into two environments of differing polarity, often one can detect this by the
splitting of one of the hyperfine lines, especially the high field
line (Fig. 7).

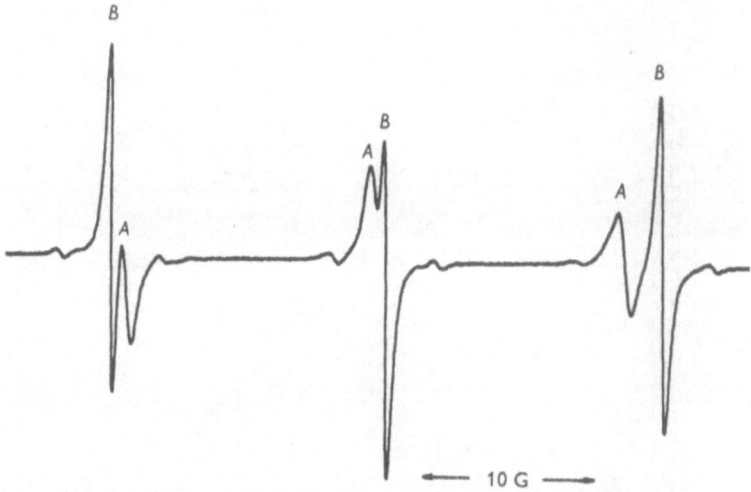

Fig. 7. Observation of spin labels in two environments. ESR
spectrum of cat sciatic nerve labeled with a keto-heterocyclic
nitroxide that has partioned into hydrophobic and hydrophillic
environments. Because increased polarity leads to greater hyperfine
splitting the lines (labeled B) from the label in the aqueous phase
are distinguishable from those in the hydrophobic environment (A)
(from Haak et al., 1976).

Changes in the partitioning of the spin label may be used as an
independent experimental parameter.

F. Reduction of Spin Labels

    While spin labeled fatty acids generally are stable in model
systems, many functional biological systems such as mitochondria and
intact cells reduce the spin labels making them not detectable by
ESR (Quintanilha and Packer, 1977). The reduction of the spin
labels probably occurs via enzymic pathways and requires energy.
The rate of reduction may provide a useful experimental parameter of
cell function. In comparing two cell systems, differences in the
rate of reduction may provide insights as to fundamental differences
between the cell lines (Fig. 8).

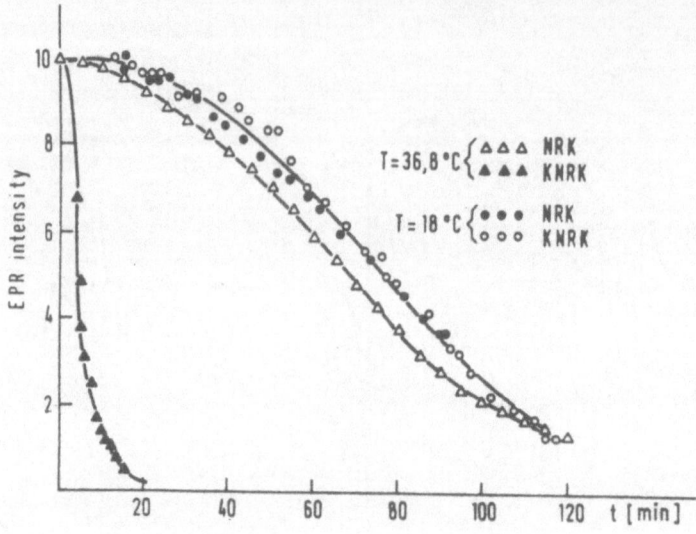

Fig. 8. Use of rate of reduction of spin labels as a parameter to study differences in cell lines. This figure indicates the rate of reduction of a fatty acid spin label by transformed (KNRK) and control (NRK) cell lines (from Schara et al., 1977).

The ability to regenerate virtually all of the reduced spin label signal by means of an extracellular mild oxidizing agent implies that the spin labels are located in the outer bilayer of the membrane. An alternative explanation would be that the spin label moves very rapidly throughout all of the membranes in the cell including the plasma membrane. In any case, it seems clear that intermediate rates of movement of the spin labels from the outer bilayer seems inconsistent with the data on regeneration of the signal by oxidizing agents.

G. Membrane Protein Spin Labels

Several investigators have spin labeled the membranes of erythrocytes and found spectral features for strongly immobilized and weakly immobilized spin labels, as indicated in Fig. 9 (Butterfield et al., 1980; Rigaud et al., 1974; Sinha and Chignell, 1979).

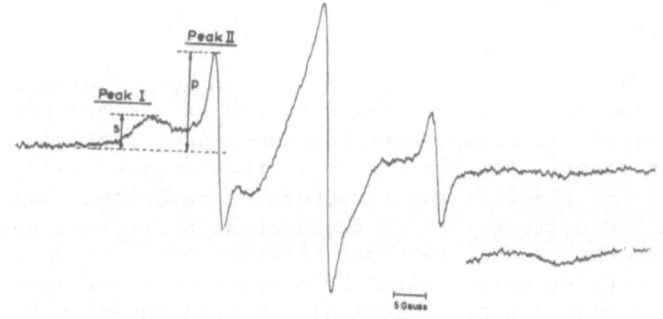

Fig. 9. ESR spectrum of protein labeled with maleimide spin label.
Peak I reflects spin labels that are strongly immobilized while
peak II reflects weakly immobilized spin labels. To measure
relative changes the ratios of the heights of the two peaks often
are compared (from Rigaud et al., 1974).

Upon different chemical and/or biological treatment of the
erythrocytes the ratio of these two types of spin label sites may
become altered (Table 4).

The W/S ratio appears to reflect structural information in proteins
and their organization in membranes and does not correlate closely
with measurements of order parameters by fatty acid spin labels.

Table 4 Effects of previous proteolytic enzyme treatment of intact
         cells on the W/S ratio of maleimide spin label attached to
         membrane proteins in control and Huntington's Disease
         erythrocyte ghost membranes

Treatment	Control	HD	N	Pb
None	5.14 ± 0.20	5.64 ± 0.22	13	<0.02
Pronase	4.75 ± 0.20	4.70 ± 0.19	13	>0.5
Chymotrypsin	4.86 ± 0.23	5.07 ± 0.26	10	>0.8
Trypsin	4.86 ± 0.21	4.67 ± 0.20	4	>0.5

W/S ratio: ratio of peak heights of components of ESR spectra due
to "weak" and "strong" immoblization.
Means ± S.E.M. are presented.
P-value calculated by a two-way analysis of variance.
(From Butterfield et al., 1980).

SUMMARY

This paper has summarized the use of a number of different
experimental parameters that are based on ESR-spin label techniques
to study membranes. Although there are some interrelations between
some of the parameters, for the most part they provide different
ways to measure the function and structure of membranes. When
applied to complex systems such as cells these parameters usually
provide relative rather than absolute information. The most useful
information appears to result when these approaches are combined
with independent measures of biological function and/or with other
physical techniques. Although there are many sources of potential
experimental difficulties and artifacts, these various spin label
techniques have the potential of providing very useful information
that cannot be obtained as readily and/or reliably by other
techniques.

REFERENCES

Berliner, L. J., ed., 1976, "Spin labeling theory and applications,"
    Academic Press, New York.
Butterfield, D. A., Chesnut, D. B., Appel, S., and Roses, A. B.,
    1976, Spin label study of erythrocyte membrane fluidity in
    myotonic and Duchenne muscular dystrophy and congenital
    myotonia, Nature, 263:159.
Butterfield, D. A., Doorley, P. F., and Marksbery, W. R., 1980,
    Evidence for a membrane surface defect in erythrocytes, in
    Huntington's Disease, Life Sciences, 27:609.
Butterfield, D. A., Whisnant, C. C., and Chesnut, D. B., 1976, On
    the use of the spin labeling technique in the study of
    erythrocyte membranes, Biochim. Biophys. Acta., 426:697.
Gaffney, B. and Chen, 1977, "Spin label studies of membranes"
    in methods in Membrane Biology 8:291-356.
Gordon, L. M., Sauerheber, R., and Esgate, J. A., 1978, Spin label
    studies on rat liver and heart plasma membranes: effects of
    temperature, calcium, and lanthanum, J. Supramolecular
    Structure, 9:299.
Haak, R. A., Kleinhans, F. W., and Ochs, S., 1976, The viscosity of
    mammalian nerve axoplasm measured by electron spin resonance,
    J. Physiol., 263:115.
Haak, R. A., Ingraham, L. M., and Baehner, R. L., 1979, Membrane
    fluidity in human and mouse Chediak-Higashi leukocytes,
    J. Clin. Inv., 64:138.
Hemminga, M. A., 1975, An ESR spin label study of structural and
    dynamical properties of oriented lecithin-cholesterol
    multibilayers, Chem. Phys. Lipids, 14:151.
Hyde, J. S. and Dalton, L. R., 1979, "Saturation-transfer
    spectroscopy," in Spin Labeling II, L. J. Berliner ed.,
    pp. 1-70.

Ingraham, L. M., Boxer, L. A., Haak, R. A., and Baehner, R. L., 1981, Membrane fluidity changes accompanying phagocytosis in normal and in chronic graulomatous disease polymorphonuclear leukocytes, Blood, 58:830.

Keith, A. D. and Snipes, W., 1974, Viscosity of cellular protoplasm, Sciences, 183:666.

Kivelson, D., 1960, Theory of ESR linewidths of free radicals, J. Chem. Phys., 33:1094.

Kornberg, R. D. and McConnell, H. M., 1971, Inside-outside transitions of phospholipids in vesicle membranes, Biochemistry, 10:1111.

Lai, C.-S., Hopwood, L. E., and Swartz, H. M., 1980, Electron spin resonance studies of changes in membrane fluidity of Chinese hamster ovary cells during the cell cycle, Biochim. Biophys. Acta., 602:117.

Marsh, D. and Watts, A., 1980, Molecular motion in phospholipid bilayers in gel phase: spin label saturation transfer ESR studies, BBRC, 94:130.

Mason, R. P., Giavecloni, E. B., and Dalmasso, A. P., 1977, Complement-induced decrease in membrane mobility: introducing a more sensitive index of spin-label motion, Biochemistry, 16:1196.

Newhall, W. J., Kleinhans, F. W., Rosenthal, R. S., Sawyer, W. D., and Haak, R. A., 1979, Neisseria gonorrhaeae membrane microenvironment studied by spin-label electron spin resonance: comparison of colony types, J. Bacteriol., 139:98.

Quintanilha, A. T. and Packer, L., 1977, Surface localization of sites of reduction of nitroxide spin-labeled molecules in mitochondria, PNAS, 74:570.

Rigaud, J. L., Gary-Bobo, C. M., and Taupin, C., 1974, Effect of chemical modifiers of passive permeability on the conformation of spin-labeled erythrocyte membranes, Biochim. Biophys. Acta., 373:211.

Sauerheber, R. D., Zimmerman, T. S., Esgate, J. A., VanderLaan, W. P., and Gordon, L. M., 1980, Effects of calcium, lanthanum, and temperature on the fluidity of spin-labeled human platelets, J. Membrane Biol., 52:201.

Schara, M., Séntjurc, M., Cotic, L., Pécar, S., Palcic, B., and Monti-Bragadin, C., 1977, Spin label study of normal and Ki-MSV transformed rat kidney cell membranes, Stud. Biophysica. 62:141.

Schreier, S., Polnaszek, C. L., and Smith, I.C.P., 1978, Spin labels in membranes--problems in practice, Biochim. Biophys. Acta., 515:375.

Schreier-Mucillo, Marsh, D., and Smith, I.C.P., 1976, Monitoring the permeability profile of lipid membranes with spin labels, Arch. Biochem. Biophys., 172:1.

Sinha, B. and Chignell, C. F., 1979, Interaction of antitumor drugs with human erythrocyte ghost membranes and mastocytoma P815: a spin label study, BBRC, 86:1051.

FREEZE-ETCHING ELECTRON MICROSCOPY : RECENT DEVELOPMENTS AND
APPLICATION TO THE STUDY OF BIOLOGICAL MEMBRANES AND THEIR
COMPONENTS

T. Gulik-Krzywicki

C.N.R.S., Centre de Génétique Moléculaire
91190 Gif sur Yvette, France

Electron microscopy, which is one of the most direct and most
powerful techniques for obtaining structural information, has been
applied only sparingly in the study of untreated samples. Examination of such samples is virtually impossible due to the damaging
effects of the high vacuum and electron beam and because of the
very weak contrast provided by the biological molecules. Generally
pretreatments (chemical fixation and staining) to permit electron
microscopic observation may modify the sample structure.[1] One procedure which seems to avoid some of the pitfalls of the other
techniques is cryofixation. Frozen hydrated samples may thus be
examined directly[2] or by freeze fracture electron microscopy.
Ideally, when cryofixation is used, the sample should be quenched
rapidly enough to avoid structural alterations due to temperature
induced structural transitions, change of the partial specific
volumes, ice crystal formation, etc. Practically, this goal may
be difficult to achieve and one must assess the extent of preservation of the sample structure after cryofixation. Among the methods that might be used for such an assessment, X-ray, neutron
or electron diffraction, which provide detailed structural information, are probably the most straightforward. We have developed
an approach based on the combined use of ambient and low temperature X-ray diffraction and freeze-etching electron microscopy to
study the effect of quenching upon sample structure. This approach
was applied first to study a variety of lipid-water phases of
known structure cryofixed by different procedures.[3-4] More recently
the same approach was used to study solutions of low density serum
lipoproteins.[5] The results of these studies provided some understanding of the factors which affect sample structure during quenching and permitted the elaboration of a more general approach for
structural investigations by freeze etching electron microscopy.

In this paper we will describe this general approach and its use in the study of biological membrane components. The use of freeze-etching electron microscropy for the study of native membranes will be illustrated by some recent studies of the molecular mechanism of synaptic transmission. In the latter case, the use of ultrarapid cryofixation procedures allow one to follow rapid morphological changes taking place within the presynaptic membrane after different types of stimulation which lead to the release of acetylcholine.

METHODS

FREEZE -ETCHING ELECTRON MICROSCOPY

Freeze-etching electron microscopy has become, at least for biological membranes, a routine method for studying structure. Well defined procedures giving highly reproducible results have been devised and applied to a variety of different materials.[6] The technique consists of four main steps. Each step, for which there are many technical variations, must be optimized to insure that the structure observed in the electron microscope is an exact representation of the ultrastructure of the original sample.

Cryofixation of the Sample

The goal of this step is to quench the sample rapidly enough so that, after fixing, the structure is the same as it was at the initial temperature. Many factors, however, such as ice crystal formation, temperature induced structural transitions, changes in partial specific volume, etc., have the potential to alter the sample structure during quenching. One approach to minimize such perturbations is to pre-treat the sample by chemical fixation or by adding cryoprotectants.[7] For lipids and some other materials, such pre-treatment itself may alter the sample structure. In these cases, optimal quenching must be achieved by one of the ultrarapid cyrofixation methods such as sandwich,[4] propane jet[8], spray[9] or copper block freezing.[10] In the first two methods, the sample is squeezed between two thin metal (copper or gold) plates and then quenched by rapid plunging into liquid propane or by projecting, at high speed, a stream of liquid propane onto the metal surfaces. The rate of cooling, about 13×10^{3}°K/s in both cases, can be increased by applying high pressure to the sample.[11] The spray freezing technique employs an artist's air brush and filtered compressed air to spray small droplets (5-15 µm diameter) of sample containing solution into liquid propane. The rate of cooling for this technique is unknown but may be surmised to be of the same order of magnitude as the other rapid quenching methods. Projection of the sample against a very clean copper surface maintained at liquid helium temperature forms the basis for the copper block technique (cooling rate about 6×10^{3}°K/s.) The relative advantages

56

and disadvantages of these procedures have been discussed recently by Costello et al.[12]

Fracture of the Cryofixed Sample

Quenched samples are fractured at low temperatures (less than 173°K) under high vacuum (better than 10^{-6} Torr) by the action of a liquid nitrogen or liquid helium cooled knife or by separation of the two metal plates of "sandwiched" samples. Deposition of re sidual water vapor resulting in contamination of the exposed surface may be reduced by operating under a higher, cleaner vacuum. The sample may or may not be deformed during the fracture process. Plastic deformation[13] may be reduced by fracturing at lower tempeture. Finally, parts of the sample situated below the fracture plane may be exposed subsequently by sublimation (freeze-etching) of the solvent prior to replication. Some samples may be very sensitive to this procedure and undergo structural changes[14].

Replication of the Fracture-Exposed Surfaces

The replication of surfaces exposed by fracture (freeze fracture) or by fracture and etching (freeze etching) is performed by evaporation of heavy metals. Platinum-carbon is commonly used but a finer grain size is obtained with tungsten or tungsten-tantalum alloys.[15-16] The metal deposit, which has a mean thickness of 1-2 nm, is too fragile to manipulate or observe in the electron microscope. To strengthen the replica, a 10-20 nm thick layer of carbon is evaporated onto the surface. The carbon layer, at this thickness does not contribute appreciably to the final electron microscope image. Since evaporation of the metals during the replication step requires high temperatures, heating and possible distortion of the exposed surface must be carefully avoided. This is particularly true when metals other than platinum are used. Surface heating can be minimized by working at the optimal conditions for a given source of evaporation and by using a multistep evaporation procedure allowing cooling of the partially shadowed surface between steps.[16] The best revelation of the various aspects of the sample ultrastructure frequently needs to be determined empirically by using unidirectional or circular (obtained by rotating the sample) shadowing or combinations thereof at different shadowing angles.

Cleaning and Observation of the Replica

After replication, the sample is thawed and then solubilized or digested with an appropriate solvent system. After elimination of the sample, the replica is washed with distilled water, deposited on an electron microscope grid and dried prior to observation in the electron microscope. Incomplete elimination of sample material giving a "dirty" replica can, on occasion, raise a problem. Finally, it must be stressed that since the replica is an object having three dimensions, stereo electron microscopy is invaluable for properly determining the sample ultrastructure.

Further, for ordered phases, optical diffraction of the electron micrographs may provide information about the lattice parameters and symmetry of the sample ultrastructure revealed by different fracture planes.

X-RAY DIFFRACTION

The X-ray scattering instrument and the experimental procedures are those routinely used in our laboratory.[3-5] The sample is placed

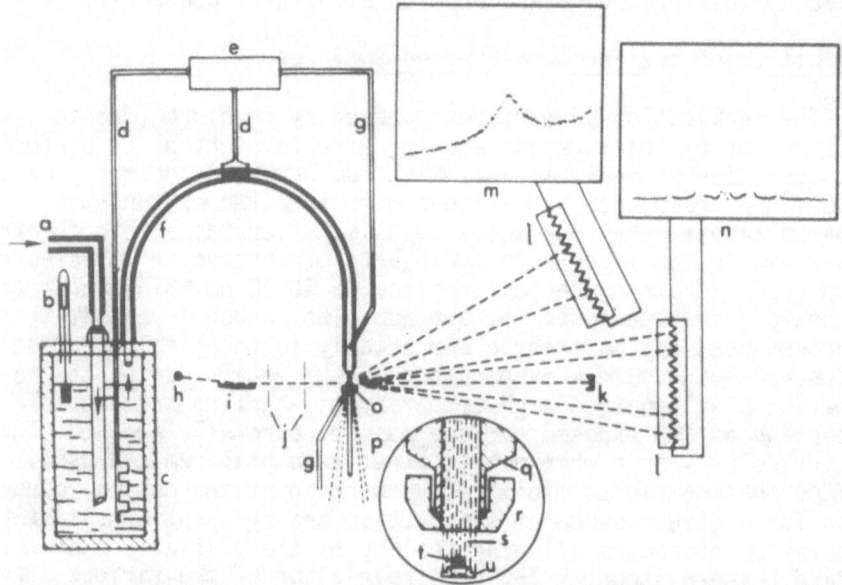

Figure 1. Schematic drawing of the low temperature X-ray diffraction set-up : (a) liquid nitrogen input ; (b) level indicator ; (c) constant level Dewar ; (d) heaters ; (e) heater control ; (f) flexible Dewar transfer tube ; (g) thermocouples ; (h) X-ray source ; (i) focusing mirror ; (j) slits ; (k) beam attenuator ; (1) linear position-sensitive detectors ; (m) wide-angle output from multichannel analyser; (n) low-angle output from multichannel analyser; (o) sample support cup; (p) glass rod; see insert for (q) cryostat output tip; (r) heated cap to eliminate frost; (s) laminar flow of gaseous nitrogen; (t) sample; (u) sample holder

on an appropriate holder which is mounted on a small metal support
(O) fixed on a glass capillary (p), whose position in the X-ray beam
is controlled by a micromanipulator (Figure 1). The X-ray diffraction
pattern of the sample is recorded with linear position sensitive
detectors.[17] This pattern is compared with that obtained from the
sample in a sealed specimen holder, to ensure that no changes occur
during preparation of the specimen for freezing. If no changes occur,
then the sample is immediately frozen and replaced on the same small
metal support (o) under a laminar stream of cold dry nitrogen gas
(133°K). After recording the low temperature spectrum, the sample is
transferred to liquid nitrogen for storage prior to subsequent analysis by freeze-etching electron microscopy. Comparisons of X-ray
diffraction patterns from samples before and after freezing permit
the assessment of the structural changes induced by the freezing
process. Low angle diffraction was used to follow changes in the
long range organization of samples, specifically, the change of
dimensions of lattices and their disordering (as judged from the
broadness of X-ray reflections). Wide angle X-ray diffraction was
used mainly to follow the perturbations in the packing of the lipid
hydrocarbon chains but it also gives information about the presence
or absence of ice crystals which display characteristic strong
reflections between 0.3 and 0.4 nm.[18].

RESULTS AND DISCUSSION

Biological membrane lipids

      Our previous combined X-ray diffraction-freeze-etching electron
microscope studies have shown that the extent of conservation, after
cryofixation, of the initial sample structure depends strongly upon
the composition of the sample, its structure and, in most cases, the
quenching procedure.[3-4] The lipid phases containing less than 10%,
by weight, water appear to be the least sensitive to the rate of
cooling. Among these, the phases displaying an ordered conformation
of the hydrocarbon chains (L_β phase) showed no changes in the low
angle X-ray scattering pattern, before and after quenching, indicating complete preservation of the lattice order and dimensions.
Broadening of the wide angle X-ray reflections for these samples
after cryofixation reflects some disordering of the initial packing
of the hydrocarbon chains. In contrast, phases which initially have
disordered hydrocarbon chains (L_α phase) become more ordered and
more closely packed after quenching. This ordering is reflected
by the narrowing and shifting to a wider angle of the 0.45 nm
broad X-ray diffraction band (a band characteristic of highly disordered hydrocarbon chains).[19] The ordering, however, is very
limited since there is no change in the lattice dimensions. That
such ordering is limited is also suggested by similar behavior for
phases having structures incompatible with the existence of highly
ordered hydrocarbon chains such as the two dimensional egg lecithin

water hexagonal phase (H_α) shown in Figure 2. This phase also can be quenched at relatively slow cooling rates without any appreciable change in lattice dimensions. The laser optical diffraction pattern of the electron micrograph of the cross fractured, hexagonally ordered water cylinders shows a hexagonal lattice whose dimensions match those observed by X-ray diffraction before quenching. This last observation also indicates that no major structural perturbations occured during fracture, replication and washing of the replica.

For phases containing 10 to 30% water, insufficiently rapid cooling rates result in an increase in lattice dimensions and segregation of different lipid species into separate domains after cryofixation (Figure 3).

When the water content of a phase exceeds 30% by weight, only ultrarapid cryofixation methods conserve, with varying degrees of success, the initial sample structure. Slower rates of quenching lead to the formation of ice crystals resulting in disordering of the lattice concomittant with an important decrease in lattice dimensions (Figure 4). Some samples may undergo phase transitions toward phases existing either at lower temperatures or lower water contents. In extrême cases, a highly ordered phase, such as a cubic phase, may be broken down into an amorphous like material.[20] The addition of cryoprotectants, such as glycerol, may partially relieve the perturbations due to quenching.[3]

Given the preceding results concerning lipid-water phases, we can now summarize the relationship of freeze fracture and freeze-etching electron microscopy to X-ray diffraction for the structural analysis of lipid phases. In particular, we can attempt to show what unique information can be provided by the electron microscopic technique.

First, the extent of this information depends upon the nature of the sample. For pure, highly ordered phases, providing rich X-ray diffraction data, electron microscopy can reveal the overall (general) morphology of the sample and shape and dimensions of some of the structural elements. This information may permit one to determine which of several molecular models (all compatible with the X-ray diffraction data) is the correct one. Further, the combination of freeze-etching with freeze fracture may give information about the location of the solvent within the structure.

For less well ordered phases, providing more limited X-ray diffraction data, the information provided by electron microscopy is of greater importance. For samples composed of a mixture of ordered phases, the freeze-etching data can indicate the number of different phases present and their structure. Further, the data may help to correctly assign X-ray reflections in, often, very complex diffraction patterns.

For phases composed of both ordered and disordered domains or disordered domains alone, freeze-etching electron microscopy permits the morphological analysis of each individual domain. Such an analysis may be very useful for following phase transitions. Replicas of samples quenched while undergoing, for example, the lamellar to cubic phase transition in the monoglycerides-water system, display many structures other than pure lamellar and/or pure cubic (Figures 5 and 6) phases.[20] These structures, some of which are shown in Figure 7, may represent intermediate steps in the lamellar to cubic phase transition. The stacked arrangement of smooth (characteristic of pure lamellar phase) and rough lamellae may correspond to an early stage of reorganization of the lipid lamellae (Figure 7A). Stacked lamellae showing a large number of holes (Figure 7B) or hexagonally arranged globules (Figure 7C) may correspond to the later stages of the transition. The latter, called "lipidic particles", have been the topic of much discussion and speculation recently.[23-25]

For the study of biological membrane lipids, several conclusions concerning the use of freeze-etching electron microscopy may be drawn. Knowledge of the phase diagram for various water contents may be useful for predicting how difficult adequate quenching (that is, quenching with minimal structural perturbation) might be. Since in some cases, even the most rapid cooling rates presently in use do not provide quenching without structural perturbations, comparison of structural parameters (lattice dimensions, lattice order, ice crystal size, etc.) before and after cryofixation are of crucial importance. One very powerful method which permits such a comparison is X-ray diffraction which may be performed before and after quenching using the same sample that will undergo analysis be electron microscopy.

Biological membrane proteins

Freeze-etching electron microscopy has been applied only rarely to the study of biological molecules in solution. Indeed, as compared to negative staining, which is the most frequently used techniques to investigate such molecules[1], freeze-etching is a more complex procedure. However, it may be quite appropriate for studying materials which are, like detergent solubilized membrane proteins, sensitive to staining agents. Freeze-etching electron microscopy can be used succesfully to study the morphology of biological macromolecules in solution when ultrarapid cryofixation and well controlled etching conditions are used.[26, 27] smaller biological molecules, such as extrinsic and intrinsic membrane proteins, can be studied by freeze-fracture electron microscopy. Extrinsic membrane proteins can be studied in buffer solutions containing

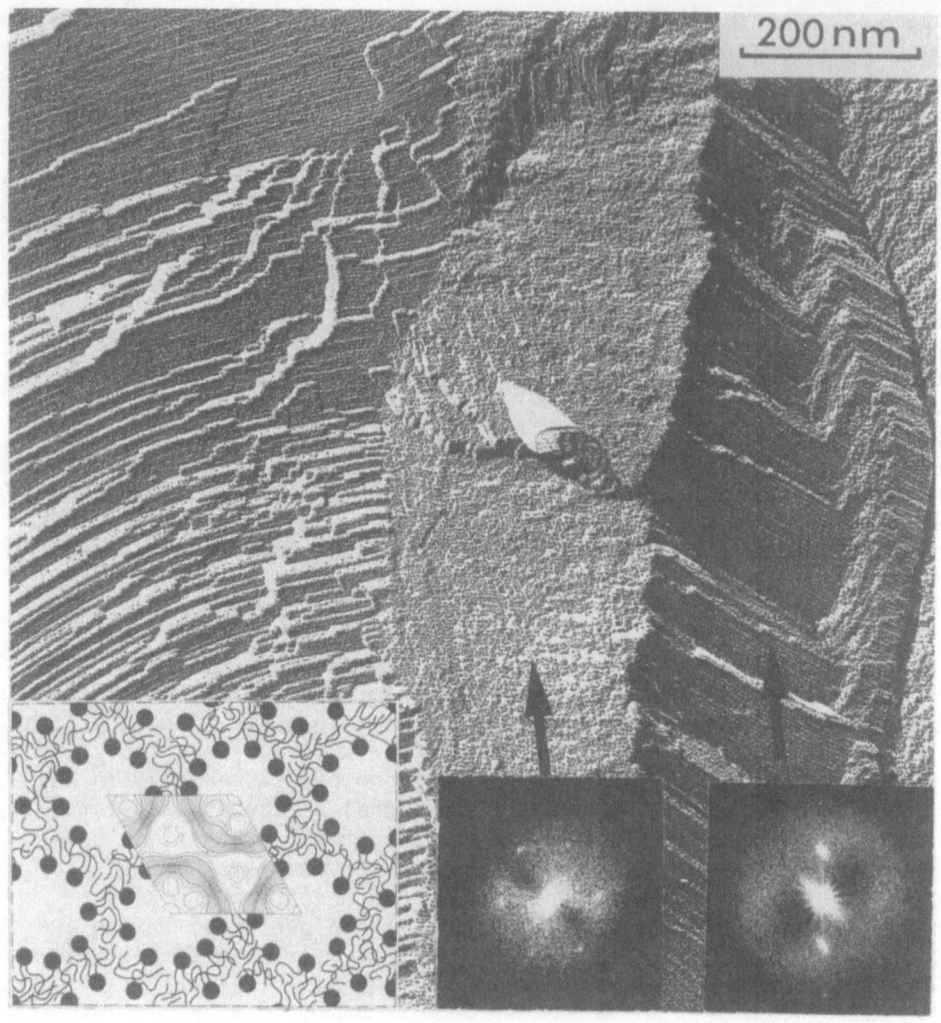

Figure 2. Freeze-fracture electron micrograph of egg lecithin-water reversed hexagonal phase (H_α). Shadowing with W-Ta (shadows are white) is unidirectional from lower right to the upper left. The structure of the pase is shown at the bottom left corner (hexagonally packed water cylinders, surronded by lipid polar head groups and separated by disordered lipid hydrocarbon chains[19]). Two laser optical diffraction pictures, shown at the bottom right corner, were obtained from the areas indicated by arrows. Notice that fracture planes propagating in between the cylinders show striations, whose period corresponds to the distance between the cylinders (according to X-ray diffraction results). the cross fractured cylinders show the hexagonal packing on the optical diffraction patterns obtained from the corresponding area.

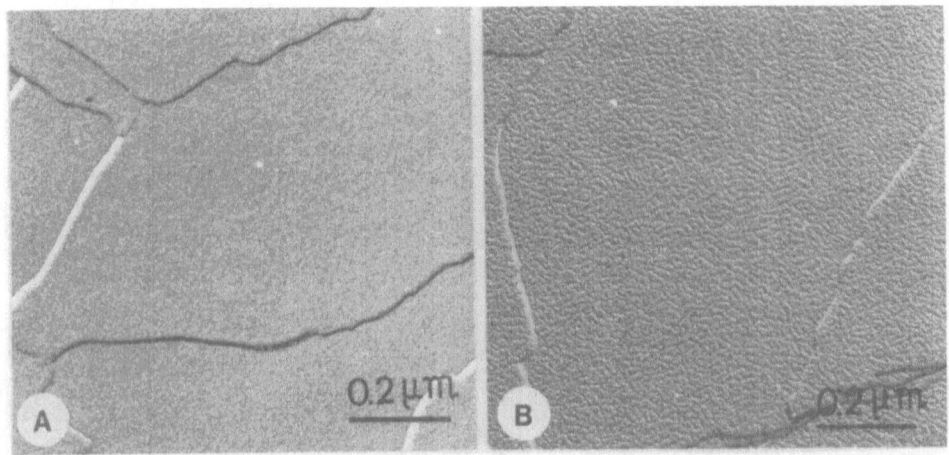

Figure 3. Freeze-fracture electron micrographs of the Lα phase of egg lecithin containing 16% water. A. Ultrarapid cryofixation using "sandwich" procedure[4]. Notice extended smooth fracture surfaces. B. Conventional cryofixation in Freon-22. Notice "worm-like" pattern on extended fracture surfaces, due to the freezing induced segregation of more saturated egg lecithin molecules[3].

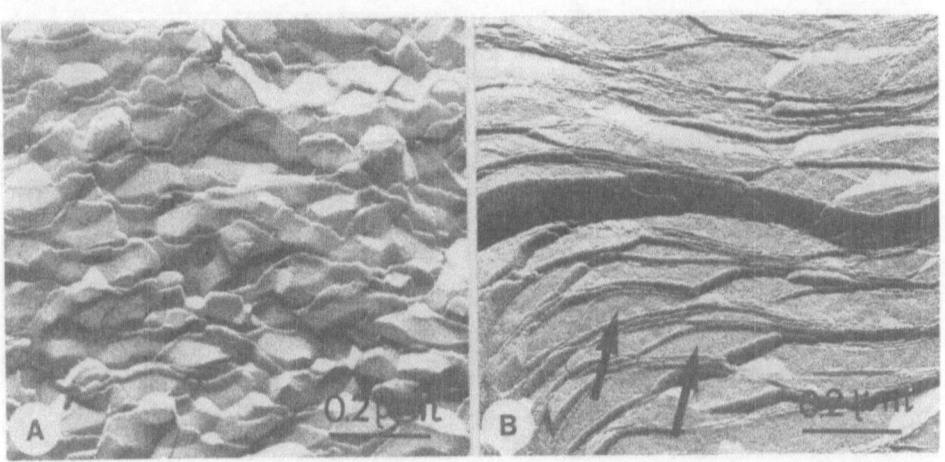

Figure 4. Freeze-fracture electron micrographs of the Lα phase of egg lecithin phosphatidyl inositol (50:50 by weight) containing 60% water. A. oblique fracture of conventionally frozen sample. Notice striking angular perturbations which reveal small patches of smooth fracture faces. B. Perpendicular fracture of conventionally frozen sample. Notice the clearly visible "ice pockets" (arrows) between stacks of closely packed lipid lamellae.

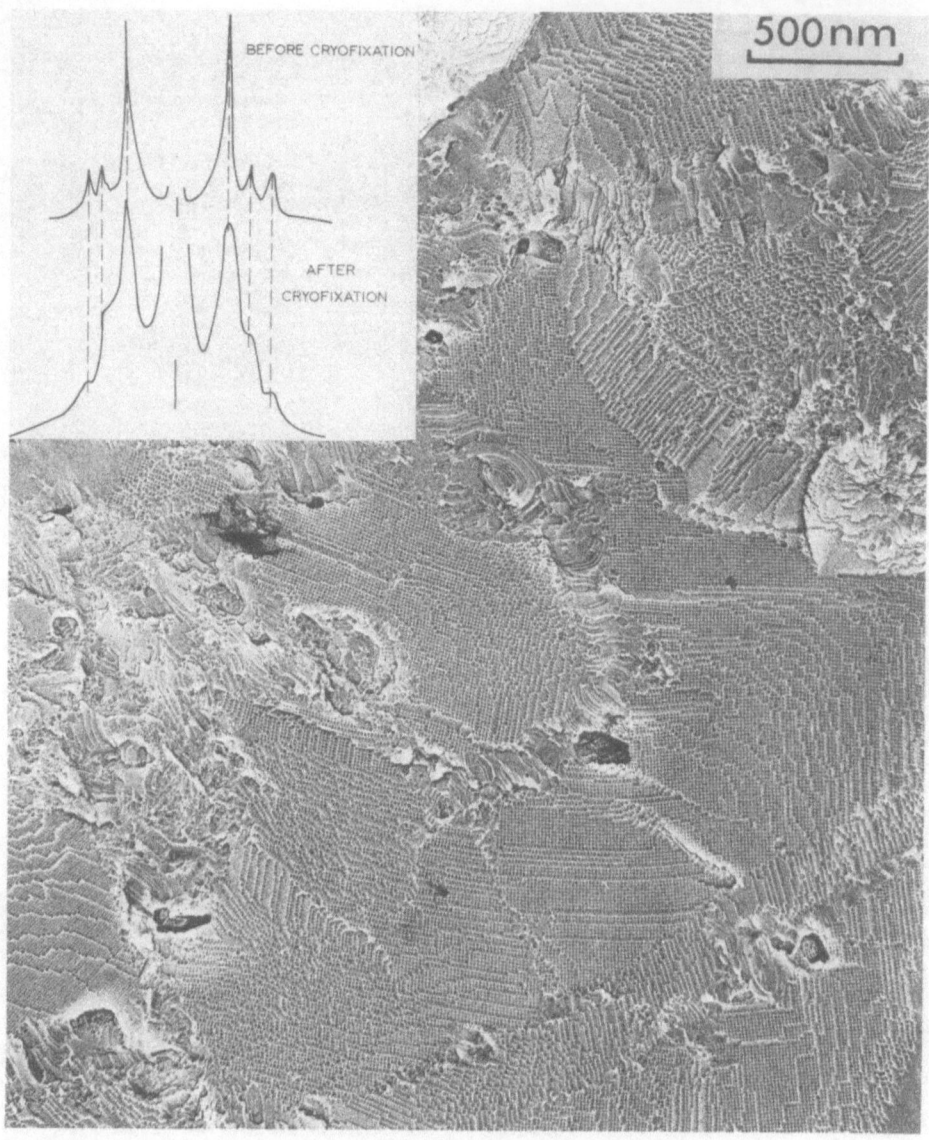

Figure 5. Low magnification freeze-fracture electron micrograph of sunflower oil monoglycerides-cytochrome c-water cubic phase. Rotary shadowed with W-Ta (shadows are white). X-ray diffraction patterns of the sample before and after cryofixation (inset) show a good preservation of the sample structure in the quenched state. The freeze-fracture replica shows extended domains of different, higly ordered, fracture planes, each compatible with a body centered cubic symmetry. In some places, very small domains of a reversed hexagonal phase can also be seen.

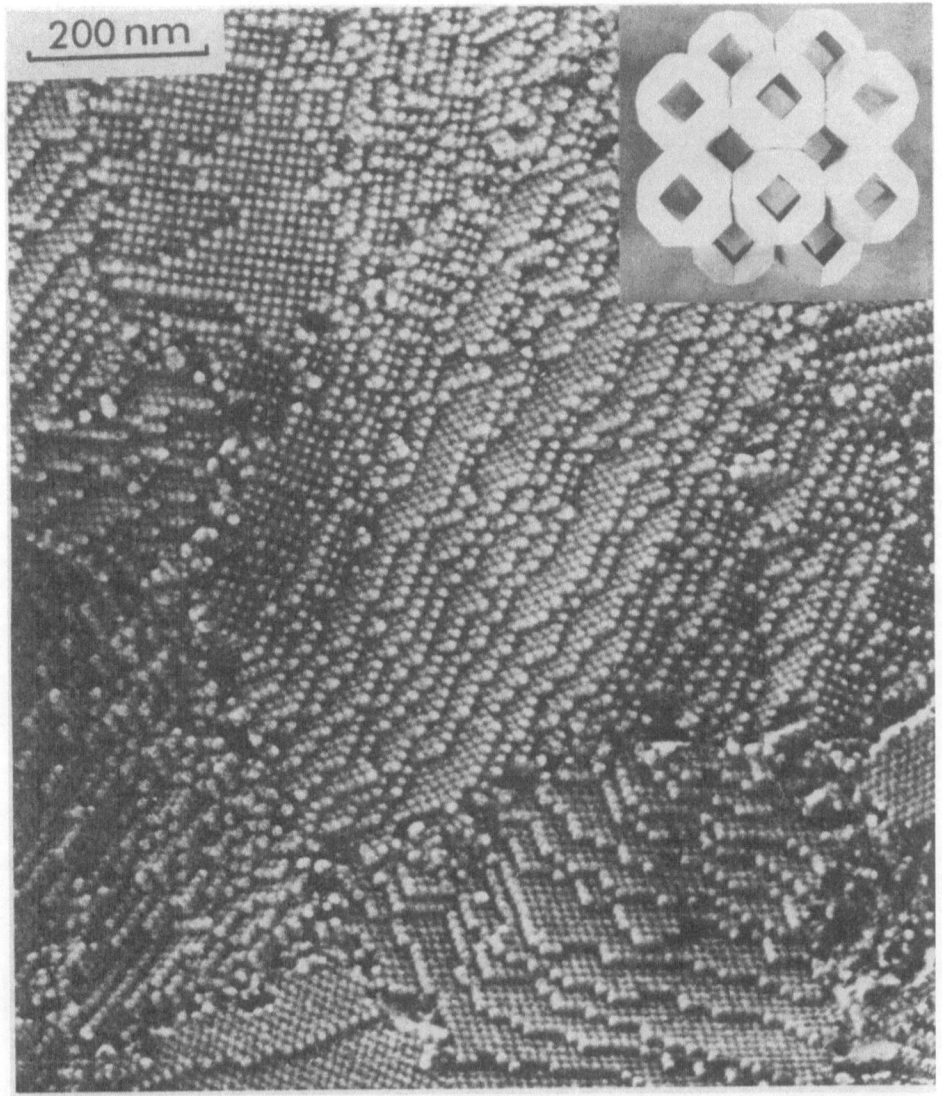

Figure 6. Freeze-etching electron micrograph of sunflower oil monoglycerides -cytochrome c-water cubic phase. Rotary shadowed with W-Ta but contrast is reversed (shadows are black). Partial sublimation of the ice (30 seconds at 173 °K, under 5.10^{-7} Torr) leads to some changes in fracture face morphology, mainly due to an enlargement of the water channels, which are clearly seen at the bottom right part of the lectron micrograph. Inset. Photograph of a model of the tetrakaidekahedral polyhedron proposed as the building block of the body centered cubic structure in monoglyceride water systems.[21,22]

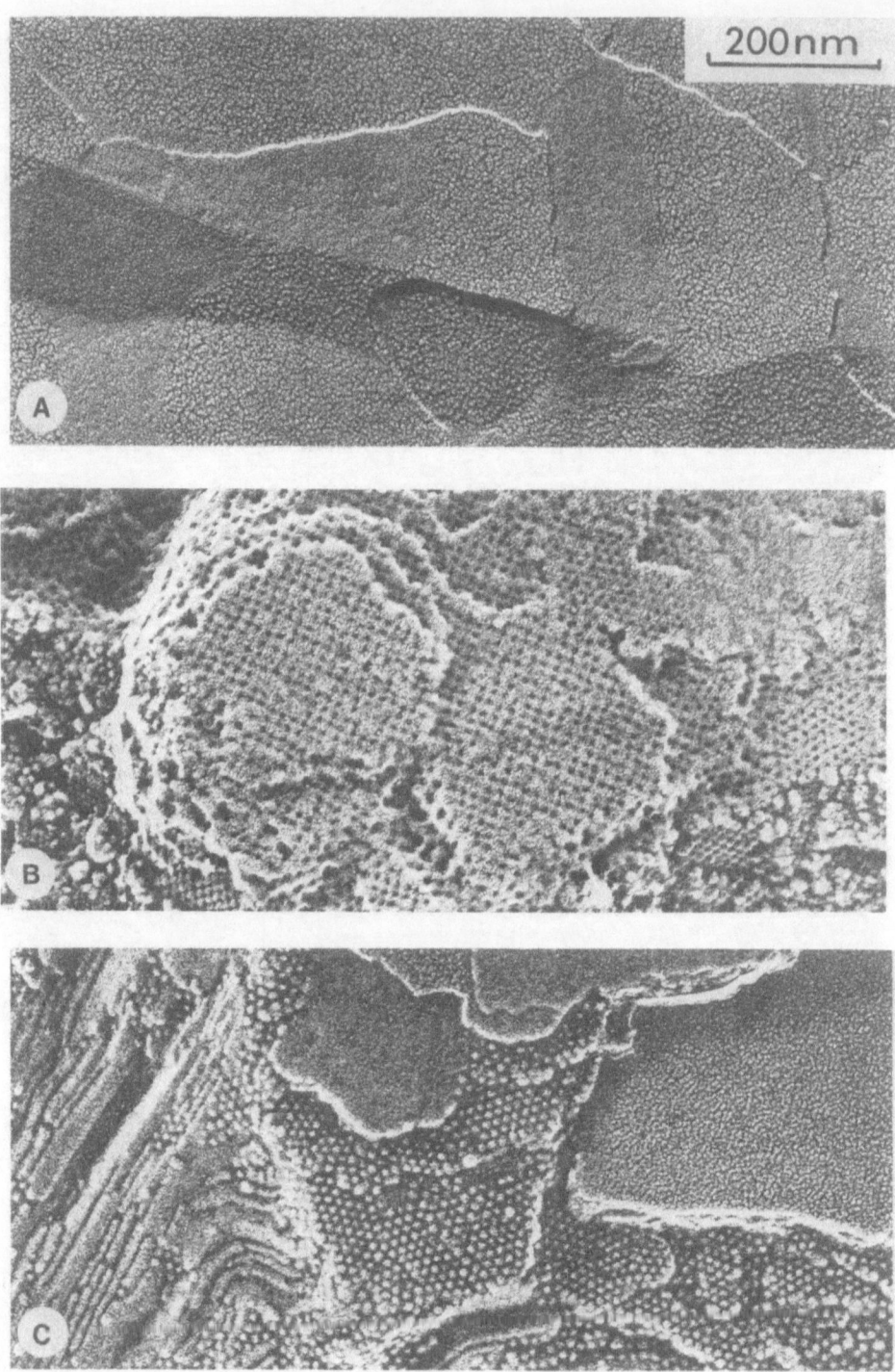

Figure 7. Freeze-etching electron micrographs of, probably, the early stages of the transition between the lamellar and the cubic phases in sunflower oil monoglycerides-cytochrome c-water system. The micrographs were obtained from the replicas of etched (1minute at 173°K, under 5.10^{-7} Torr vacuum) surfaces of the sample in which the formation of cubic phase was not complete. A. Stacked lamellae, parts of which are very smooth (typical of corresponding pure lamellar phase) and parts of which are rough. Unidirectional shadowing with W-Ta from the bottom toward the top. B. Large holes in stacked lamellae, some of which form square lattices whose dimensions are very close to that of the final cubic lattice dimension. Rotary shadowing with W-Ta. C. Stacked lamellae, parts of which are smooth and parts of which are arranged in small globules (particles) forming hexagonal lattices with interglobular distances approximating the dimensions of the cubic phase. Rotary shadowing with W-Ta. The contrast is reversed for all 3 photographs (shadows are black).

---

cryoprotectants and intrisic membrane proteins in detergent solubilized forms.[16, 28] The presence of proteins is indicated, in both cases, by the appearance of dispersed particles on the fracture surfaces of the frozen solutions. If one assumes that the freeze-fracture particles represent the portions of the protein molecules exposed above the surface of the ice by the fracture, then the distribution of these particle sizes should reflect the varying degrees of exposure and orientation of the proteins and is thus related to the morphology of the protein molecules. A direct interpretation of histograms of the freeze-fracture particle sizes in terms of molecular dimensions requires detailed knowledge of the phenomena occuring during the fracture of frozen protein solutions. In the absence of such knowledge, we decided to use, as a first stage for our work, molecules of known dimensions and shapes, in order to establish some general correlations between the morphology of the molecules and their corresponding freeze-fracture histograms. A good correlation was found between the general shape of the molecule and the form of its histogram and between the maximal dimension of the molecule and the maximal size of the corresponding freeze-fracture particles.[16] Further data and analysis are needed to better understand the histograms of freeze-fractured solutions of biological molecules. It appears, nevertheless, that freeze-fracture electron microscopy is a promising technique for the study of biological membrane proteins, especially those sensitive to stains.

Protein-lipid associations

The electronmicrographs of freeze-fractured extrinsic membrane protein-lipid associations are very similar to the corresponding lipid-water associations.[29] This is exactly what is expected if one admits that the plane of fracture upon cleavage propagates in between the lipid hydrocarbon chains, as in biological membranes.

Consequently, freeze-fracture electron microscopy alone is of very little use to study extrinsic membrane protein-lipid associations. Intrinsic membrane protein-lipid associations show entirely different fracture plane images. Fracture planes are, in this case, covered with particles, yielding images which are qualitatively very similar to those obtained with intact biological membranes.[29] The density and dimensions of these particles are related to the protein-lipid ratio in the reconstituted material and to the size of the protein used. Very small integral membrane proteins or those which are anchored in the lipid vesicles by only a small hydrophobic tail, may not form freeze-fracture particles until the formation of multimers of appropriate size. It has been shown, indeed, that small hydrophobic polypeptides or proteins may undergo a micellar-type association within the lipid bilayers, leading to the formation of multimers which can then be seem on the fracture planes of the protein-lipid associations. Freeze-fracture electron microscopy can be, furthermore, used for analyzing the nature and composition of reconstituted protein-lipid vesicles, and particularly for assessing the presence of pure lipid vesicles and of symmetrically and asymmetrically distributed proteins in integral membrane protein-lipid vesicles.[31] Such an analysis is straightforward when a double replication procedure is used. Pure lipid vesicles (or the vesicles associated with the peripheral membrane proteins) give rise to two complementary smooth fracture planes (convex and concave) while symmetrically distributed proteins in lipid-protein vesicles give rise to two equally particulate fracture planes. Totally asymmetric protein-lipid vesicles show one particulate and one smooth fracture plane[31] (Figure 8).

Biological membranes

Freeze-etching electron microscopy is the only morphological method which allows one to investigate the organization of intrinsic membrane proteins in native biological membranes. These proteins, or their oligomers, appear as intramembrane particles on both halves of fractured membranes. The organization, size and density of these intramembrane particles are characteristic for each membrane but my undergo rapid changes under different physiological conditions. With the most recent improvements in freeze-etching electron microscopy, and particularly the development of novel ultrarapid cryofixation procedures, such a rapid morphological changes may now be followed.[10, 32-35] The most important and exciting advantage of the ultrarapid cryofixation method, in comparison with conventional ultrastructural approaches, is that living specimens may be almost instantaneously frozen with good ultrastructural preservation at known intervals after a biological stimulus. This permits the use of the electron microscope for kinetic studies of transient biological events which are completed within a few seconds or even, with certain favourable specimens,

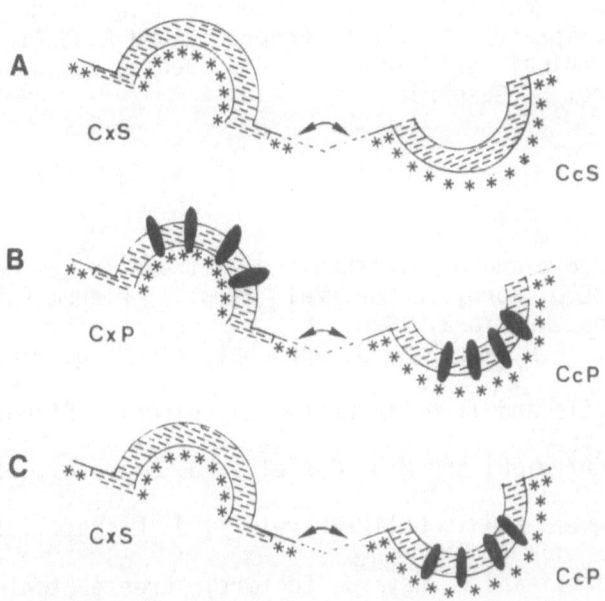

Figure 8. Schematic representation of different types of fractured vesicles. Three kinds of vesicles are considered (see the text) : pure lipid vesicles (A), protein-lipid vesicles with proteins distributed symmetrically on both sides (B), protein-lipid vesicles with proteins anchored only on one side of the vesicles (C). Any fractured vesicle produces one convex (Cx) and one concave (Cc) fracture plane. Depending on the nature of the vesicle the fracture planes can appear as smooth (S) or particulate (P). The analysis of a large number of fracture planes allows estimation of the percentage of each type of vesicle present in the mixture.

within a few milleseconds. The best example of such an approache is the recent study of the morphological changes taking place in the presynaptic membrane during synaptic activity.[10, 33-35] The realization of such direct kinetic studies represents a significant breakthrough in membrane biology and will, certainly, be applied in the near future to other membrane processes.

ACKNOWLEDGEMENTS

   The author gratefully acknowledge the help in the preparation
of the manuscript from Drs. L.P. Aggerbeck and A. Gulik and the
excellent technical assistance from J.C. Dedieu and secretarial
assistance from B. Gascard.

REFERENCES

1. H.P. Zingsheim and H. Plattner, in "Methods in Membrane
   Biology," E.D. Korn, Editor, Vol. 7, p. 1, Plenum Publishing
   Corporation, New York, 1976.
2. J. Lepault, F.P. Booy and J. Dubochet, (1982) J. Microsc. in
   press.
3. M.J. Costello and T. Gulik-Krzywicki, Biochim. Biophys. Acta,
   455, 412 (1976).
4. T. Gulik-Krzywicki and M.J. Costello, J. Microsc., 112, 103
   (1978).
5. L.P. Aggerbeck and T. Gulik-Krzywicki, J. Microsc., 126, 243
   (1982).
6. E.L. Benedetti and P. Favard, Editors, "Freeze-Etching, Technique and Applications, "Société Française de Microscopie Electronique, Paris, 1973.
7. F. Franks, J. Microsc., 111, 3 (1977).
8. M. Mueller, N. Meister and H. Moor, Mikroscopie (Wein), 36, 129,
   (1980).
9. L. Bachmann and W.W. Schmitt-Fumian, in "Freeze-Etching Technique and Applications", E.L. Benedetti and P. Favard, Editors,
   p. 73, Société Française de Microscopie Electronique, Paris,
   1973.
10. J.E. Heuser, T.S. Reese, M.J. Denis, Y. Jan, L. Jan and L. Evans,
    J. Cell. Biol., 81, 275 (1979).
11. H. Moore, in "Proceedings of Electron Microscopy Society of
    America, "G.W. Bailey, Editor, p. 334, Claitor's Publishing
    Division, Baton Rouge, 1977.
12. M.J. Costello, R. Fetter and M. Höchli, J. Microsc., 125, 125
    (1982).
13. V.B. Sleytr and A.W. Robards, J. Microsc., 110, 1 (1977).
14. J. Lepault and J. Dubochet, J. Ultrastruct. Res., 72, 223
    (1980).
15. R. Abermann, M.M. Salpeter and L. Bachman, in "Principles and
    Techniques of Electron Microscopy", M.A. Hayat, Editor, Vol. 2,
    p. 197, Van Nostrand Reinhold, New York, 1972.
16. A. Gulik, L.P. Aggerbeck, J.C. Dedieu and T. Gulik-Krzywicki,
    J. Microsc., 125, 207 (1982).
17. A. Gabriel and Y. Dupont, Rev. Sci. Instrum., 43, 1600, (1972).

18. L.G. Dowell, S.W. Moline and A.P. Rinfret, Biochim. Biophys. Acta, 59, 158 (1962).
19. V. Luzzati, in "Biological Membranes", D. Chapman, Editor, p. 71, Academic Press, New York, 1968.
20. T. Gulik-Krzywicki, L.P. Aggerbeck and K. Larsson, in "Symp. Surfactants in Solution", Editor, K.L. Mittal, in press.
21. G. Lindblom, K. Larsson, L. Johansson, K. Fontell and S. Forsen, J. Am. Chem. Soc., 101, 5465, (1979).
22. K. Larsson, K. Fontell and N. Krog, Chem. Phys. Lipids, 27, 321 (1980).
23. S.W. Hui and L.T. Boni, Nature, 296, 175 (1982).
24. W.P. Williams, A. Sen, A.P.R., Brain, P.J. Quinn and M.J. Dickens, Nature, 296, 175 (1982).
25. A.J. Verkleij, C. Monbers, J. Bijvelt-Leunissen and P.J.J.Th. Ververgaert, Nature, 279, 162 (1979).
26. T. Gulik-Krzywicki, M. Yates and L.P. Aggerbeck, J. Mol. Biol. 131, 475 (1979).
27. L.P. Aggerbeck, M. Yates and T. Gulik-Krzywicki, Ann. N.Y. Acad. Sci. 348, 352 (1980).
28. M. Le Maire, J.V. Møller and T. Gulik-Krzywicki, Biochim. Biophys. Acta, 643, 115 (1981).
29. T. Gulik-Krzywicki, Biochim. Biophys. Acta, 415, 1 (1975).
30. J.P. Segrest, T. Gulik-Krzywicki and Ch. Sardet, Proc. Natl. Acad. Sci. U.S., 71, 3294 (1974).
31. D. Brethes, N. Averet, T. Gulik-Krzywicki and J. Chevallier, Arch. Biochem. Biophys., 210, 149 (1981).
32. T. Gulik-Krzywicki, M. Balerna, J.P. Vincent and M. Lazdunski, Biochim. Biophys. Acta, 643, 101 (1981).
33. N. Morel, R. Manaranche, T. Gulik-Krzywicki and M. Israel, J. Ultrastr. Res., 70, 347 (1980).
34. M. Israel, R. Manaranche, N. Morel, J.C. Dedieu, T. Gulik-Krzywicki and B. Lesbats, J. Ultrastr. Res. 75, 162 (1981).
35. M. Israel, R. Manaranche, B. Lesbats and T. Gulik-Krzywicki, in "Advances in the Biosciences", P. Lechat, Editor, vol. 35, p. 173, Pergamon Press, Oxford and New York, 1982.

THE NATURE OF INTRAMEMBRANEOUS PARTICLES

A.J. Verkleij

Institute of Molecular Biology
State University of Utrecht
8 Padualaan, 3584 CH Utrecht, The Netherlands

INTRODUCTION

Freeze-fracturing electron microscopy has become a very popular method in membrane biology over the last decade and it is also of great value in the study of the structural details of aqueous lipid dispersions. This can simply be attributed to a faithful physical fixation method and the replica method, both of which exclude the artifacts frequently introduced in this section (positive staining) and negative staining electron microscopy (Zingsheim, 1972). At present freeze fracturing is technically straightforward and the reproducibility is high. Moreover, the fracturing process intrinsic to the method is now reasonably well understood (Verkleij and Ververgaert, 1978).

THE TECHNIQUE OF FREEZE FRACTURING

As many reviews have been written about the freeze-etch technique (Mühlethaler, 1971; Branton, 1971; Moor, 1971; Zingsheim, 1972; Bullivant, 1974; Verkleij and Ververgaert, 1975; Zingsheim and Plattner, 1976; Sleytr and Robards, 1977; Verkleij and Ververgaert, 1978) I shall only describe the crucial points of the method, which are essential for a better understanding of the micrographs. I shall also discuss the pit falls and limitations relevant for the study of biological and artificial membranes.

Three important steps can be distinguished in the freeze etch or freeze-fracture etch method: (i) rapid freezing or quenching; (ii) fracturing and, if required, etching (sublimation of ice); (iii) shadowing and replication.

In the first step the specimen is quenched in freon, propane
or a mixture of liquid and solid nitrogen. Ideally, fixation should
proceed in such a way that no redistribution of molecules takes place
during the quenching process. In a combined X-ray diffraction and
freeze-fracture study Deamer et al. (1970) found that several lipid-
water phases do not transform during the quenching. On the other
hand, it is clear from other studies (Dupont et al., 1972; Gulik-
Krzywicki and Costello, 1978) that, especially with pure phospholipids, disorder-order transitions may not be preserved with the
current quenching methods. This has been confirmed experimentally in
a study with pure saturated phosphatidylcholines by Ververgaert et
al. (1973b). They found that the L_α phase of dimyristoyl- and dipalmitoylphosphatidylcholine, which have transitions to the $P_{\beta'}$ phase at
23ºC and 41ºC, respectively, could not be preserved if quenched
from above the phase transition temperature with the current quenching procedures. However, with the ultra-rapid quenching method of
'spray-freezing', introduced by Bachmann and Schmidt (1971), the L_α
phase can be preserved (Ververgaert et al., 1973b). A second example
is the preservation of hexagonal II phases in dispersions of unsaturated phosphatidylethanolamines, which exhibit a temperature-dependent transition from a hexagonal II to a lamellar orientation. It has
been found extremely difficult to preserve the hexagonal II phase
when such a transition occurs above 20ºC. In this case the spray-
freezing method (Bachmann and Schmidt, 1971) cannot be applied, as
hexagonal II phase lipids form a sticky precipitate which cannot
pass through the diaphragm of the spray gun. However, with the jet-
freezing method introduced by Müller et al. (1980) this problem can
be solved (Van Venetië et al., 1981).

Next to internal alterations in the lipid organization current
quenching methods do not completely prevent ice crystal formation,
which can introduce gross morphological damage by squeezing the lipid
vesicles together. When studying sizes of small vesicles such phenomena are particularly destructive. In these cases one may employ
cryoprotectants, such as glycerol or dimethylsulphoxide (DMSO). However, care must be taken that these agents do not introduce changes
in the physical properties of lipids (Papahadjopoulos et al., 1976;
Boni et al., 1981) and consequently in the structure of the lipids.

In conclusion, it is recommended to use as a control one of the
ultra-rapid quenching methods currently available, e.g. spray freezing (Bachmann and Schmidt, 1971), the 'sandwich' method (Gulik-
Krzywicki and Costello, 1978), the copper-block method (Heuser et al.,
1979) or the jet-freezing method (Müller et al., 1980).

In the second step of the procedure the specimen is fractured
under high vacuum (< 10^{-6} Torr). Fracturing at a specimen temperature
lower than -120ºC, or fracturing at a higher temperature followed by
immediate replication, will generate two fracture faces of a membrane. At present, it is widely assumed that these fracture faces

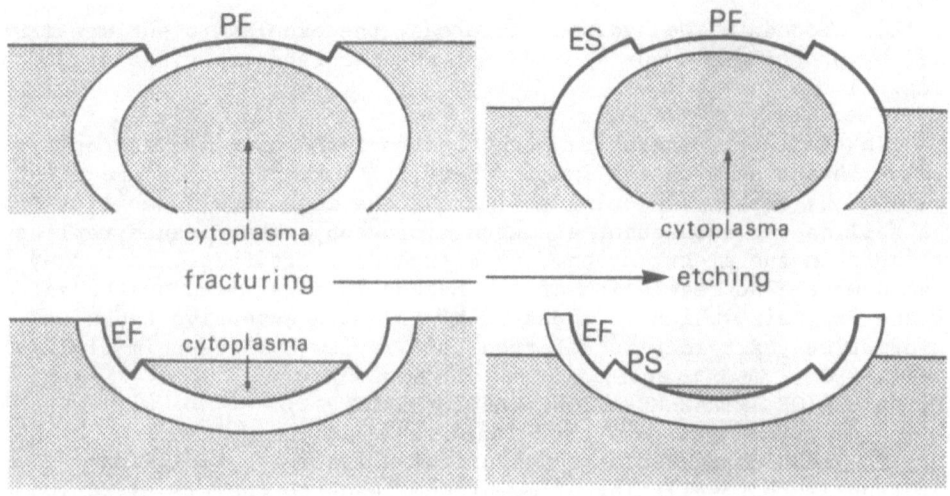

Fig. 1. Rationale of the fracturing and etching of a membrane with the nomenclature: protoplasmic fracture face (PF), exoplasmic fracture (EF), exoplasmic surface (ES) protoplasmic surface (PS).

are the result of a splitting of the membrane into two monolayers (Branton, 1966; Pinto da Silva and Branton, 1970). The fracture plane is determined by the hydrophobic interface, i.e., it runs between the terminal CH_3 groups of the acyl chains of the phospholipids (Fig. 1). One fracture face reveals the hydrophobic side of the outer mono-layer and the other fracture face the hydrophobic side of the inner monolayer, i.e., the exoplasmic fracture face (EF) and the protoplas-mic fracture face (PF), respectively, according to the proposed nomenclature (Branton et al., 1975).

A crucial requirement for accurate replication is that no con-tamination takes place, i.e., there must be no condensation of gaseous components onto the fracture faces before shadowing. It ap-pears that the main contaminating component is water (Gross et al., 1978). To prevent water contamination a moisture trap is required adjacent to the specimen during the fracture procedure. This has been achieved in the Denton apparatus by a cooled shroud at -196°C and in the Balzers machine by a cooled knife at -196°C. Alternatively, con-densation of water vapour can be prevented by fracturing of the spe-cimen in ultra-high vacuum (Gross et al., 1978).

Etching or sublimation of frozen water occurs by fracturing at -100°C and leaving an interval between fracturing and replication.

By this procedure the hydrophilic areas, the exoplasmic surface (ES) and the protoplasmic surface (PS) of the membrane are visualized too (Fig. 1).

In the last preparative step the fractured specimen is replicated by shadowing with platinum/carbon (Pt/C) at angles of about 45°C, followed by carbon shadowing to improve the mechanical stability of the replica for subsequent electron microscopy. The cleaned replica is inert in the electron beam, thus excluding artifacts induced by beam damage. The resolution of the method is determined by the size of the Pt grain, which is about 30 Å. For more extensive technical information the reader is referred to a review by Zingsheim (1972).

THE NATURE OF INTRAMEMBRANEOUS PARTICLES (IMP's)

The most characteristic feature of a fractured biological membrane is that splitting of membranes leads to the visualization of particles on a smooth background. It is generally assumed that the smooth surface reflects lipids which are organized in a bilayer configuration. This may be deduced from the facts that liposomes exhibit generally smooth fracture faces and that partial lipid removal from membranes by for instance ether extraction leads to an almost removal of the smooth areas between the IMP's (Cullis and Grathwohl, 1977). However, the absence of complementarity - smooth areas are found complementary to particles - asks for caution in interpreting the smooth areas merely on the acyl moiety of a lipid monolayer. In this respect attention should be paid to the finding that on the exoplasmic fracture face of the erythrocyte membrane large holes can be etched (Pinto da Silva et al., 1974). Only frozen water has a vapour pressure sufficient to account for sublimation at -100°C under convential vacuum conditions.

At present the topography of the IMP's, i.e., the number of particles on PF and EF or density, their lateral distribution and their organization in specialized structures have been reported and proposed as being important in many biomembrane processes (see review Verkleij and Ververgaert, 1978). In fact, each membrane has a specific freeze-fracture topography. Number, size and shape of the IMP and their distribution over both fracture faces as well as laterally can vary dramatically. This already leads us to the conclusion that each membrane is morphologically different as has been shown biochemically. It is even clear that in differentiated cells like the epithelial cell, the apical membrane is different from the basolateral membrane both in freeze-fracture morphology, biochemically and functionally. These different membranes are laterally separated by tight junctions (Steahelin and Hull, 1978). Moreover, a membrane can exhibit distinct areas in the membranes which have been called membrane specialization. Well-known examples of such specialized areas are the purple membrane of Halobacterium halobium, and the gap junction which can be isolated and analysed.

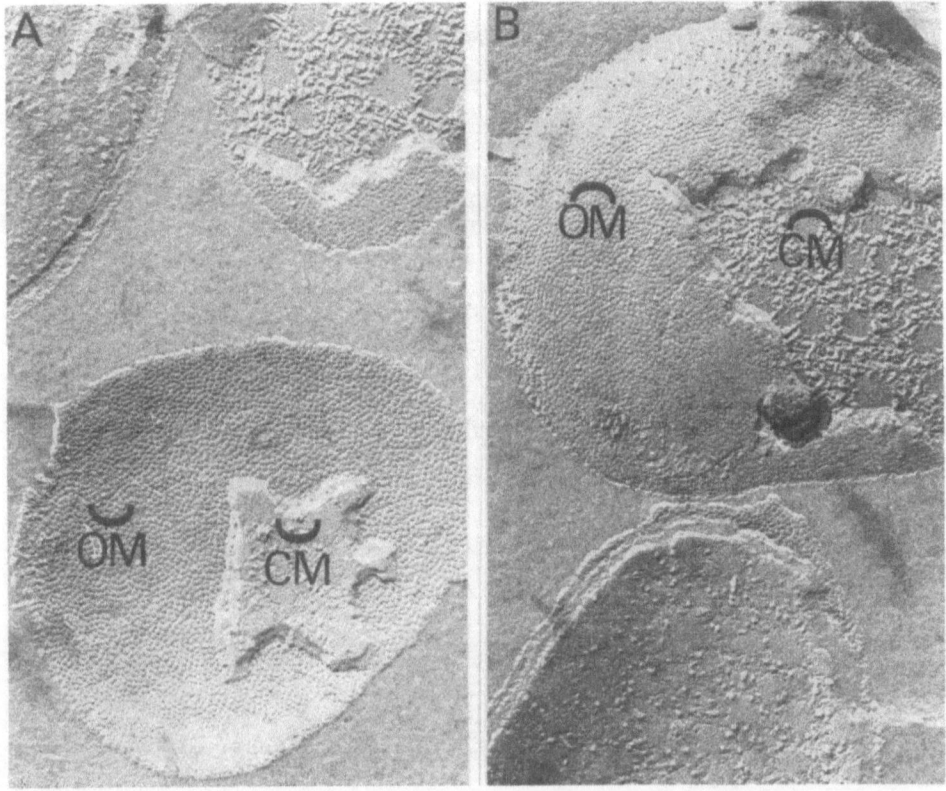

Fig. 2. Freeze-fracture morphology of the outer membrane of an ompA
mutant of E. coli strain B^E. The concave (outer) fracture
face of the outer membrane (OM) is almost covered with particles (A), whereas the convex (inner) fracture face (OM)
contains numerous pits (B), presumably complementary to the
particles. Cells were quenched at 4°C (x 120,000).

With respect to the nature of the intramembrane particles (IMP's)
it is generally assumed that they represent membrane intrinsic (membrane spanning) proteins. This can also be deduced from the fact that
recombination of pure lipids and intrinsic proteins exhibit fracture
faces with IMP's similar as in the native membrane. Also labelling
studies at the surface with antibodies linked to ferritin suggest
that particles of the erythrocyte membrane are proteins in nature,
i.e., band III and glycophorin (Pinto da Silva and Nicolson, 1974).
However, it has become clear that not all IMP's consist exclusively
of proteins but can be complexes of protein plus lipid, as found for
the IMP's of the outer membrane of Escherichia coli. It has been
shown on the basis of a variety of mutants of this gram-negative bacteria that aggregates of lipopolysaccharides and the pore protein are

responsible for the framework of the IMP's (Verkleij and Ververgaert, 1978). Of particular interest is the fact that these particles have complementary pits on the opposite fracture face (Fig. 2) in contradiction to the particles which are almost exclusively protein. These latter particles, which do not show complementarity, are the result of plastic deformation during the fracturing. From this complementarity aspect it was hypothesized that IMP's may partially reflect lipid in case complementary pits have been found (Verkleij and Ververgaert, 1978). Recently we have shown in fact that IMP's of pure lipid can occur in artificial lipid systems, the so-called "lipidic particles" (Verkleij et al., 1979). Such particles indeed have complementary pits on the opposite fracture face. Whether such particles indeed occur in biomembranes is still the question, but the phenomenon of lipidic particles has certainly attributed to the alternative concept that tight junctions are pure lipidic structures (Kachar and Reese, 1982).

REFERENCES

Bachmann, L. and Schmidt, W.W., 1971, Weniger Artefacten in der Gefrierätzung durch erhöhte Einfriergeschwindigkeit, Proc. Natl. Acad. Sci. USA, 68: 2149.

Boni, L.T., Stewart, T.P., Aldorfer, J.L. and Hui, S.W., 1981, Lipid polyethylene glycol interactions. -I. Induction of fusion between liposomes, J. Membr. Biol., 62: 65.

Branton, D., 1966, Fracture faces of frozen membranes, Proc. Natl. Acad. Sci. USA, 55: 1048.

Branton, D., 1971, Freeze etch studies of membrane structure, Phil. Trans. Roy. Soc. Lond. B, 261: 133.

Branton, D., Bullivant, S., Gilula, N.B., Karnovsky, M.J., Moor, H., Mühlethaler, K., North, D.H., Parker, L., Satir, B., Satir, P., Speth, V., Staehelin, J.A., Steere, R.L. and Weinstein, R.S., 1975, Freeze-etching nomenclature, Science, 190: 54.

Bullivant, S., 1974, Membranes, freeze-etching techniques applied to biological membranes, Phil. Trans. Roy. Soc. Lond. B, 268:

Cullis, P.R. and Grathwohl, A., 1977, Hydrocarbon phase transitions and lipid-protein interactions in the erythrocyte membrane. A ^{31}P NMR and fluorescence study, Biochim. Biophys. Acta, 471: 213.

Deamer, D.W., Leonard, R., Tardieu, A. and Branton, D., 1970, Lamellar and hexagonal lipid phases visualized by freeze etching, Biochim. Biophys. Acta, 219: 47.

Dupont, Y., Gabriel, A., Chabre, M., Gulik-Krzywicki, T. and Schechter, E., 1972, Use of new detector for X-ray diffraction and kinetics of the ordering of the lipids in E. coli membranes and model systems, Nature, 238: 331.

Gross, H., Bass, E. and Moore, H., 1978, Freeze-fracturing in ultrahigh vacuum at -196°C, J. Cell Biol., 76: 712.

Gulik-Krzywicki, T. and Costello, M.J., 1978, The use of low temperature X-ray diffraction to evaluate freezing methods used in freeze-fracture electron microscopy, J. Microsc., 112: 103.

Hauser, J.E., Reese, T.S., Dennis, M.J., Jan, Y., Jan, L. and Evans, L., 1979, Synaptic vesicle exocytosis captured by quick freezing and correlated with quantal transmitter release, J. Cell Biol., 81: 275.

Kachar, B. and Reese, T.S., 1982, Evidence for the lipidic nature of tight junction strands, Nature, 296: 464.

Moor, H., 1971, Recent progress in the freeze etching technique, Phil. Trans. Roy. Soc. Lond. B, 261: 121.

Mühlethaler, K., 1971, Studies on freeze etching of cell membranes, Int. Rev. Cytol., 31: 1.

Müller, M., Meister, N. and Moor, H., 1980, Freezing in a propane-jet and its application in freeze fracturing, Mikroskopie (Wien) 36: 129.

Papahadjopoulos, D., Hui, S., Vail, W.J. and Poste, G., 1976, Studies on membrane fusion. -II. Induction in pure phospholipid membranes by calcium ions and other divalent metals, Biochim. Biophys. Acta, 448: 245.

Pinto da Silva, P. and Branton, D., 1970, Membrane splitting in freeze-etching: covalently bound ferritin as a membrane marker, J. Cell Biol., 45: 598.

Pinto da Silva, P. and Nicolson, G.L., 1974, Freeze-etch localization of concanavalin A receptors to the membrane intercalated particles in human erythrocyte membrane, Biochim. Biophys. Acta, 363: 311.

Pinto da Silva, P., Moss, P.S. and Fudenberg, H.H., 1974, Anionic sites on the membrane intercalated particles of human erythrocyte ghost membranes, Exp. Cell Res., 81: 127.

Sleytr, U.B. and Robards, A.W., 1977, Understanding the artefact problem in freeze fracture replication, J. Microsc., 110: 1.

Steahelin, L. and Hull, B.E., 1978, Junctions between living cells, Sci. Am., 238: 140.

Van Venetië, R., Hage, W.J., Bleumink, J.G. and Verkleij, A.J., 1981, Propane jet-freezing. A valid rapid freezing method for the preservation of temperature dependent lipid phases, J. Microsc., 123: 287.

Verkleij, A.J. and Ververgaert, P.H.J.Th., 1975, The architecture of biological and artificial membranes as visualized by freeze fracturing, Ann. Rev. Phys. Chem., 26: 101.

Verkleij, A.J. and Ververgaert, P.H.J.Th., 1978, Freeze fracture of biological membranes, Biochim. Biophys. Acta, 515: 303.

Verkleij, A.J., Mombers, C., Leunissen-Bijvelt, J. and Ververgaert, P.H.J.Th., 1979, Lipidic intramembraneous particles, Nature, 279: 162.

Ververgaert, P.H.J.Th., Verkleij, A.J., Verhoeven, J.J. and Elbers, P.F., 1973, Spray freezing of liposomes, Biochim. Biophys. Acta, 311: 651.

Zingsheim, H.P., 1972, Membrane structure and electron microscopy. The significance of physical problems and technics (freeze-etching), Biochim. Biophys. Acta, 265: 339.

Zingsheim, H.P. and Plattner, H., 1976, Electron microscopic methods
in membrane biology, in: "Methods in Membrane Biology", Vol. 7,
(Korn, E.D., ed.), Plenum Publ. Co., New York, p. 1-146.

LATERAL AND POLYMORPHIC PHASE TRANSITIONS IN RELATION

TO THE BARRIER FUNCTION OF A LIPID MEMBRANE

J. de Gier, C.J.A. van Echteld, J.A. Killian, B. de
Kruijff, J.G. Mandersloot, P.C. Noordam, A.T.M. van der
Steen and A.J. Verkleij

Department of Biochemistry, State University of Utrecht
Padualaan 8, 3584 CH Utrecht, The Netherlands

INTRODUCTION

The lipid matrix of a biomembrane is built with a complex mixture of lipids showing defined variations in polar headgroups and
paraffin chains. For membranes with different functions different
combinations of lipid species are used. The significance of this lipid
diversity and specificity is not yet well understood. It is the aim
of this contribution to discuss properties of individual lipid molecules and to show how these properties work out in the barrier functions of membrane model systems. An important aspect will be to see
how these barrier functions can be modified by environmental conditions and by intrinsic protein elements. As a consequence of the
combination of a polar and an apolar part in the same molecule membrane lipids spontaneously form aggregates when they are dispersed
in water. The molecular organization of such aggregates is dependent
on the molecular shape of the constituting molecules. Lipids which
have a good balance between the size of the polar group and the size
of the hydrophobic part easily form bilayer structures (e.g. phosphatidylcholines, phosphatidylglycerols, sphingomyelins). This bilayer
organization is in accordance with the lipid organization in the
generally accepted molecular models of biomembranes[1]. However, it is
relevant to emphasize that in each biological membrane also large
fractions of lipids occur which in pure form in water adopt other
types of molecular arrangements. Figure 1 illustrates that lipid
molecules with a polar head of limited size and hydration (e.g. unsaturated phosphatidylethanolamines, monoglucosyldiglycerides, cardiolipin in the presence of Ca^{2+}) prefer an organization of hexagonal
packed cylinders, in which the polar heads of the molecules are
directed to the inside of the cylinders. Lipid molecules in which

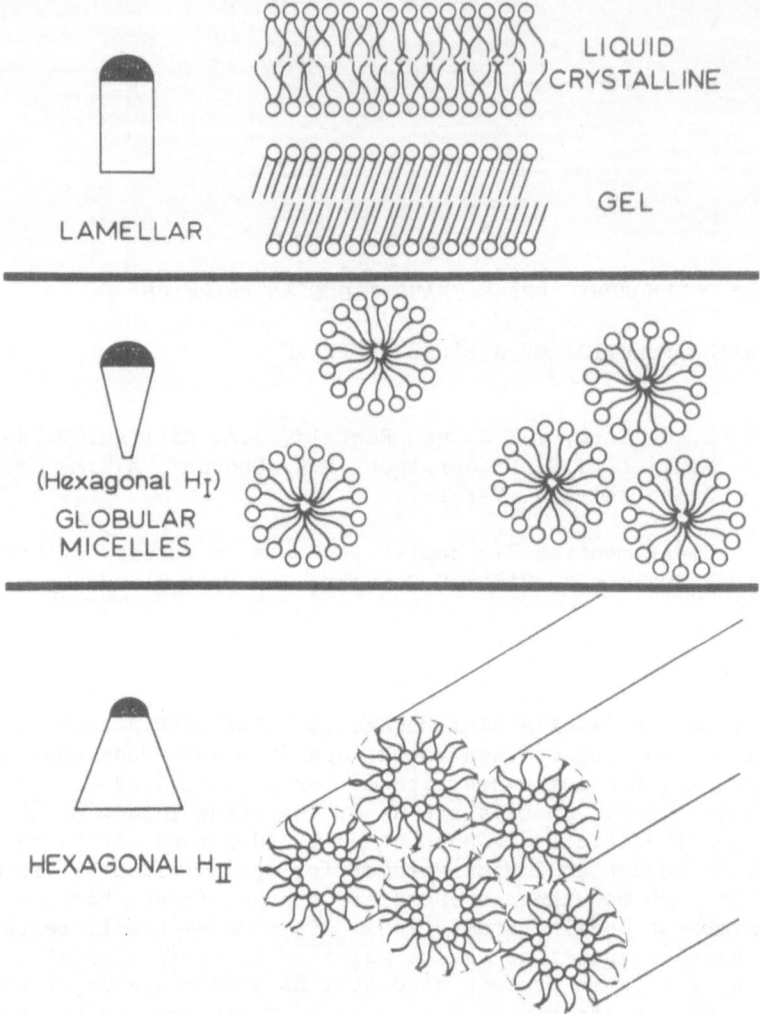

LIQUID CRYSTALLINE

GEL

LAMELLAR

(Hexagonal H$_I$)
GLOBULAR
MICELLES

HEXAGONAL H$_{II}$

Fig. 1. Molecular association structures of membrane lipids in water
as determined by the balance in size of polar and apolar part
of the molecules.

the polar headgroups are dominating (e.g. lysolecithins, gangliosides
form a reversed hexagonal (H$_I$) organization upon hydration with a
limited amount of water, which is converted into a system of globular
micelles in excess of water.

THE LIPID BILAYER A SELECTIVE PERMEABILITY BARRIER

When phosphatidylcholines are dispersed in a buffer solution
thermodynamically stabilized bilayer structures (liposomes) are

formed. Depending on the dispersion technique multilayered structures or one-walled vesicles of different size can be obtained. From osmotic volume changes of the liposomes in response to concentration changes in the outside medium it can be concluded that the bilayer is a selective permeability barrier; impermeable to ions and large polar molecules, but permeable to water and small non-electrolytes[2]. The molecular motion which enables this selective permeability is to be found in the fluid character of the paraffin chains in these bilayers. Trans-gauche conformation changes cause kink formation in the paraffin chains and result in the formation of small open cavities which can be filled with single water (or non-electrolyte) molecules. As the kinks are mobile along the chains, this process acts as an intrinsic carrier system for the passage of water molecules through the membrane[3]. Factors like chain length, degree of unsaturation and the presence or absence of sterols in the bilayer, which are known to affect the fluidity, influence the extent of the permeability in a predictable way, such that increase in fluidity increases the rate of permeation. An illustration experiment[4] is given in Fig. 2. Osmotic sensitive multilayered liposomes of dipalmitoylphosphatidylcholine liposomes are prepared by dispersing the pure lipid with an addition of 4 mole % phosphatidic acid at a temperature of 50°C in 50 mM KCl.

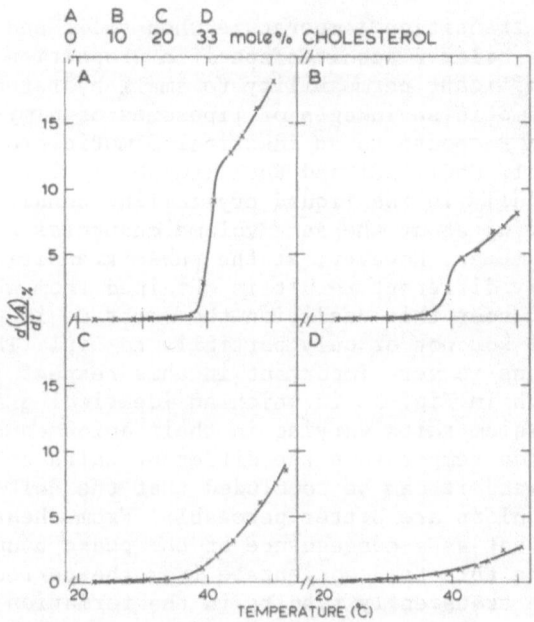

Fig. 2. Temperature dependence of the osmotic shrinkage of dipalmi-toylphosphatidylcholine liposomes, following the addition of hypertonic glucose. The liposomes contained 0 (A), 10 (B), 20 (C) and 33 (D) mole % of cholesterol.

A relative measure of the water permeability can be obtained by measuring the rate of osmotic shrinkage in response to an applied glucose gradient. Above the temperature of the phase transition, which occurs at about 40°C, the liquid crystalline bilayer is well permeable to water but upon transition to the gel state this permeability is largely reduced, although a slow but significant osmotic response also below the transition indicates that the immobilization of the chains is not complete. When cholesterol is incorporated into the dipalmitoyllecithin bilayer considerable change in water permeability can be noticed which reflects the effect of the cholesterol on the membrane fluidity. Below 40°C an increase in permeation rate of water can be noticed which indicates an increase in membrane fluidity but above the transition temperature a large decrease in water permeability is found which points to a large decrease in membrane fluidity. The result is a gradual disappearance of the abrupt change at the phase transition temperature when the amount of cholesterol increases. These observations are in agreement with conclusions about membrane viscosity which have been made from spectroscopic and calorimetric measurements[4].

IONIC PERMEABILITY AT THE GEL TO LIQUID CRYSTALLINE PHASE TRANSITION

At the phase transition temperature when solid and liquid patches are coexisting the bilayer acquires specific properties which become apparent as a significant permeability to small hydrated ions[5]. Figure 3A shows osmotic shrinkages of liposomes of dimyristoylphosphatidylcholine in response to an identical osmotic gradient obtained by addition of LiCl, NaCl, KCl and RbCl, respectively. At 35°C when the bilayers are fully in the liquid crystalline condition, the different salts bring about the same volume change as can be expected for a perfect osmometer. However, at the phase transition temperature (24°C) a completely different result is obtained from which it can be concluded that under this condition the membrane is permeable to NaCl, KCl and RbCl but not or only partially to LiCl. That the size of the hydrated ions is very important in this respect is confirmed by the result shown in Fig. 3B in which an identical gradient is applied with potassium salts varying in their anion constituent. At the phase transition temperature the different salts cause different osmotic responses and it can be concluded that the salts with the smaller hydrated anions are better permeable. From these results it can be suggested that as a consequence of the phase boundary, the lipid molecules are able to disorientate from the perfect bilayer arrangement, which transiently results in the formation of pore-like structures large enough for small hydrated ions to pass through. Disorientation from the bilayer orientation is also indicated by trans-bilayer (flip-flop) movement of the lipid molecules which occurs at the phase transition temperature, but is extremely low in the liquid crystalline state[6].

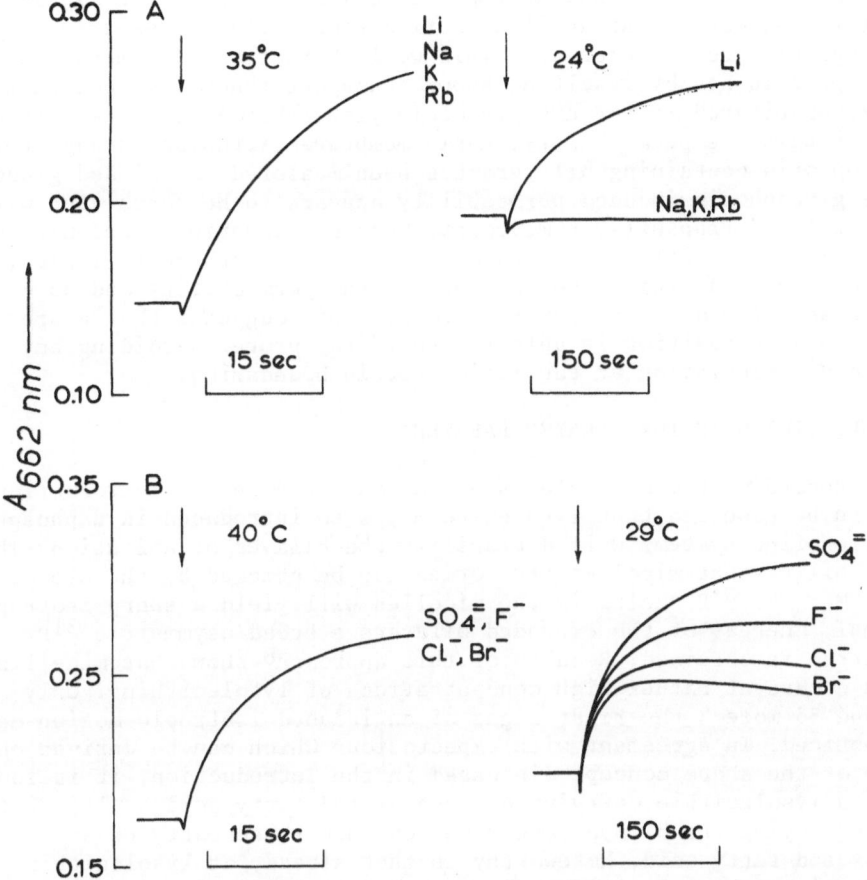

Fig. 3. Osmotic shrinkage of liposomes in response to the same osmotic gradient formed by different salts. (A) Liposome system:
dimyristoylphosphatidylcholine (96 mole %) and phosphatidic
acid (4 mole %); phase transition at 24°C. Salts used for the
gradient: LiCl, NaCl, KCl and RbCl. (B) Liposome system: dimyristoylphosphatidylcholine (72 mole %),dipalmitoylphospha-
tidylcholine (24 mole %) and phosphatidic acid (4 mole %);
phase transition at 29°C. Salts used for the gradient: KBr,
KCl, KF and K_2SO_4.

EFFECT OF RECONSTITUTION OF GLYCOPHORIN INTO A LECITHIN BILAYER

When proteins are reconstituted into liquid models this often
results in an increase of membrane permeability. This increase can
be due to the formation of polar channels by the proteins themselves,
comparable with those induced by the gramicidin, but it can also be

85

the consequence of transient lipid disorientations at the lipid-protein boundaries. As an illustration reconstitution experiments of glycophorin are of interest[7]. This well-characterized membrane spanning protein has by itself no known transport functions, but when it is reconstituted in for example dioleoylphosphatidylcholine bilayers, it enhances the permeability. Rapid membrane diffusion through the glycophorin containing bilayers has been measured for K^+ and glucose. This glycophorin-induced permeability appears to be correlated with a fast lipid transbilayer movement. When glycophorin is reconstituted into bilayer vesicles composed of total human erythrocyte lipids, there is only a minimal enhancement of the permeability and no measurable lipid transbilayer movement. This suggests that a specific lipid composition is able to "seal" the protein avoiding any lipid disorientation at the lipid-protein boundaries.

LYSOLECITHINS IN THE BILAYER BARRIER

Regarding the molecular shape as outlined in the Introduction, it can be expected that lysolecithins, when introduced in a phosphatidylcholine system, will destabilize the bilayer organization. The possibility that micelles are formed can be checked by the use of ^{31}P NMR. The ^{31}P nuclei in the micelles will yield a sharp isotropic signal, whereas of the extended bilayers a broad asymmetric ^{31}P spectrum is obtained. Results of this approach[8] show that micellization occurs at rather high concentrations of lysolecithins. Only around 35 mole % the first signs of an induced isotropic motion can be noticed. In agreement with expectations which can be derived on base of the shape concept discussed in the Introduction, it is found that a lysolecithin carrying an unsaturated fatty acid is less potent in destroying the bilayer organization than those carrying an saturated fatty acid. Noteworthy is that already at lysolecithin concentrations about 5 mole % below the concentration required to induce a measurable isotropic ^{31}P NMR signal the K^+ permeability barrier is affected. Figure 4 shows that also with respect to this permeability effect palmitoyllysolecithin is more potent than the unsaturated oleoylphosphatidylcholine, at least in liquid crystalline bilayers. In the gel state bilayer of dipalmitoylphosphatidylcholine a reversed effect can be noticed. The unsaturated oleoyllysolecithin is very potent in destroying the permeability barrier for K^+ in this gel state bilayer such in contrast to the palmitoyllysolecithin. This can be explained by a random distribution of the palmitoyllysophosphatidylcholine also in the gel state, whereas in case of oleoylphosphatidylcholine phase separation occurs. It is likely that the segregated lysolecithin easily forms polar channels in the bilayer organization already at very low concentration.

HEXAGONAL H_{II} PHASE PREFERRING LIPIDS IN THE BILAYER

When dioleoylphosphatidylethanolamine is dispersed in a cooled buffer liposomal structures are formed, but when the temperature is

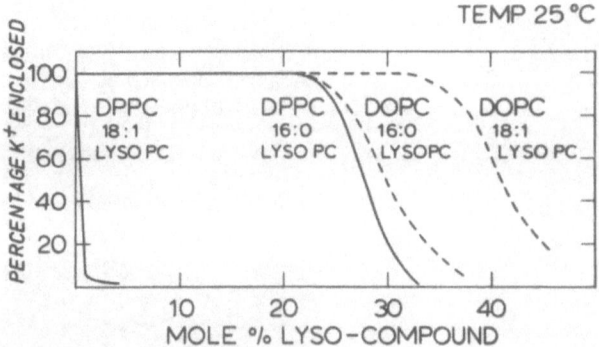

Fig. 4. Ability of liposomal structures with increasing concentrations of lysolecithins to maintain a permeability barrier
against enclosed K^+ when dialyzed in isotonic $MgSO_4$. The
liposomes are prepared of dipalmitoylphosphatidylcholine
(DPPC) or dioleoylphosphatidylcholine (DOPC) and the lysolecithins incorporated are respectively palmitoyllysophosphatidylcholine and oleoyllysophosphatidylcholine.

increased, the lamellar phase is converted to a hexagonal H_{II} phase.
Saturated phosphatidylethanolamines form stable bilayer organizations
at all temperatures, whereas dilineoylphosphatidylethanolamine gives
a stable hexagonal phase up to temperatures as low as -15°C. These
observations are easily explained in terms of the form concept discussed in the Introduction.

It will be obvious that a lamellar to hexagonal phase transition
affects the permeability barrier. Impermeable components enclosed in
liposomes of dioleoylphosphatidylethanolamine are completely released
upon transition to the hexagonal H_{II} phase[9]. The temperature at which
a membrane reorganization occurs can be affected by a number of factors. When dioleoylphosphatidylethanolamine is mixed with dioleoylphosphatidylcholine the bilayer organization is stabilized and the
temperature at which the permeability barrier for K^+ ions disappears,
shifts to higher values. When also cholesterol with its small polar
group and large hydrophobic body is added to the mixture, this temperature goes down again in agreement with the shape concept. However,
it is noteworthy that in the mixed systems the disappearance of the
permeability barrier is not correlated with a lamellar to a hexagonal H_{II} transition, but with a change from a smooth lamellar phase

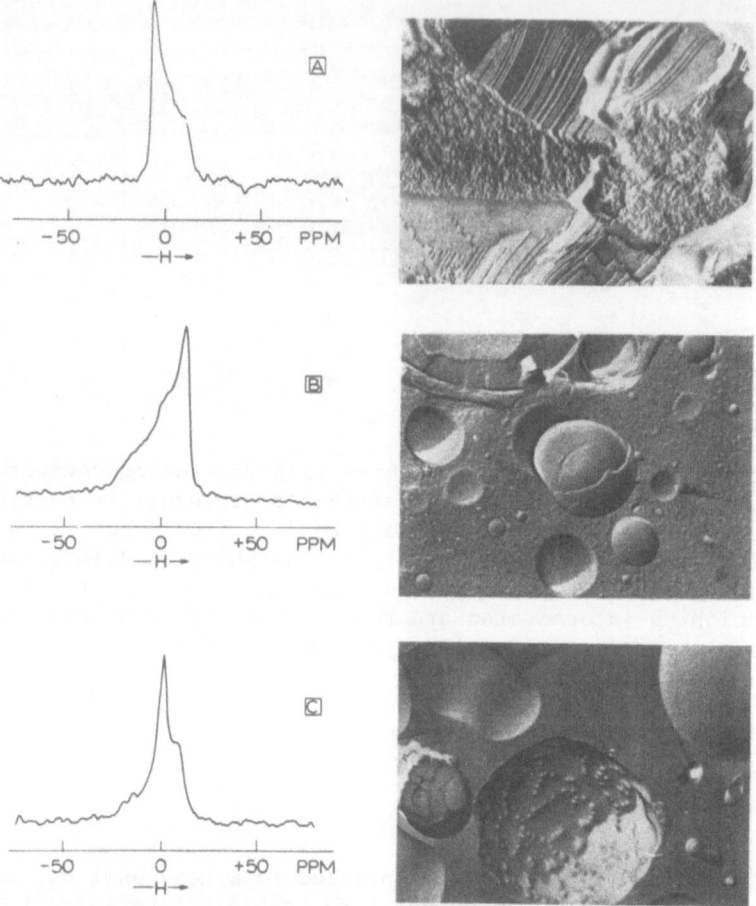

Fig. 5. ^{31}P NMR spectra and freeze-fracture electron microscopic
images of phospholipid dispersions. (A) Pure cardiolipin in
the presence of Ca^{2+} (hexagonal H_{II} phase). (B) Cardiolipin/
dioleoylphosphatidylcholine (1:1) in the absence of Ca^{2+}
(lamellar phase). (C) Cardiolipin/dioleoylphosphatidylcholine
(1:1) after addition of 2.5 mM Ca^{2+} (bilayer organization
with lipid particles).

to a lamellar phase with "particles". The membrane organization maintains but an increased dynamic state allows the formation of particles which can be noticed on electron microscopic pictures of freeze-
fracture faces of the mixed liposomes. The increased dynamic state
also must be responsible for the fact that ions can pass the barrier.
As a comparable example we discuss cardiolipin systems[10] of which the
structural information is given in Fig. 5. Cardiolipin, well-known as
one of the major lipid constituents of inner mitochondrial membranes,
easily forms bilayers when dispersed in water. This bilayer organi-

zation is transferred into a hexagonal H_{II} phase when Ca^{2+} is added (compare Fig. 5A). The Ca^{2+} ions form salt bridges between the two phosphate groups of the lipid molecule and reduce the hydration, thus inducing the conical shape. When an equimolar mixture of cardiolipin and dioleoylphosphatidylcholine is dispersed in water a multilayered liposome system is formed (compare Fig. 5B). When to this system 2.5 mM Ca^{2+} is added the barrier function of the bilayers is selectively affected. They become permeable to K^+ ions whereas the barrier with respect to large dextrane molecules is maintained[10]. In addition a rapid transbilayer movement of the lipid constituents can be measured. Freeze-fracture faces of the Ca^{2+}-treated liposomes reveal particles with a diameter of about 70 Å and ^{31}P NMR measurements show a relative sharp isotropic signal superimposed on a characteristic bilayer spectrum (compare Fig. 5C). This structural change has been taken as a strong indication for the formation of inverted micelles embedded in the hydrophobic interior of a bilayer organization[11]. The mechanism by which these inverted micelles can be formed is still subject of debate. In Fig. 6 two possibilities are indicated. The formation of an inverted micelle could be the result of a fusion event between

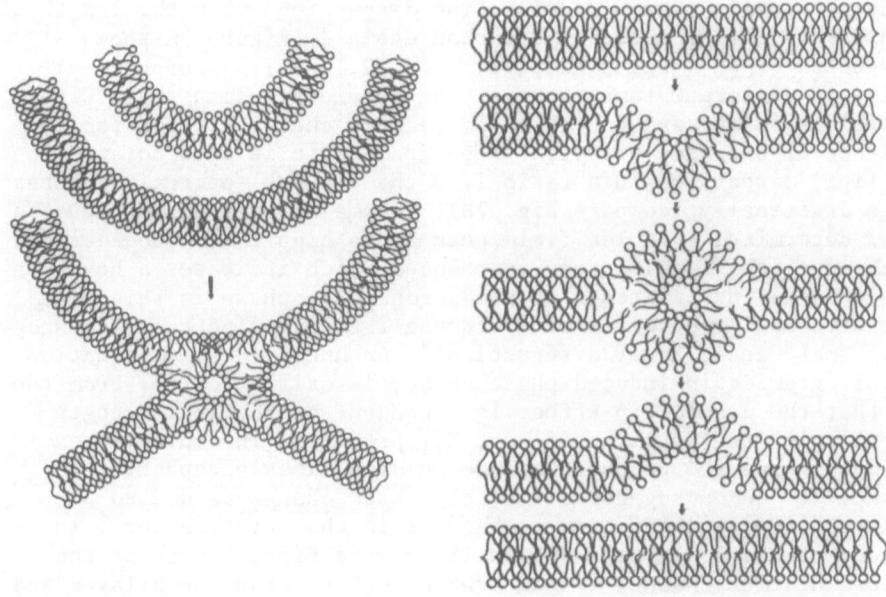

Fig. 6. Schematic presentation of possibilities by which inverted micelles may be formed.

two bilayers but it is also conceivable that the inverted micelles are spontaneously formed in the bilayer as a consequence of statical accumulation of cone shaped lipid molecules in the fluid bilayer. As small polar solutes could be enclosed in the interior of the inverted micelle a dynamic formation and resolution of such particles could offer a mechanism for ion translocation and transmembrane diffusion of lipid molecules.

PROTEIN CONTROLLED LIPID POLYMORPHISM

The foregoing experiments clearly illustrate that membrane lipids can adopt a variety of association structures. Although it appears that in mixtures a lamellar organization is often dominant, it is found that local and transient disorientations can have considerable effect on the barrier properties of such a membrane. In this respect it is important that intrinsic membrane proteins can have considerable influence on the lipid organization in that they can stabilize and destabilize the bilayer organization. We discussed already the ability of glycophorin to induce ion permeability and flip-flop diffusion of the lipid molecules in bilayers of dioleoylphosphatidylcholine. The same protein reconstituted in a system of phosphatidylethanolamine has been found to stabilize the bilayer organization and to inhibit a transition to the hexagonal H_{II} phase[12]. On the other hand it has been shown that gramicidin lowers the bilayer to hexagonal H_{II} phase transition temperature of phosphatidylethanolamines. This hexagonal H_{II} phase promoting ability of gramicidin also can be noticed in phosphatidylcholine dispersions when the length of the paraffin chains exceeds 16 carbon atoms[13]. Figure 7A shows ^{31}P NMR spectra of an aqueous dispersion of $22:1_c/22:1_c$-phosphatidylcholine at different temperatures. The asymmetric line shape with a low field shoulder and a high field peak is characteristic for the existence of extended bilayers. When gramicidin is incorporated in this lipid dispersion in a ratio 1:10 the ^{31}P NMR spectral features change drastically (compare Fig. 7B). The major part of the spectra is now determined by a low field peak and a high field shoulder and the chemical shift anisotropy is reduced which indicates a hexagonal organization. The existence of a hexagonal H_{II} phase in this mixed dispersion has been confirmed by freeze-fracture electron microscopy and by small angle X-ray diffraction[13]. An unequivocal explanation for this gramicidin-induced phase change is still lacking. From the fact that the gramicidin effect is dependent on the chain length of the lecithins (less significant H_{II} promotion in dioleoylphosphatidylcholine and no phase change in dipalmitoylphosphatidylcholine) it can be suggested that the phase change is due to a "mismatch" of gramicidin dimer channels in the paraffin core. In the bilayers of the long chain phospholipids the fixed length of the gramicidin dimer channel is not long enough to span the bilayer and could give rise to meniscus formation resulting into the established phase change. On the other hand it can be argued that the gramicidin monomer has a conical shape as bulky amino acids (tryptophans) are

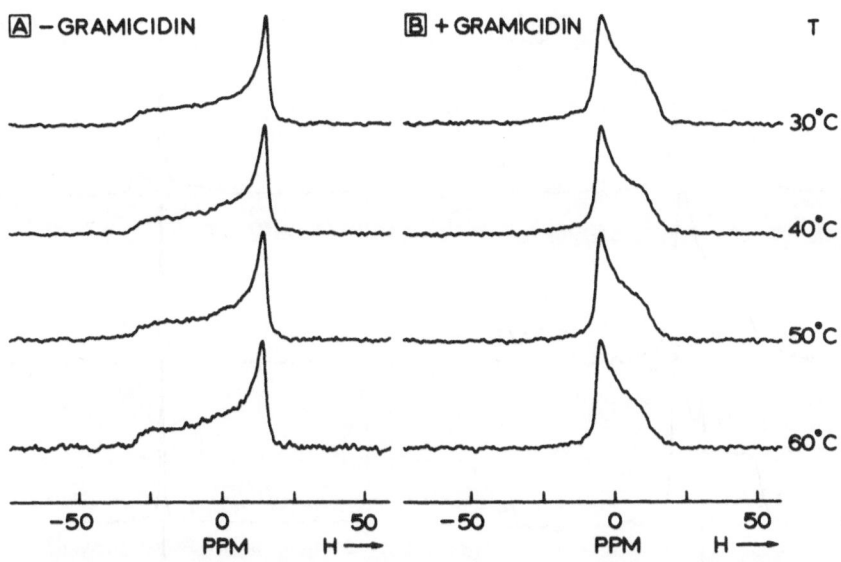

Fig. 7. ^{31}P NMR spectra of aqueous dispersions of 22:1$_c$/22:1$_c$-phosphatidylcholine and a mixture of gramicidin: 22:1$_c$/22:1$_c$-phosphatidylcholine = 1:10 at various temperatures.

concentrated to the carboxyl end of the peptide chain. Therefore if the gramicidin dissociates into monomers and orientates with the N-formyl end to the surface the hexagonal H$_{II}$ phase promoting effect could be explained according to the shape concept. That indeed a shape effect is involved in the gramicidin-lipid interaction is also supported by the finding that a mixture of gramicidin and lysophosphatidylcholine forms bilayers when dispersed in aqueous solutions. This is shown in Fig. 8 which presents ^{31}P NMR spectra of mixtures of palmitoyllysolecithin and increasing concentrations of gramicidin. Next to an isotropic signal which originates from the lysolecithin molecules in a micellar organization a typical broad asymmetric bilayer signal can be noticed which increases linearly with the increasing gramicidin concentration. From the results it can be concluded that each gramicidin molecule causes 4 lysolecithin molecules to adopt the bilayer organization.

CONCLUSION

The findings that each biomembrane contains large amounts of "non-bilayer" lipids and that protein elements can control (polymorphic) phase transitions suggest that the membrane permeability barrier is not the stable and unchangeable bilayer organization as

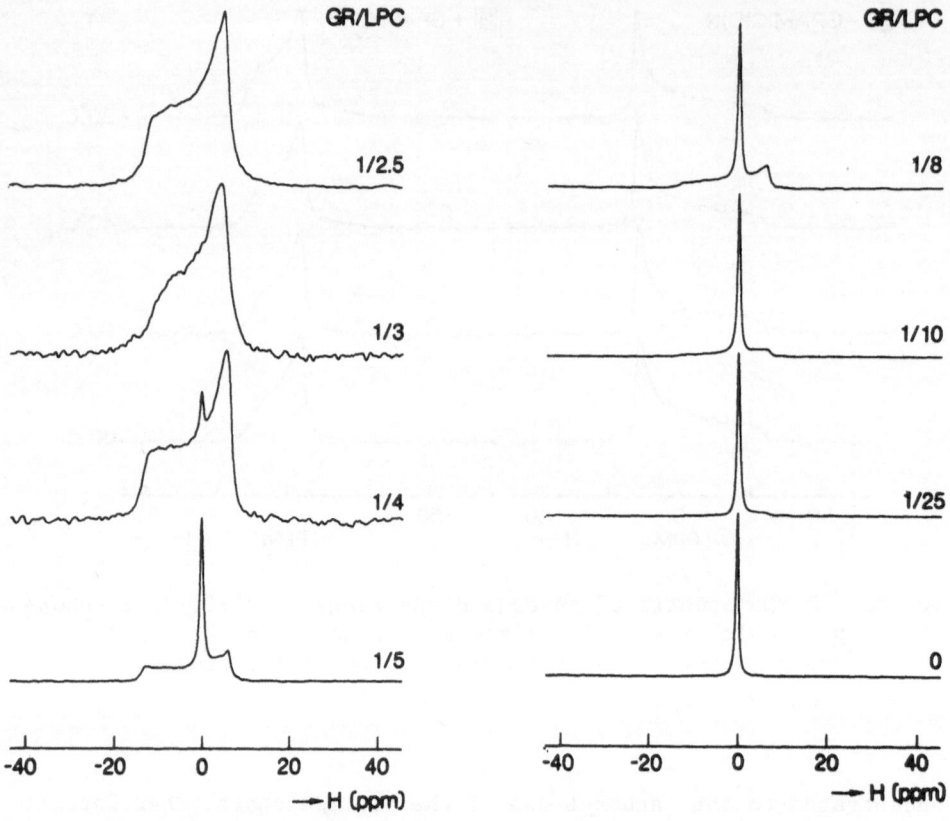

Fig. 8. ^{31}P NMR spectra of aqueous dispersions of gramicidin and palmitoyllysophosphatidylcholine at various molar ratios as indicated in the figure.

generally accepted, but that in response to environmental fluctuations and conformation changes of intrinsic membrane proteins local destabilizations of such an organization may occur which have implications for important membrane phenomena such as membrane fusion and membrane translocations.

REFERENCES

1. S.J. Singer and G.I. Nicholson, The fluid mosaic model of the structure of cell membrane, Science 175: 720 (1972).
2. J. de Gier, J.G. Mandersloot, J.V. Hupkes, R.N. McElhaney and W.P. van Beek, On the mechanism of non-electrolyte permeation through lipid bilayers and through biomembranes, Biochim. Biophys. Acta, 233: 610 (1971).
3. H. Träuble, The movement of molecules across lipid membranes. A molecular theory, J. Membrane Biol., 4: 193 (1971).

4. M.C. Blok. L.L.M. van Deenen and J. de Gier, The effect of cholesterol incorporation on the temperature dependence of water permeation through liposomal membranes prepared from phosphatidylcholines, Biochim. Biophys. Acta, 464: 509 (1977).

5. M.C. Blok, L.L.M. van Deenen and J. de Gier, Effect of the gel to liquid crystalline phase transition on the osmotic behaviour of phosphatidylcholine liposomes, Biochim. Biophys. Acta, 433: 1 (1976).

6. B. de Kruijff and E.J.J. van Zoelen, Effect of the phase transition on the transbilayer movement of dimyristoyl phosphatidylcholine in unilamellar vesicles, Biochim. Biophys. Acta, 511: 105 (1978).

7. A.T.M. van der Steen, B. de Kruijff and J. de Gier, Glycophorin incorporation increases the bilayer permeability of large unilamellar vesicles in a lipid-dependent manner, Biochim. Biophys. Acta, 691: 13 (1982).

8. C.J.A. van Echteld, B. de Kruijff, J.G. Mandersloot and J. de Gier, Effects of lysophosphatidylcholines on phosphatidylcholine and phosphatidylcholine/cholesterol liposome systems as revealed by ^{31}P NMR, electron microscopy and permeability studies, Biochim. Biophys. Acta, 649: 211 (1981).

9. P.C. Noordam, C.J.A. van Echteld, B. de Kruijff, A.J. Verkleij and J. de Gier, Barrier characteristics of membrane model systems containing unsaturated phosphatidylethanolamines, Chem. Phys. Lipids, 27: 221 (1980).

10. J.G. Mandersloot, W.J. Gerritsen, J. Leunissen-Bijvelt, C.J.A. van Echteld, P.C. Noordam and J. de Gier, Ca^{2+}-induced changes in the barrier properties of cardiolipin/phosphatidylcholine bilayers, Biochim. Biophys. Acta, 640: 106 (1981).

11. B. de Kruijff, A.J. Verkleij, C.J.A. van Echteld, W.J. Gerritsen, C. Mombers, P.C. Noordam and J. de Gier, The occurrence of lipidic particles in lipid bilayers as seen by ^{31}P NMR and freeze-fracture electron microscopy, Biochim. Biophys. Acta, 555: 200 (1979).

12. T.F. Taraschi, B. de Kruijff, A.J. Verkleij and C.J.A. van Echteld, Effect of glycophorin on lipid polymorphism, Biochim. Biophys. Acta,685: 153 (1982).

13. C.J.A. van Echteld, R. van Stigt, B. de Kruijff, J. Leunissen-Bijvelt, A.J. Verkleij and J. de Gier, Gramicidin promotes formation of the hexagonal H_{II} phase in aqueous dispersions of phosphatidylethanolamine and phosphatidylcholine, Biochim. Biophys. Acta, 648: 287 (1981).

FLUORESCENCE ENERGY TRANSFER AS A STRUCTURAL

PROBE IN MEMBRANES AND MEMBRANE-BOUND PROTEINS

W. E. Blumberg

Bell Laboratories
Murray Hill, New Jersey 07974, USA

INTRODUCTION

In 1972 Radda and Vanderkooi (Radda 72) wrote "Fluorescence has been used in biochemical studies for many years, yet it is only fairly recently that its potential to the study of problems associated with membrane-related phenomena has been realized," and in 1976 Badley (Badley 76) added "... it is only recently that fluorescence methods have been used in a more problem-oriented way in contrast to the, perhaps inevitable, initial rush of data primarily concerned with establishing the method as one of 'high potential'." I might begin by noting that as of 1982 only a handful of fluorescence experiments involving membranes have made full and rigorous use of the physical phenomena accessible to the fluorescence spectroscopist: quantum yield, spectral shifts, emission anisotropy, and resonance energy transfer including their multiple interactions. The first three of these phenomena have been treated in other lectures of this Advanced Study Institute. I will concentrate on energy transfer, but, as we shall see, it is quite useless and even misleading to try to analyze it without considering quantum yield and anisotropy, both in the steady state and time-resolved domains. Therefore, in the first few sections I will lay the groundwork necessary for the analysis of resonance energy transfer experiments.

A probe of a macromolecular system is defined to be a molecular group small relative to that system that provides information about the region in close proximity to its location. Fluorescent probes are able to reflect characteristics of their environments by various characteristics of the emission produced upon optical excitation. This localized nature of the information obtained has the advantage that one has a good chance of understanding it but the disadvantage that one can only ever obtain information about a limited part of the system. For the most part I will concentrate on the role played by fluorescent probes introduced into a system by the experimenter (so-called extrinsic probes), but there are many examples of membranes containing intrinsic probes which are already fluorescent. Applications of both types of probes to the study of membrane bound proteins and to the structure of membranes will be discussed in due course.

EMISSION DECAY RATES AND ENERGY TRANSFER RATES

The fluorescence quantum yield is a measure of the rate of photon emission k_f compared with the rate of photon absorption or

$$\phi_f = \frac{k_f}{k_f + k_{nr}},$$

where k_{nr}, the rate constant for nonradiative decay of the optically excited state, is the sum of all the internal conversion and intersystem crossing rate constants.

The decay of fluorescence following a very short pulse is found to be exponential in sufficiently simple systems and therefore can be given a characteristic lifetime. The lifetime may be defined as

$$\tau_f = 1/(k_f + k_{nr}).$$

This brief discussion of rates of dissipation of optical energy serves to introduce yet another rate: the rate of resonance energy transfer between a donor and an acceptor. Singlet-singlet resonance energy transfer involves the exchange of excited state energy over distances of 10-100Å from a donor D to a nearby acceptor chromophore A whose absorption spectrum overlaps the fluorescence spectrum of the donor. The phenomenon has been described by considering dipole-dipole near zone resonance interactions (Förster 48, 51, 59). The transfer rate k_T is given by

$$k_T = 8.71 \cdot 10^{23} \, \kappa^2 \, J_{\bar{\nu}} \, k_f \, / \, n^4 \, R^6$$

where

$$J_{\bar{\nu}} = \int_0^\infty \bar{\nu}^{-4} \, F_D(\bar{\nu}) \, e_A(\bar{\nu}) \, d\bar{\nu},$$

$\bar{\nu}$ is the wave number, $F_D(\bar{\nu})$ is the relative number of quanta emitted

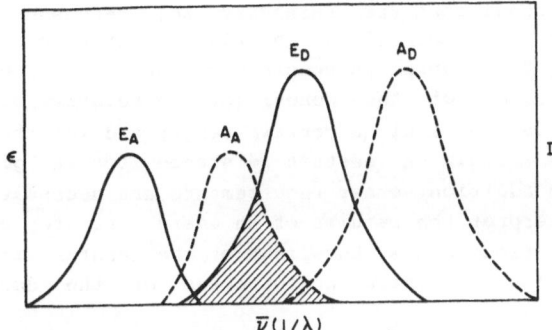

Fig. 1. Schematic diagram of the absorption and emission spectra of a typical donor-acceptor pair vs. $\bar{\nu}$ or $1/\lambda$. The overlap region used in the computation of the Förster radius for resonance energy transfer is shaded. The acceptor is shown to have a fluorescence spectrum; while a fluorescent acceptor makes analysis of the results simpler, it is not essential to the energy transfer process.

over wave number interval $d\bar{\nu}$ by the donor, $e_A(\bar{\nu})$ is the acceptor molar absorptivity at $\bar{\nu}$, n is the refractive index, R is the donor-acceptor separation, $J\bar{\nu}$ is the designated overlap integral (see Fig. 1), κ^2 is the orientation factor defined as $(\cos\theta_T - 3\cos\theta_D\cos\theta_A)^2$. As in Fig. 2, θ_T is the angle between transition moments in D and A, and θ_D and θ_A are the angles between the transition moments of the acceptor and the donor and the line joining them. An examination of this equation reveals the conditions necessary for energy transfer and also the sources of information. The inverse sixth power distance dependence indicates that energy transfer may be useful for measuring inter-chromophore distances in the nanometer range.

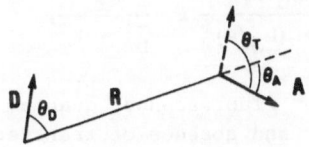

Fig. 2. Diagram of the angular parameters needed in the definition of κ^2. D and A represent the transition dipole moments of the donor and acceptor, and R is the separation vector between them.

97

Several conditions are necessary for resonance transfer to be observed: (a) the donor must be fluorescent with a sufficiently long lifetime; (b) the absorption spectrum of the acceptor must overlap the emission spectrum of the donor; (c) the relative orientations of the oscillators must be within a certain range; and (d) the donor and acceptor must be within a certain distance for a given efficiency of transfer. In addition, other requirements are necessary if one is to be able to interpret the results of an energy transfer experiment. These include, in various cases, knowledge of the quantum yields, fluorescence lifetimes, and emission anisotropies of the donor and acceptor molecules.

In order to analyze energy transfer experiments it is necessary to obtain values for $J_{\overline{V}}$ and the physical constants (straightforward) and for k^2 (difficult). For systems showing rapid isotropic motion (in comparison with τ_f) of the chromophores, it can be shown that the average value of k^2 is 2/3. For a rigid random solution an average value of $k^2 = 0.496$ has been given (Steinberg 71). Since this value applies to a rigid solution, its use is not justified for situations in which both the orientations and the distribution of chromophores are not random (Eisinger 74). Neither condition is likely to be met in the case of specifically located donors and acceptors in biological systems. In membranes, there may be situations approaching randomness and also with sufficiently rapid rotational reorientations to allow the use of $k^2 = 2/3$, but all such situations should be experimentally verified.

Resonance energy transfer is characterized by a critical transfer distance R_0, the so-called Förster radius, at which the rate constants for transfer and fluorescence are equal.

$$R_O^6 = 8.71 \cdot 10^{23} \, k^2 \, k_f \, \tau_D^O \, J_{\overline{V}} \, / \, n^4$$

R_O can be calculated from the spectral overlap integral, the physical constants, and τ_D^O, the lifetime of the donor in the absence of the acceptor. The experimental quantity usually measured is the transfer efficiency T, which is given by

$$T = 1 - \phi_D/\phi_D^O = \frac{(R_O/R)^6}{1 + (R_O/R)^6} = \frac{k_T}{\tau_D^{-1} + k_T},$$

where ϕ_D and ϕ_D^O are the fluorescence quantum yields for the donor molecule in the presence and absence of transfer, respectively.

It only remains to obtain the appropriate value for k^2 in order to solve for R. It is the average value of k^2 over the whole donor-acceptor ensemble, denoted by $\langle k^2 \rangle$ that is the relevant average. This is not in general trivial to obtain, and the remainder of the lecture will be concerned primarily with this problem. It has been shown (Dale 79) that k^2 can <u>never</u> be determined exactly except in trivial limiting cases. In order to maximize our information about k^2, it is of importance to measure the emission anisotropies of the donor and acceptor separately and the anisotropy of the transfer process itself.

What type of averaging must be carried out in order to obtain the correct value of $\langle k^2 \rangle$ depends on the geometrical nature of the biophysical situation itself. Three geometrical cases will be examined: point-to-point transfer from donors and acceptors located at specific points on macromolecules; transfer from a fixed donor to a surface array of acceptors; and transfer from donors to a semi-infinite array of acceptors. The first is applicable to fixed associations of proteins, whether membrane-bound or not. The second is applicable to donors attached to membrane-bound proteins and acceptors distributed in the membrane. The third is applicable to membrane-associated donors, protein-bound or not, and acceptors distributed in either the cytosol or medium. Examples of each type will be given.

EMISSION ANISOTROPIES

All the intensity information from a fixed fluorescent molecule can be summarized by the polar and azimuthal angles θ and ϕ of the molecule's transition moment with respect to the observer and the emission anisotropy r. In cases composed of molecules with random orientations, of course, the angles will be averaged over. Since it is often convenient in experimental spectrofluorometers for the plane of polarization of the exciting light to be vertical and for the emission to be observed at right angles to it, the subscripts "vertical" and "horizontal" are useful to denote two directions of polarization of the exciting and emitted light. The fluorescence emission anisotropy can then be calculated according to

$$r = \frac{I_{VV} - I_{VH}}{I_{VV} + 2I_{VH}}.$$

This is related to the emission polarization parameter sometimes computed: p = 3r(2+r). The anisotropy has the advantage of being additive for mixtures and multiplicative for sequential processes and of containing a denominator proportional to the total emitted intensity. The maximum value r can take in a random rigid solution is 0.4, assuming parallel absorption and emission vectors, no motion during τ_f, and no energy transfer, all of which can only decrease r further. Since each absorption band in a fluorophore is associated with a particular direction and since emission is normally from the lowest singlet excited state, the angle between absorption and emission will in general not be zero, and the limiting anisotropy in rigid solution will be less than 0.4. If molecular motion can take place on a time scale faster than τ_f, the initially nonrandom collection of excited dipoles can become completely randomized, i.e., r = 0. If the rotation is a little slower (on the same time scale as τ_f), only partial depolarization occurs. If τ_f is known, the rotational relaxation time ρ, correlation time $\tau_c = \rho/3$, or the rotational diffusion coefficient Θ of a spherical molecule can be determined by

$$\rho = \frac{3\tau_f}{r_o/r - 1}.$$

ρ can be interpreted as

$$\rho \equiv \frac{1}{2\Theta} = \frac{3V\eta}{RT},$$

where V is the molar volume of solute in ml; η is the solution viscosity; R is the gas constant; and T is the absolute temperature. For energy transfer experiments, overall rotational diffusion of the donor-acceptor system merely stands in the way of the analysis, and, if present, a correction must be allowed for it. Fortunately, for experiments in membranes, rotational diffusion is rarely a complication.

AVERAGING REGIME

When _every_ donor and _every_ acceptor can take up its entire range of orientations during the energy transfer process, the system is said to be in the dynamically averaged regime. In this case $\langle \kappa^2 \rangle$ is the same for every D-A pair, and the energy transfer may be written as

$$\langle T \rangle_d = \langle \kappa_T \rangle / (1/\tau_D^o) = \langle k_T \rangle = \langle \kappa^2 \rangle / (C^{-1}R^{-6} + \langle \kappa^2 \rangle).$$

Thus knowing $\langle T \rangle_d$ and $\langle \kappa^2 \rangle$ would permit a determination of R. On the other hand, the statically averaged energy transfer efficiency between a donor and an acceptor separated by a fixed distance R is

$$\langle T \rangle_s = \langle k_T / (1/\tau_D^o + k_T) \rangle = \langle \kappa^2 / (C^{-1}R^{-6} + \kappa^2) \rangle.$$

This formulation cannot readily be adapted, in general, to evaluate R for an ensemble of D,A pairs since the R and κ^2 dependence cannot be separated, each D,A pair having a different and unknown k_T and κ^2. The only exceptions occur when (a) $\langle T \rangle_s$ is vanishingly small so that it approaches the dynamic average efficiency $\langle T \rangle_d$ or (b) both D and A have isotropic orientational distributions ($\kappa^2 = 2/3$). The latter situation has been considered by Dale et al. (Dale 79) as it is amenable to an analytic solution. The result is displayed in Fig. 3, which also shows for comparison the corresponding dynamically averaged efficiency $\langle T \rangle_d$

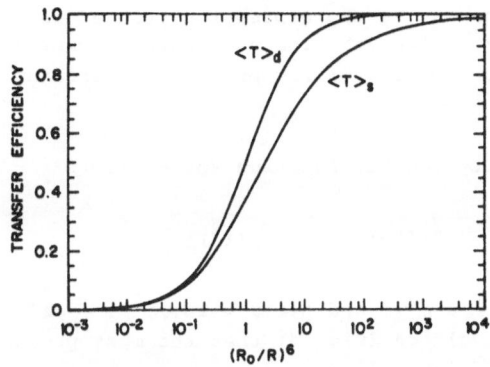

Fig. 3. The dynamically and statically averaged transfer efficiencies, $\langle T \rangle_d$ and $\langle T \rangle_s$, as functions of $(R_o/R)^6$, calculated under the isotropic assumption for the orientational freedom of both luminophores. The averaging regime is seen to make little difference if the transfer efficiency is low. From Dale et al. (Dale 79).

with $k^2 = 2/3$ as appropriate for the isotropic case considered.

Since in the dynamic limit all D,A orientations, including those favorable for rapid transfer, are sampled during the transfer time, $\langle T \rangle_d$ always exceeds $\langle T \rangle_s$ for a given intramolecular separation. The error in the intramolecular separation R, introduced by assuming the wrong averaging regime in deriving R from energy transfer efficiencies, in this case is readily seen to be about 15% when $\langle T \rangle_d$ is 0.5 and about 50% when $\langle T \rangle_d$ is 0.9. At low transfer efficiencies the error becomes negligible. In the following sections the dynamic averaging regime will be assumed.

THE ORIENTATION FACTOR k^2

The orientation factor k^2 can in principle have any value between 0 and 4, although in a random assembly, values at or near the lower limit occur with high frequency. In a real situation, a considerable amount of angular averaging of k^2 may be expected (Dale 79). As will be demonstrated later, the extent of angular averaging is indeed of paramount importance in the interpretation of a given energy transfer efficiency in terms of intramolecular separations.

I digress here to give the exact expression for the probability density function for k^2 and its probability distribution function, as it seems quite difficult to get the correct expressions in print. Expressions given in two sources (Jones 70, Tompa 79) are marred by typographical errors as is the expression printed in the article by Dale et al. (Dale 79) although these authors attempted unsuccessfully to correct these errors on the galley proofs. An erratum (Dale 80) gives the correct expressions.

The probability density function for k^2 is

$$p(k^2) = 1/(\ 2\sqrt{3k^2})\,[\ln(2+\sqrt{3}) - g(k^2)].$$

where $g(k^2) = 0$ for $0<k^2<1$ and $g(k^2) = \ln(\sqrt{k^2-1} + \sqrt{k^2})$ for $1<k^2<4$. It can be seen at a glance (Fig. 4) that the most probable value of k^2 is zero.

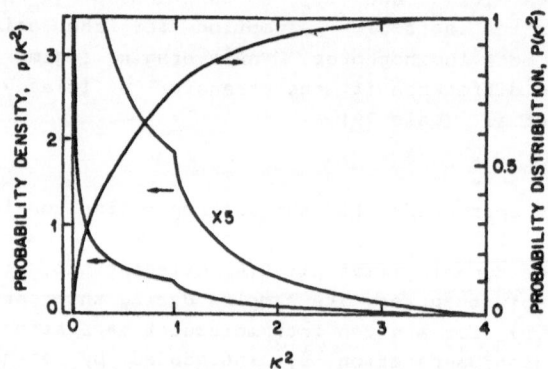

Fig. 4. The probability density $p(k^2)$ as a function of k^2 (left scale). $P(k^2)$ is the probability distribution for k^2 being between zero and k^2 (right scale). The most probable value for k^2 is zero, while its average value is 2/3. From Dale et al. (Dale 79).

The probability distribution of this density function is obtained by integrating from zero to any arbitrary value of k^2. This gives

$$P(k^2) = \sqrt{k^2/3}\ \ln(2+\sqrt{3}) \qquad (0<k^2<1)$$

$$P(k^2) = \sqrt{k^2/3}\,[\ln(2+\sqrt{3}) - \ln(\sqrt{k^2} + \sqrt{k^2-1}) - \sqrt{k^2-1}] \qquad (1<k^2<4).$$

Inspection of the latter curve shows that, for instance, there is approximately a 60% probability that k^2 has a value outside the limits 1/3 — 4/3. There is approximately a 20% probability that k^2 has a value less than 0.1 of its average value of 2/3 and almost 12% probability that it is less than 0.02. The corresponding values derived for the D-A separation in the latter two cases are $0.1^{1/6}$ and $0.03^{1/6}$, i.e., about 0.68 and 0.56, of that obtained under the assumption of an average value of 2/3 for k^2. It has in the past been argued that the "isotropic" value of 2/3 has statistical validity for D-A pairs with unknown relative orientations, since extreme values for k^2 (0 or 4) are unlikely to obtain for a pair of independent transition dipoles. The mathematical expression dispels the credibility of this purely intuitive argument. Indeed, it shows that the opposite is true, since values of k^2 near zero are seen to predominate in a random distribution. There exists, moreover, a fundamental objection to obtaining an estimate for the orientation factor on statistical grounds. A fluorescent label, be it intrinsic or extrinsic, is found at a specific site of a macromolecule because of the unique structural or functional properties of its interaction with the site. It is therefore likely that the resultant distribution of dipole orientations, while perhaps somewhat heterogeneous, will be characterized by a narrow range of geometries.

Some authors (Hillel 76) have made use of k^2 distributions to assess an error in a determination of R caused by a particular choice of k^2. At best, such considerations have statistical validity for cases in which D and A have a large number of entirely independent possible orientations on a common substrate. It is, however, inappropriate to use the k^2 distribution to draw conclusions about the k^2 value in a case for which some particular D,A geometry obtains, since the experimenter has no way of knowing whether his particular case is a "probable" or an "improbable" one. Thus, when considering D,A pairs on a protein or a protein-protein complex, statistical arguments are invalid, but when considering D (or A) to be on a membrane-bound protein, for example, and A (or D) to be distributed in the membrane, cytosol, or medium, statistical distributions are indeed appropriate.

DEPOLARIZATION FACTORS

The most promising strategy for delimiting the possible values of the orientation factor makes use of the results of depolarization measurements to quantitate the model illustrated in Fig. 5. The upper part of the figure depicts the sequence of events which occurs in the course of a typical energy transfer experiment: the donor is excited, its transition moment direction $\underline{D}_i$ reorients to $\underline{D}_j$, transfer occurs to the acceptor with orientation $\underline{A}_i$, which then reorients to the orientation $\underline{A}_j$.

For application to an ensemble of identical D,A pairs attached to identical substrate, the analysis of this model is greatly facilitated by expressing the reorientational averaging which occurs in terms of depolarization factors.

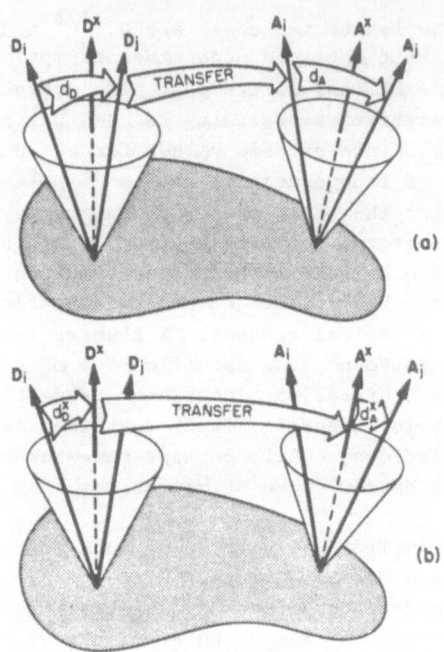

Fig. 5. (a) Schematic representation of the depolarizing steps following the absorption of excitation energy by the donor: donor depolarization, transfer depolarization and acceptor depolarization. (b) An equivalent representation to (a), in terms of axial depolarization of $\underline{D}$ and $\underline{A}$ and transfer between the symmetry axes of the donor and acceptor transition dipole distribution ($\underline{D}^x$ and $\underline{A}^x$). From Dale et al. (Dale 79).

When the direction of the transition moment of a fluorophore is changed by an angle θ (over random azimuths), the emission anisotropy, initially r_i, is related to its final value r_f by

$$r_f = dr_i$$

in which d, the depolarization factor, is given, according to Perrin's law of isotropic depolarization (Perrin 26), by

$$d = (3\cos^2\theta - 1)/2.$$

It can be shown (Soleillet 29) that, for a series of n independent depolarizing events, the overall depolarization is related to the individual ones by

$$d = \prod_{i=1}^{n} d_i$$

where d_i is the depolarization factor for the ith step.

If the orientation of a transition moment is defined by a unit vector $\underline{M}_i$, its reorientation, say, $\underline{M}_i \to \underline{M}_j$, is independent of any intermediate steps in the reorientation process. If, in particular, the vectors $\underline{M}_i$ are distributed with axial symmetry about the axis $\underline{M}^x$, this depolarization is equivalent to the two-step path $\underline{M}_i \to \underline{M}^x$ followed by $\underline{M}^x \to \underline{M}_j$. It is then convenient to define an average depolarization factor $\langle d \rangle$ and an average axial depolarization factor $\langle d^x \rangle$. Dale et al. (Dale 79) have shown that

$$\langle d \rangle = \langle d^x \rangle^2 .$$

The axial depolarization factor is the same as the (axial) order parameter S used in several other of these lectures.

The reorientational averaging which is indicated by the angular brackets is assumed to belong to the dynamic regime, as is discussed above and in greater detail elsewhere (Dale 79), i.e., it is assumed that reorientational relaxation is fast compared with the fluorescence and transfer rates.

The events discussed above are described pictorially in the lower part of Fig. 5, which represents a donor and acceptor pair D,A with limited orientational freedom about the axes $\underline{D}^x$ and $\underline{A}^x$, attached to an immobile substrate and separated by a fixed distance. The observed average depolarization may then be written

$$\langle d_T \rangle = \langle d_D^x \rangle \, d_T^x \, \langle d_A^x \rangle$$

where

$$d_T^x = (3\cos^2\theta_T - 1)/2$$

is the depolarization factor which corresponds to the reorientation $\underline{D}^x \to \underline{A}^x$, and θ_T is the angle between $\underline{D}^x$ and $\underline{A}^x$. Experimentally one determines the depolarization factors of D and A from separate measurements of r, one in the absence of the other. Then $\langle d \rangle = r_\infty / r_{max}$, where r_∞ is the anisotropy of the time-resolved fluorescence at long times and r_{max} is 0.4. If dynamic averaging is known to be very well satisfied, one may approximate $\langle d \rangle$ as r_{ss}/r_{max}, where r_{ss} is the steady state value. If the acceptor is fluorescent also, $\langle d_T \rangle$ can be determined in a similar manner from the limiting anisotropy $r_{\alpha T}$ of the emission of the acceptor when the donor is excited.

When both donor and acceptor are immobile and attached to fixed points on a well-defined protein or association of proteins, all the distances and angles will be defined. Then the appropriate average $\langle k^2 \rangle$ is just the value of k^2. Unfortunately, in no real case is this obtainable from fluorescence data. The best one can do is to measure as many of the depolarization factors as possible in addition to the transfer efficiency. Then upper and lower bounds can be placed on $\langle k^2 \rangle$ and thus on R.

Depolarization measurements generally permit the evaluation of $\langle d_D \rangle$, $\langle d_A \rangle$ and $\langle d_T \rangle$. If the possible ambiguity with respect to sign in deriving $\langle d^x \rangle$ from $\langle d \rangle$ is ignored for the moment (see Dale 79), such experiments yield a minimum of three angular parameters related to the model: θ_T, and the angular widths of the D and A orientational distributions. $\langle k^2 \rangle$ can be seen, however, to depend on at least five angular parameters: the three angles appearing in its definition, plus at least two more (depending on the orientational distribution model invoked (Dale 79)), those characterizing the widths of the (assumed axially symmetric) D and A distributions. Thus, while polarization experiments cannot, except in special situations, yield unique values for $\langle k^2 \rangle$, they can provide bounds for $\langle k^2 \rangle$ which are consistent with the measured depolarization factors. This is accomplished by expressing $\langle k^2 \rangle$ in terms of the angular parameters which define the model illustrated in Fig. 6 and then solving numerically for the maximum and minimum values of $\langle k^2 \rangle$ consistent with the observed depolarization factors for the D, A, and T processes. Dale et al. (Dale 79) have presented contour maps representing many such computer solutions.

In the absence of an experimental determination of $\langle d_T \rangle$, d_T^x cannot be determined, and a somewhat wider range of possible $\langle k^2 \rangle$ values is obtained from the single contour plot shown in Fig. 6 reproduced from Dale et al. (Dale 79). Contour plots for d_T^x values in intervals of 0.1 have also been published there.

It may be instructive to examine in some detail the use of such contour plots in the determination of the D-A separation R. If other sources of error (e.g., imperfectly dynamic orientational averaging, uncertainty in probe attachment site) are negligible, the uncertainty in $\langle k^2 \rangle$ introduces a negligible error in R only when both D and A have nearly isotropic distributions. In this case depolarization is complete, $\langle d \rangle$ and therefore $\langle d^x \rangle$, become vanishingly small for both donor and acceptor, and the $\langle k^2 \rangle$ value is indeed close to 2/3. Since R_o is proportional to $\langle k^2 \rangle^{1/6}$, a factor of two between the minimum and maximum

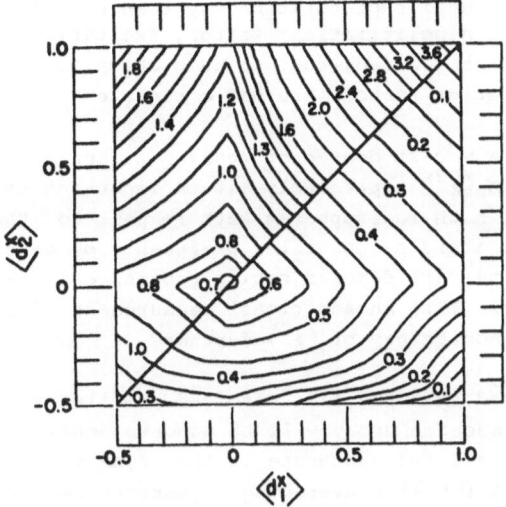

Fig. 6. Contour plot for obtaining extreme values of κ^2 consistent with donor and acceptor depolarization factors when d_T is unknown. Contours for the minimum and maximum values appear in the lower right and upper left halves of the diagram, respectively. For finding the minimum, $\langle d_1^x \rangle$ is the larger of $\langle d_D^x \rangle$ and $\langle d_A^x \rangle$, while for the maximum $\langle d_2^x \rangle$ is the larger of the two. This and more detailed contour plots have been given by Dale et al. (Dale 79).

values of $\langle \kappa^2 \rangle$ produces an uncertainty of only some ±6% in R. Fig. 6 shows that such precision is readily obtainable for most situations in which the two axial depolarization factors are in the range ±0.4, corresponding to $\langle d \rangle$ values of less than 0.16. Serious uncertainties in R obtain when both fluorophores are relatively immobile, with axial depolarization factors near unity, e.g., $\langle d^x \rangle$ > 0.8. The condition, $\langle d_D^x \rangle$, $\langle d_A^x \rangle$ > 0.8 corresponds to a factor of more than 22.6 between the upper and lower bounds of $\langle \kappa^2 \rangle$, or an error greater than ±25% in R, and this error is not greatly reduced when $\langle d_T \rangle$ is also measured. This limit still corresponds to relatively small measured depolarization factors, $\langle d_D \rangle = \langle d_A \rangle = 0.64$. If both the donor and the acceptor have axial depolarization factors as large as about 0.9 (corresponding to $r_o = 0.32$, which would by no means constitute an unexpectedly high value), the resulting uncertainty in R approaches a factor of 2. Dale et al. (Dale 79) have provided protocols for analyzing energy transfer experiments when various combinations of the depolarization factors are known.

When extrinsic labels are used, they should be closely approximated by the limits of either linear or planar transition moment. The

107

chromophores' rotational reorientation may then be determined. To measure the transfer depolarization factor, excitation into higher donor transitions should be avoided, for, in such cases, the orientation and transfer depolarization factors are not correlated (Dale 79).

Because even a modest degree of reorientational freedom can reduce the uncertainty in $\langle k^2 \rangle$ appreciably, it is important that the depolarization factors for both luminophores with respect to the substrate be measured as accurately as possible, preferably by determining the time dependence of the overall depolarization process directly. As discussed by Dale et al. (Dale 79), an additional measurement of d_T serves to narrow the range of uncertainty of $\langle k^2 \rangle$ further.

It may be noted that the time resolved profile of the total intensity of the fluorescence decay will be monoexponential in the case of a unique D-A pair on a fixed substrate in the dynamic averaging regime. Partial (or complete) static averaging, substrate reorientation, or multiple D-A interactions will in each case destroy the monoexponential nature of the time profile.

An Example

The 11S form of acetylcholinesterase (AchE) isolated from Torpedo electric organs is a tetrameric enzyme composed of apparently equivalent, independent subunits of 80,000 mol wt. Each subunit possesses an active center at which acetylcholine is hydrolyzed, as well as a peripheral anionic site which is spatially distinct from the active center. Cationic ligands can alter catalysis by association at either the active center or the peripheral site, and formation of a ternary complex with active-center- and peripheral-site-selective ligands can be demonstrated. Cationic ligands that bind at the peripheral or active sites in a 1:1 stoichiometry with each subunit have been identified.

Berman et al. (Berman 80) employed pyrene and dansyl fluorescent phonphonate labels of the active center as energy-transfer donors, and propidium, which is specific for the peripheral anionic site, served as an acceptor. The pyrene ligand, pyrenebutyl methylphosphonofluoridate (PBMPF), and the dansyl ligand, (dansylamido)pentyl methylphosphonofluoridate (DC$_5$MPF), react solely with the active site serine on AchE to form the respective fluorescent conjugates PBMP-AchE and DC$_5$MP-AchE.

The steady state fluorescence spectra of the two donors and the optical absorption of the acceptor were measured in order to compute the overlap integrals necessary to determine R_o. All the other measurements were done in the time domain.

108

In the presence of propidium, the lifetimes and respective amplitudes of PBMP-AchE and DC_5MP-AchE fluorescence are dramatically altered.

The axial depolarization factors $\langle d^x \rangle$ for the pyrene, dansyl, and
propidium fluorophores were found to be 0.55, 0.87, and 0.91, respectively. The transferred excitation between pyrene and propidium was
found to be virtually depolarized; i.e., $\langle d_T \rangle$ = 0, and, hence, d_T^x, the
transfer depolarization factor also is zero.

The contour maps of Dale et al. (Dale 79) relating $\langle d_D^x \rangle$ and $\langle d_A^x \rangle$
for the case when d_T = 0 afford limits on the range of values for the
orientation factor characteristic of the pyrene-propidium D-A pair:
$0.25 \leq k^2 \leq 2.2$. In the absence of knowledge of the degree of polarization
of the transferred excitation, a broader range of values is obtained for
k^2 based solely on D and A anisotropies: $0.2 \leq k^2 \leq 2.7$. For the dansyl-
propidium D-A pair, estimates of k^2 were obtained solely from analysis
of $\langle d_D^x \rangle$ and $\langle d_{2A}^x \rangle$ and lead to a wide range of values for the orientation
factor, $0.08 \leq k^2 \leq 3.2$.

The former D-A pair leads to an intersite distance of 19-28Å and
the latter to a distance of 20-37Å. The rather large range in each case
arises because the chromophores are relatively immobile, especially the
propidium. A systematic error of the experiment, over and above the
uncertainty in distance between the fluorophores, is the fact that each
of them is joined to the binding site by chains several bonds in length,
leading to a greater uncertainty in the distance between the active
sites. See Berman et al. (Berman 80) for some other considerations not
discussed here.

DISTRIBUTIONS ON SURFACES (e.g., IN MEMBRANES)

I now consider resonance energy transfer among a distribution of
donors and acceptors which are confined to a plane or the surface of a
sphere. Planar distributions may be viewed as the limiting cases
approached by spheres of large radius and are appropriate to oriented
planar bilayers as well as to cell membranes. The curvature of cell
surfaces is always small compared with that of a sphere with a radius on
the order of Förster distances and may be considered a plane. When
acceptors are distributed in the bilayer surface of small vesicles, the
radius of the vesicle may become important.

Energy transfer experiments for such systems may be useful for
determining local acceptor densities around donor sites. These may then
be compared with the density calculated for uniform or random surface
distributions and can therefore elucidate surface topological relationships between specifically labeled membrane components, such as antibody
binding sites, hormone receptors and ion transport proteins.

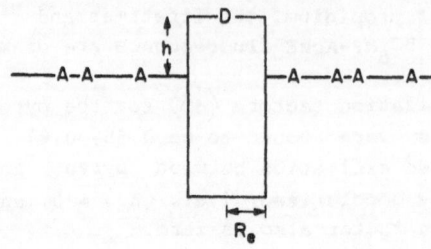

Fig. 7. Schematic representation of acceptors randomly distributed on a surface. The donor may be located on the same
surface or at a height h(±) above or below it. In addition,
there may be a region of radius R_e from which acceptors are
excluded, if, for example, the donor is attached to a bulky
membrane-bound protein.

Consider the following model: donors and acceptors, considered as
point dipoles, are randomly distributed over the surface, as shown in
Fig. 7 with h and R_e considered zero for the moment. All donors have
equivalent depolarization factors, as do all acceptors. The acceptors
have an average surface concentration c (mol per cm^2), and the critical
Förster concentration c_o corresponds to one acceptor per disc of radius
R_o, i.e., $c_o = 1/\pi NR_o^2$.

The decay of the excited donor population has been calculated by
Tweet et al. (Tweet 64).

$$\bar{I}(t) = \exp\{-[(t/\tau_D^o) + (c/c_o)(t/\tau_D^o)^{1/3}\Gamma(2/3)]\}$$

where the bar indicates an average over all acceptor distributions and
τ_D^o is the donor decay time in the absence of acceptors. This derivation
assumes that both donor and acceptor have zero depolarization factors,
but the result holds whether the averaging regime is static or dynamic.
If the time dependence $\bar{I}(t)$ of the fluorescence decay is measured, then
this expression may be used as the kernel of a fitting algorithm in
order to determine c. It should be noted that, because of the $t^{1/3}$
dependence in the exponent, the time profile (plotted as log(I) vs. t)
will not be linear but curved upward. One should resist the temptation
to fit the profile with a sum of exponentials.

Wolber and Hudson (Wolber 79) have expressed the donor quantum
yield relative to its yield in the absence of acceptors analytically as

110

$$\langle \phi \rangle / \phi_0 = \frac{\int_0^\infty \overline{I}(t)\, dt}{\int_0^\infty I_0(t)\, dt}$$

$$= \sum_{j=0}^\infty \{ [-(c/c_0)\Gamma(2/3)]^j \Gamma[(j/3)+1]/j! \}$$

where $I_0(t)$ is the donor decay function in the absence of acceptors. The efficiency of two-dimensional transfer is then given by

$$\langle T \rangle = 1 - \langle \phi \rangle / \phi_0.$$

Wolber and Hudson (Wolber 79) also have treated the case in which acceptors are excluded from a circular region centered on the donor. This would be appropriate to an intrinsic donor, such as tryptophan, embedded in a membrane-bound protein. The expression involves integrals over the incomplete gamma functions and is too lengthy to be reproduced here. As in most of the complicated cases involving planar distributions, evaluation must be done numerically. Unfortunately this formulation assumes isotropy for both donor and acceptor (unlike tryptophan embedded in a protein).

Koppel et al. (Koppel 79) have derived expressions for both the time dependence of the fluorescence decay and the energy transfer efficiency in the case in which the donor is somewhere in the interior of a sphere (vesicle) on the surface of which the acceptors are randomly distributed, except for a circle, just above each donor, from which the acceptors are excluded. They took into account the depolarization factors $\langle d \rangle$ of the donors and acceptors and also a computation of $\langle d_T \rangle$ (in a different, somewhat less informative notation). In deriving the transfer efficiency, they separated the acceptors into the "nearest acceptor" and "secondary acceptors," leading to somewhat less general expressions than could be given. The formulation is not very easy for an experimentalist to use in analyzing data from a different experiment.

Eisinger et al. (Eisinger 81) have given a more general treatment of donors and acceptors distributed on vesicular surfaces in which the radius of the vesicle was considered and the explicit dependence of $\langle \kappa^2 \rangle$ on the depolarization factors was emphasized. They did not include the cases $h \neq 0$ and $R_e \neq 0$. When the vesicular radius is large, its surface is

111

effectively planar and the result, also applicable to cell membranes and lipid bilayer preparations, reduces to

$$\left\langle k^2 \right\rangle = \left\langle d_D^x \right\rangle\left\langle d_A^x \right\rangle + \frac{1 - \left\langle d_D^x \right\rangle}{3} + \frac{1 - \left\langle d_A^x \right\rangle}{3}.$$

These authors presented their results is terms of graphs which are easy for the experimenter to use if the case at hand is appropriate for this analysis.

Recently Dale and co-workers (R. E. Dale, personal communication) have formulated a more general problem for planar distributions of acceptors in which both $h \neq 0$ and $R_e \neq 0$ are considered as well as the explicit inclusion of the depolarization factors of both donors and acceptors. Unfortunately, calculations based on this formulation have not yet been published.

An Example

Cytochrome b_5 is an amphipathic molecule composed of a hydrophilic heme peptide segment that is joined by a short, apparently flexible sequence to a nonpolar segment. The primary structures of both the heme peptide segment and the nonpolar peptide segment have been determined. This latter amino acid sequence reveals three tryptophanyl residues in a predominantly nonpolar structure. It has been shown that only Trp-109 if fluorescent. This single excitation energy donor within the cytochrome has been utilized by Fleming et al. (Fleming 79) to determine the distance between this residue and trinitrophenyl or dansyl groups placed at the head group region of a phospholipid membrane with quantitative excitation energy transfer experiments.

Cytochrome b_5 was enzymically cleaved by addition of trypsin to make the nonpolar peptide segment (residues 91-133). Vesicles containing 95% dimyristoylphosphatidylcholine and 5% dimyristoylphosphatidylethanolamine were then prepared and either the holoprotein or the peptide segment was added at lipid/protein ratios of 200-600, along with various amounts of either of the two acceptors.

The fluorescence emission spectrum of each sample was recorded at an excitation wavelength of 285nm. Energy transfer efficiencies were calculated from the fluorescence intensities at 335nm vs. a standard sample containing no acceptor groups. Steady state anisotropies of the tryptophan and dansyl fluorophores were measured at the red end of their emission bands, although values from the time-dependent decays would have been better. As the trinitrophenyl chromophore is non-fluorescent, its reorientational properties cannot be determined in this way. In fact, Fleming et al. assumed that $\left\langle d_A^x \right\rangle$ for trinitrophenyl would be the same as for an entirely unrelated fluorophore — a completely unjustified assumption. Axial depolarization factors used for tryptophan, dansyl, and trinitrophenyl were 0.42, 0.31, and 0.42 (assumed), respectively. They then fitted the energy transfer efficiency for various concentrations of acceptor to the theory of Koppel et al. (Koppel 79) assuming various values of h and R_e, yielding a best fit of h=21$\mathring{A}$ and $R_e \cong 0$ for all the D-A combinations except for the case of tryptophan in the holoprotein transferring to trinitrophenyl, in which case R_e=16$\mathring{A}$ (with a large uncertainty). Heuristic arguments why this difference is reasonable have been discussed by Fleming et al. Using the analysis method of Koppel et al., there is a trade-off between h and R_e so that they can seldom be determined accurately independently of one another. The most relevant parameter in the analysis is the distance of closest approach of the nearest acceptor to the donor.

Another Example

The plasma membrane of the retinal rod outer segment (ROS) portion of the retinal rod cell encloses a stack of separated disc membranes whose primary protein component is rhodopsin, the photopigment. The absorption of a single photon by rhodopsin leads to a reduction of the sodium influx and consequent hyperpolarization of the plasma membrane. It has been suggested that membrane lipids may regulate these and other membrane-mediated events by controlling, in part, protein diffusion, association, and enzymatic activity; solute permeability; the surface binding and release of electrolytes; and membrane fusion.

Sklar et al. (Sklar 79) examined the thermally effected reorganization of retinal ROS membranes and phospholipids using the fluorescence properties of cis-parinaric acid (cis-PnA, 9,11,13,15-cis,trans,trans,cis-octadecatetraenoic acid) and trans-parinaric acid (trans-PnA, 9,11,13,15-all-trans-octadecatetraenoic acid). It is known that there is a preferential partition of the trans probes into phospholipid solid phases where they exhibit a strongly enhanced quantum yield and polarization; the polarization of trans-PnA fluorescence is responsive to as little as 1% solid phase. cis-PnA distributes more equally between co-existing solid and fluid phospholipid phases. Much of this study was devoted to effects of temperature on quantum yield and polarization of these probes, but experiments demonstrating fluorescence energy transfer from the probes to the visual pigment were also performed as well as experiments using intrinsic tryptophan as the donor to the probes.

The emission spectra of cis-PnA and trans-PnA overlap the absorption spectra of retinal and rhodopsin. If retinal is sufficiently close to PnA in a membrane, energy transfer between PnA as donor and retinal as acceptor can occur although neither free retinal nor rhodopsin-bound retinal fluoresce. In ROS phospholipids with added retinal, the energy transfer efficiency from trans-PnA to retinal decreases strongly over the temperature range associated with the phospholipid lateral phase separation. This result can be explained by the partition of trans-PnA into solid phase lipid as it forms and the exclusion of retinal (due to its bulky β-ionine ring) from solid phase. Preliminary calculations suggest that solid phase patches must be at least twice the Förster radius for this system or about 70Å in diameter, and probably much larger, in order to account for the reduced energy transfer at low temperature.

Tryptophan is a fluorescence donor to either cis-PnA or trans-PnA (R_o equals 25 to 30Å). In ROS membranes, the energy transfer efficiency between opsin or rhodopsin tryptophan residues and trans-PnA is reduced over the temperature range ascribed to the lateral phase separation of several PS/PE molecular species in the membrane outer monolayer; the energy transfer to cis-PnA is virtually temperature independent.

Sklar et al. conclude that the reduced fluorescence energy transfer to trans-PnA and the relatively constant energy transfer to cis-PnA are consistent with the lateral phase separation of a small fraction of solid phase lipids (≈15%) in ROS membranes, and in particular that trans-PnA is included, opsin and rhodopsin excluded and cis-PnA distribution virtually unaffected by that phase.

DISTRIBUTIONS IN SEMI-INFINITE MEDIA

A semi-infinite medium may be considered to be all of space sliced by some plane dividing it into a region in which there is a uniform (or random) distribution of acceptors and a region in which there are none. A biological membrane would be a good example of such a dividing plane. Then the acceptors could be distributed in the extracellular medium or in the cytosol or vesicle interior. Cases can be analyzed in which donors are in the plane or at some fixed distance above or below it (either into the acceptor region or away from it).

Dobretsov et al. (Dobretsov 82) have considered the former case. (Incidentally, the reference list of this paper gives the reader an entree to the Russian literature on energy transfer, not often cited in the West.) The geometry of this case is shown in Fig. 8. The region denoted as filled with acceptors is the interior of a large lipid particle, but it could just as well be the aqueous phase of a cytosol or the external medium. The donors are all assumed to be at some depth h from the interface. Dobretsov et al. integrate over all the semi-infinite space by considering partial spherical shells of radius R and thickness dR.

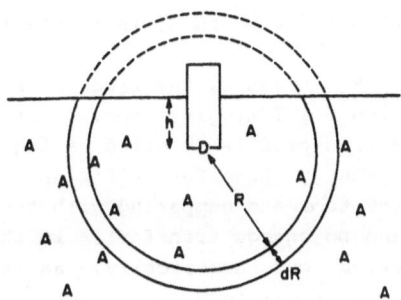

Fig. 8. Schematic representation of acceptors randomly distributed in a semi-infinite medium. The donor may be located
on the boundary or at a height h($\pm$) above or below it. In
either case the computation of the energy transfer proceeds by
constructing spherical shells about the donor with radius R
and thickness dR and having an accessibility coefficient
dependent on h and R.

In an infinite medium the time profile of the fluorescence decay in
the presence of energy transfer is described by the following expression
(Förster 49, Antonov 57, Rozman 58)

$$I(t) \propto \exp(-t/\tau_D^o - 4\pi cJ),$$

where c denotes the acceptor concentration, and

$$J = \int_0^\infty E(t,R) \; R^2 \; dR,$$

$$E(t,R) = 1 - \exp\left[-\frac{t}{\tau_D^o}\left(\frac{R_o}{R}\right)^6\right].$$

These expressions can be generalized to the semi-infinite case by constructing the following accessibility coefficients B

$$B(h,R) = dV_a(R)/dV(R)$$

$$= \begin{cases} 1 & \text{for } R<h \\ (1+h/R)/2 & \text{for } R>h. \end{cases}$$

Then

$$J = \int_0^\infty B(h,R) \; E(t,R) \; R^2 \; dR.$$

By inserting the expressions for B and E, this integral can be computed.

For given values of R_o, h and c, the fluorescence decay can be computed using the expression for $\bar{I}(t)$ given above. As in the case of surface distributions, the time profile (plotted as log(I) vs. t) will be concave upward. The energy transfer efficiency can be computed by integrating $\bar{I}(t)$ from t=0 to ∞ and comparing with the case in which c=0 (i.e., no acceptors and no energy transfer). In this formulation, the calculation must be carried out numerically, as there does not at present seem to be any analytical formulation for $\int \bar{I}(t)dt$.

Dobretsov et al. have shown that the efficiency is well approximated by an exponential when the quenching due to energy transfer is not too large (T<0.8).

$$T \cong 1 - \exp(-4.8 \left\langle B(h) \right\rangle cR_o^3),$$

where $\left\langle B(h) \right\rangle$ is the average accessibility coefficient for donors located at h. If c and/or R_o is small, the exponential may be expanded as $T \cong 4.8 \left\langle B(h) \right\rangle cR_o^3$, valid for small T.

These authors also formulated the case in which the donor is located outside the region containing the acceptors, although not in as great detail. Quite independently of them, Eisinger and co-workers (Eisinger 82a,b, 83) have considered the latter case. They have not derived an expression for the time dependence of the fluorescence decay but have gone directly to the energy transfer efficiency. They have also formulated the problem in terms of partial spherical shells centered on the donor. They used reduced units in which the unit of length is R_o and therefore $1/k_T = \tau_D^o R^6$. They equated this rate with the ensemble rate $\left\langle k_T \right\rangle$ for acceptors at constant R. The consequences of this approximation are unknown and left undiscussed, but surely it becomes intolerable for large values of T. In the isotropic dynamic averaging regime the ensemble average rate from D to any partial shell is then

$$\left\langle k_T(R) \right\rangle = 2\pi c \ (1+\frac{h}{R}) \ R^2 \left(\frac{1}{\tau_D^o R^6} \right) dR.$$

Integration over all shells with radii between h and ∞ gives the transfer rate from D to all the acceptors in the semi-infinite region.

$$\left\langle k_T \right\rangle = \frac{2\pi c}{\tau_D^o} \int_0^\infty (R^{-4} - hR^{-5}) dR$$

$$= \frac{\pi}{6\tau_D^o} \frac{c}{h^3} .$$

116

This may be compared with Dobretsov's approximation (of an approximation) valid for small T, which is also linear in concentration. Rearranging to express k_T and τ_D^o as a transfer efficiency, one obtains

$$\frac{c}{h^3} = \frac{6}{\pi} \frac{T}{1-T}.$$

The ratio c/h^3 has been separated in the last two equations to emphasize that only this combination results from Eisinger's analysis in the case of every energy transfer experiment in this geometry. In no way can either quantity be obtained independently of the other without additional knowledge. However, one can envision cases in which the concentration is well known and h is to be determined or vice versa.

It is assumed that this analysis will be applied to cases in which the semi-infinite region is isotropic, whether it is the cytosol or the medium. On the other hand, the donor can be in an anisotropic medium (e.g., a membrane) and its depolarization factor may not be zero in the dynamic averaging regime. In the case that the transition moment of the donor lies preferentially in the membrane plane (the only case treated), the energy transfer rate is modified by a factor of $(1 - \langle d_D^x \rangle/4)$, and the expression for c/h^3 is similarly altered.

The case in which the boundary is not planar but spherical, as would be appropriate to small unilamellar vesicles or to micelles, has not been treated but would not be too difficult in the formulation of either Dobretsov or Eisinger. Such an analysis should be undertaken only in the case in which the boundary radius of curvature is well known by other means; it is not advisable to try to use fluorescence energy transfer to determine it.

An Example

Serum low density lipoproteins (LDL) are more or less spherical particles with a diameter of about 200-300Å. Triacylglycerols and cholesterol esters are located inside the particles, whereas most or all of the phospholipids are situated on the surface. The LDL apolipoproteins are located mainly (or entirely) on the outer surface. However, there have been many viewpoints regarding their conformation. On the basis of small-angle X-ray scattering techniques, neutron analysis, ^{1}H-NMR and estimation of the areas occupied by the proteins on the lipoprotein surfaces two models have been suggested: apolipoprotein located on the LDL surface in thin layers and apolipoprotein spikes of 25-80Å protruding from the LDL sphere surface, as discussed by Luzzati in another of these lectures.

117

Dobretsov et al. (Dobretsov 82) made use of fluorescent energy transfer in an attempt to determine the location of the apolipoprotein in LDL. Apolipoprotein contains tryptophan residues as intrinsic fluorescence energy donors. An acceptor was incorporated into the LDL lipid phase, and the efficiency of energy transfer from the apolipoprotein to the acceptor was used to estimate the relative positions of the proteins and lipids.

The fluorescence spectrum of LDL showed a maximum at 333nm, due primarily to the fluorescence of tryptophan residues. This spectrum overlaps fairly well the absorption spectrum of pyrene. Therefore, one might hope that with the introduction of pyrene into the LDL particle, a protein — pyrene energy transfer would take place. Such a quenching of the intrinsic fluorescence did take place in these experiments, its efficiency depending on the concentration of pyrene.

Analysis of the spectral and energy transfer data gives $R_o = 31\text{Å}$ and $h = 15\text{Å}$. Of course, since there are 13 ± 2 tryptophan residues per LDL particle, this value of h is an average. As a small portion of the protein fluorescence (about 12%) is not quenched by pyrene, perhaps one or two tryptophan residues out of the 13 may be withdrawn out of the lipid surface by more than R_o. The remainder, however, must be embedded into the lipids of the particle to an average depth of about 15Å. Certainly one can conclude that the proteins are located at or near the surface. These authors are of the opinion that the thin layer hypothesis fits better with the energy transfer data, but I see no conflict with the picture presented in these lectures by Luzzati provided that a significant portion of the protein is embedded in the lipid as well as protruding from it.

Another Example

When phospholipids or fatty acids are added to intact erythrocytes, they enter the outer monolayer of the cell membrane and remain there for many hours without mixing into the inner monolayer. Eisinger et al. (Eisinger 82a,b, 83) have taken advantage of this fact in order to label erythrocyte membranes with (9-anthroyloxy)-stearic acids in which the fluorescent group was attached variously at the 2,6,9, and 12 carbon positions on the hydrocarbon chain. The emission spectrum of this fluorophore (λ_{max}=440nm) overlaps the Soret absorption band of hemoglobin, and one would expect resonant energy transfer provided hemoglobin is able to approach the inner surface of the membrane in the regions in which the labeled stearates are dispersed. The Förster radius for these probes transferring to hemoglobin varies from 33-36Å, thus a layer of hemoglobin just at the inner membrane surface is about one Förster radius from the donors in the outer monolayer.

After labeling a sample of cells with one of the donors and measuring the time profile of the fluorescence decay by front-face methods, the cells were lysed by osmotic shock and washed free of all hemoglobin. The fluorescence decay was again measured. Eisinger et al. did not make use of the shape of these decay curves; in fact, the time profile expression given there is incorrect. In addition, there is an experimental complication: there are reported to be (at least) two components to the decay of all the donors in the ghosts, and the presence of these various components makes the non-exponentiality of the time profile in the presence of energy transfer (see Dobretsov 82, discussed above) less obvious. Instead, the mean lifetime of each fluorophore was computed using an expansion of its decay into multiple exponentials, which were then solved for by the method of moments. This procedure is not only inappropriate but entirely unnecessary, as the mean lifetime is given exactly by the centroid of the time profile minus the centroid of that of the exciting lamp. Having obtained the mean lifetime in the presence and absence of energy transfer (i.e., in the intact cells and in their ghosts) the energy transfer efficiency was computed from

$$\langle T \rangle = 1 - \langle \tau_c \rangle / \langle \tau_g \rangle.$$

In Eisinger's approximation the computation of c/h^3 is straightforward when the depolarization factors of both the donors and acceptors are zero and dynamically averaged. For heme in hemoglobin this is only partially true; each heme, being a planar absorber, has $\langle d_A^x \rangle = 0.5$, but the four hemes in each tetramer are all at different orientations, and the hemoglobin molecules are assumed to be randomly oriented. Thus, part of the averaging is static, leading to slightly smaller values of $\langle T \rangle$ than if the averaging were completely dynamic. The donors, although dynamically averaged, are not isotropically disposed in the lipid. Their transition moments have a preferential orientation parallel to the membrane plane with a $\langle d_D^x \rangle$ of about 0.5. The way this preferential orientation affects $\langle T \rangle$ has been given in the main text. The quantity c/h^3 can now be computed from $\langle T \rangle$ as previously discussed.

There are two ways of proceeding from here. One can take given conditions (say pH 7.4, ionic strength 300mosm) and compute c/h^3, or better $h/\sqrt[3]{c}$, for the different donors. Since c is constant, a scale of proportionality can be found such that these values represent the distances along the hydrocarbon chain at which the fluorophores are attached. This gives c_b (heme concentration near the membrane-cytosol boundary) of 15mM, a value considerably lower than the average cell concentration, 21mM.

As conditions are varied, however, either c or h (or both) can change. At pH 7.4 the value of c/h^3 is a strong function of ionic strength. If all the dependence is ascribed to changes in concentration, then c_b varies up to about the close packed crystalline density (34mM) or to the mean cell concentration of 30mM at 750mosm as the intact cell shrinks in response to increasing osmolarily. On the other hand, as the pH is lowered to 6.1 (at 300mosm) the value of c/h^3 increases by more than a factor of three. Clearly this cannot be ascribed entirely to a change in c_b — h must have become smaller, i.e., the bilayer structure must have become thinner. Just which proportion of these changes are due to c and which to h cahnot in general be determined from this type of experiment using this analysis. This is an excellent example in which information about the cell architecture from a cell biologist would guide the interpretation of the physical experiment.

CONCLUSION

We have seen over the course of the previous sections on the analytical theory and in the several chosen examples that fluorescence energy transfer experiments can be used to determine certain structural parameters of membranes, of proteins with respect to membranes, and of membrane-bound proteins. It has been pointed out in many places that the data analysis of a given experiment may be quite complex, with many different subtle considerations having to be taken into account. Nonetheless, with proper experimental planning and careful analysis, useful information about many systems of physical-chemical or biological interest can be obtained.

Acknowledgements

I am indebted to my colleagues R. E. Dale, J. Eisinger, and A. A. Lamola for stimulating discussions on all aspects of fluorescence spectroscopy which have taken place over many years. The development of this lecture is based in large part on two published articles (Dale 79, Eisinger 81). I thank J. Eisinger for permission to quote results on energy transfer experiments in erythrocyte membranes which are at this time in press.

120

REFERENCES

Antonov-Romanovsky, V. V., and Galanin, M. D. (1957) Theoretical derivation of the law of luminescence decay for the case of resonance
   quenching. Optika i Spectroscopia 3:389-391.

Badley, R. A., (1976) in "Modern Fluorescence Spectroscopy," (E. L.
   Whery, ed.), Plenum Press, New York, pp. 91-168.

Berman, H. A., Yguerabide, J., and Taylor, P. (1980) Fluorescence energy
   transfer on acetylcholinesterase: spatial relationship between
   peripheral site and active site. Biochemistry 19:2226-2235.

Dale, R. E., Eisinger, J., and Blumberg, W. E. (1979) Orientational
   freedom of molecular probes — the orientation factor in
   intramolecular energy transfer. Biophysical J. 26:161-194.

Dale, R. E., Eisinger, J., and Blumberg, W. E. (1980) Correction: Orientational freedom of molecular probes — the orientation factor in
   intramolecular energy transfer. Biophysical J. 30:365.

Dobretsov, G. E., Spirin, M. M., Chekrygin, O. V., Karmansky, I. M.,
   Dmitriev, V. M., and Vladimirov, Y. A. (1982) A fluorescence study
   of apolipoprotein localization in relation to lipids in serum low
   density lipoproteins. Biochim. Biophys. Acta 710:172-180.

Eisinger, J., and Dale, R. E. (1974) Interpretation of intramolecular
   energy transfer experiments. J. Mol. Biol. 84:643-647.

Eisinger, J., Blumberg, W. E., and Dale, R. E. (1981) Orientational
   effects in intra- and intermolecular long range excitation energy
   transfer. Ann. New York Acad. Sci. 366:155-175.

Eisinger, J., Flores, J., and Salhany, J. M. (1982a) Association of
   cytosol hemoglobin with the membrane in intact erythrocytes. Proc.
   Natl. Acad. Sci. USA 79:408-412.

Eisinger, J., and Flores, J. (1982b) The relative locations of intramembrane fluorescent probes and of the cytosol hemoglobin in erythrocytes, studied by transverse resonance energy transfer. Biophysical J. 37:6-7.

Eisinger, J., and Flores, J. (1983) The cytosol-membrane interface of
   human erythrocytes — a resonance energy transfer study. Biophysical J. (in press).

Fleming, P. J., Koppel, D. E., Lau, A. L. Y., and Strittmatter, P.
   (1979) Intramembrane position of the fluorescent tryptophanyl residue in membrane-bound cytochrome b_5. Biochemistry 18:5458-5464.

Förster, T. (1948) Zwischenmolekulare energiewanderung und fluoreszenz.
   Ann. Physik 2:55-75.

Förster, T. (1949) Experimentelle und theoretische untersuchung des
   zwischenmolekularen Übergangs vom electronenanregungsenergie. Z.
   Naturforsch. 4a:322-327.

Förster, T. (1951) "Fluoreszenz Organischer Verbindung," Vandenhoeck and
   Ruprecht, Göttingen.

Förster, T. (1959) Transfer mechanisms of electronic excitation. Disc.
   Faraday Soc. 27:7-17.

Hillel, Z., and Wu, C.-W., (1976) Statistical interpretation of fluorescence energy transfer measurements in macromolecular systems. Biochemistry 15:2105-2113.

Jones, R. E. (1970) Nanosecond fluorimetry. Ph.D. thesis, Stanford University.

Koppel. D. E., Fleming, P. J., and Strittmatter, P. (1979) Intramembrane positions of membrane-bound chromophores determined by excitation energy transfer. Biochemistry 18:5450-5457.

Perrin, F. (1926) Polarisation de la lumière de fluorescence. Vie moyenne des molécules dans l'etat excité. J. Phys. (Paris) 7:390-401.

Radda, G. K., and Vanderkooi, J. (1972) Can fluorescent probes tell us anything about membranes? Biochim. Biophys. Acta 265:509-549.

Rozman, I. M. (1958) Theory of quenching of luminescence in solutions. Optika i Spectroscopiya 4:536-538.

Sklar, L. A., Miljanich, G. P., Bursten, S. L., and Dratz, E. A. (1979) Thermal lateral phase separations in bovine retinal rod outer segment membranes and phospholipids as evidenced by parinaric acid fluorescence polarization and energy transfer. J. Biol. Chem. 254:9583-9591.

Soleillet, P. (1929) Sur les paramètres caractérisant la polarisation partielle de la lumière dans les phénomènes de fluorescence. Ann. Physique 12:23-97.

Steinberg, I. Z. (1971) Long range non-radiative transfer of electronic excitation energy in proteins and polypeptides. Ann. Rev. Biochem. 40:83.

Tompa, H., and Englert, A. (1979) The frequency distribution of the orientation factor of dipole-dipole interaction. Biophys. Chem. 9:211-214.

Tweet, A. G., Bellamy, W. D., and Gaines, G. L. (1964) Fluorescence quenching and energy transfer in monomolecular films containing chlorophyll. J. Chem. Phys. 41:2068-2977.

Wolber, P. K., and Hudson, B. S. (1979) An analytic solution to the Förster energy transfer problem in two dimensions. Biophysical J. 28:197-210.

THE INTERACTION OF INTRINSIC PROTEINS AND LIPIDS IN BIOMEMBRANES

D. Chapman, J.C. Gomez-Fernandez and F.M. Goni

Royal Free Hospital School of Medicine
8 Hunter Street
London WC1N 1BP

The concept that biomembranes are built on a lipid bilayer
matrix is now commonly accepted. Suggestions have been made that
non-lamellar regions may also occur within the structure of natural
biomembranes[1], although this view is presently somewhat controversial[2,3]. The proteins associated with biomembranes are located
either within the lipid matrix (intrinsic proteins) or attached
to the hydrophilic faces of the bilayer (extrinsic proteins). In
some instances, intrinsic proteins may be held in position by a
cytoskeleton.

An important question is the degree to which the presence of
the intrinsic protein modulates the dynamics of the lipid matrix
and conversely how the lipid may modulate the dynamics of the
protein[4].

THE PERTURBATION OF LIPID DYNAMICS BY INTRINSIC PROTEINS

Early experiments with model systems showed that cholesterol
molecules within a lipid matrix order the lipid chains above the
phase transition temperature (Tc) but how do intrinsic proteins
and neighbouring lipids interact? Do intrinsic proteins form
stoichiometric complexes with the lipids where a specific number
of lipid molecules are 'bound' to the protein? These questions
have received much attention in recent years. From experimental
evidence we can conclude that intrinsic proteins do indeed perturb
fluid lipid bilayer dynamics. This perturbation usually extends
beyond the first lipid boundary, and decays rapidly, within a few
layers. Considering the phospholipid:protein ratios in most native
membranes it is likely that no unperturbed lipid exists in such
systems. With respect to the proposed formation of protein-lipid

complexes, we can say that on a timescale of $\sim 10^{-4}$s all the phospholipids form a homogeneous phase, hence the existence of a long-lived
shell surrounding the proteins is precluded, i.e. there is no
evidence for a stoichimetric complex.

Spin-label measurements by Jost and co-workers[5] on lipid
bilayer systems containing intrinsic proteins, i.e. cytochrome
oxidase, show two signals in the EPR spectra, one of which was
attributed to the presence of an immobilized spin label, whilst the
other was attributed to a more mobile spin label in the bulk lipid.
This led to the view that an immobilized lipid layer, termed the
boundary lipid, occurred around each protein. The amount of
immobilized lipid was said to correspond to the number of lipid
molecules in direct contact with the protein. Later workers[6] with
other intrinsic protein-lipid systems had similar results with spin
label studies, i.e. they observed two signals in the spectra, which
they interpreted as indicating the presence of an immobilized lipid
layer. They extended the concept of a boundary layer surrounding
an intrinsic protein to correspond with a long-lived shell of lipid
bound to and surrounding the protein with a lifetime in some cases
as long as minutes. This view was generally accepted for some time
by biochemists as being a further sophistication of what is sometimes
referred to as the fluid-mosaic model for biomembranes.

Recently, NMR studies have been applied to this problem. For
example, proton NMR of rod outer segment membranes[7] was interpreted
to show that although the rotational motion of the hydrocarbon chains
is affected by the proximity of an intrinsic protein, all the phospho
lipids can diffuse rapidly in the plane of the membrane. This
suggests that the translational diffusion rates of the phospholipids
are not altered by the protein during the measurement process, and
thus that the lipid phase is homogeneous. Deuterium magnetic resonance of various reconstituted systems where deuterated lipids have
been used[8-10], reveal a number of effects: (a) unlike the studies
using spin labelled lipids where two separate signals occur, the
NMR spectra show only one signal, i.e. the deuterium NMR studies
show that only one homogeneous lipid environment exists in the fluid
lipid bilayer state; there is no evidence in favour of a stoichiometric complex of protein and 'bound' lipid. (b) ^{2}H-NMR signals
are 'split', because of the quadrupolar moment of the nucleus which
has a spin value of higher than one half, namely $1 = 1$. The separation between the two peaks of ^{2}H-NMR signals is called 'the
quadrupole splitting' and from it an order parameter can be deduced.
It therefore provides information about the static order of the
system. The deuterium quadrupole splitting of the lipid shows only
slight changes due to the presence of intrinsic proteins in the
fluid lipid matrix and at high protein concentrations can even show
a decrease of splitting, in contrast to what is observed with
cholesterol where the splitting is markedly increased. This shows
that the fatty acyl lipid chains which are already disordered in

the fluid state may even increase slightly their <u>disorder</u> in the presence of the protein. (c) Another important parameter that can be retrieved from NMR spectra is the 'spin-lattice T_1 relaxation time', a measurement which refers to the speed with which nuclei in the upper energy state lose energy externally to the 'lattice', i.e. the molecular environment; the higher the molecular mobility of the environment, the less likely the 'coupling' between spin and 'lattice', and the longer the T_1 relaxation time. The relaxation times provide information about dynamic order. Deuterium T_1 relaxation times are shorter in the presence than in the absence of proteins, i.e. the rate of lipid chain segment re-orientation in the presence of protein is decreased. For some proteins and polypeptides, e.g. gramicidin A, close examination of the deuterium NMR data reveals that the quadrupole splitting, and hence the static order, increases, and then decreases at higher protein:lipid ratios[11].

The viewpoints based upon results from the two different techniques, ESR and NMR spectroscopy, the ESR method indicating two types of lipid population and the NMR only one, have led to some confusion.

With the ESR technique, we note that (a) whilst it is assumed that the spin label will mimic a natural lipid molecule, the probe will have its own inherent perturbation characteristics. (b) The time-scale of the ESR measurements is 10^{-8}s; when a perturbation occurs on a time-scale $\sim 10^{-7}$s, an immobilization of the spin probe will be seen to occur as a separate signal. (The time for a lipid to move its own diameter in pure fluid phospholipid bilayers[12] is about 10^{-7}s.) (c) The interpretation of the data assumes that the number of lipid molecules which are immobilized by the intrinsic protein is constant over the whole range of protein to lipid ratios. (d) The perturbation is often considered to extend only to the first lipid layer around the protein.

With the deuterium magnetic resonance method we note that (a) there should be no problems of inherent perturbation; (b) the time-scale for the appearance of two separate signals is $\sim 10^{-4}$s; (c) the quadrupole splitting provides information about the static order or disorder of the lipid chains; and (d) the relaxation times provide information about the dynamic disorder.

To some extent, the discrepancies between the conclusions derived from these two techniques may be rationalized in terms of the different time-scales involved. On a short time-scale, a molecule may appear to be immobile and yet mobile on a longer time-scale. However the ^{2}H-NMR technique is more reliable than the EPR because of its non-perturbing nature. ^{2}H-NMR gives a rather different view of the way intrinsic proteins can perturb the lipid dynamics, with its emphasis on non-stoichiometric complexes of protein and lipid and the slight disordering effects on lipid chains.

Further suggestions have been made to refine the interpretation of the deuterium magnetic resonance data. The data show a slight increase in the quadrupole splitting at a particular protein:lipid ratio and a decrease in splitting at higher protein concentrations. This can be interpreted to indicate that a single contact of a protein with a lipid chain causes a slight ordering of the lipid chain whilst lipid chains squeezed between intrinsic proteins become disordered[13]. If this is the case, the deuterium magnetic resonance results depend on the lipid:protein molar ratios of the samples considered. At high protein concentrations, the populations of the lipid squeezed among the protein are predominant.

An interesting line of evidence concerning the problem of protein-lipid perturbations comes from the use of a different probe molecule, i.e. fluorescent probes. Recent fluorescence polarization studies of diphenylhexatriene (DPH) included in reconstituted systems of varying protein:lipid ratios have been made. In this technique, a fluorescent probe is excited with plane-polarized light. Provided the absorption and emission dipoles are parallel or at a small angle to one another then the emitted radiation will also be polarized. If the probe changes its orientation between the excitation and emission processes then the depolarization of the emitted radiation will occur. The polarization value

$$P = \frac{I_{||} - I_{\perp}}{I_{||} - I_{\perp}}$$

where $I_{||}$ and $I_{\perp}$ are respectively the intensities observed parallel and perpendicular to the excitation plane, provides an index of the polarization characteristics of the emitted radiation. This technique, as applied to the study of protein-lipid recombinants, shows that an increase of protein concentration in the bilayer causes the steady state polarization values to increase and reach a maximum saturation value. Therefore, this probe apparently indicates that the order of the lipid system is _increased_ by the presence of the intrinsic protein. A naive interpretation of this result would be that the microviscosity of the lipid matrix is increased or that the chains are becoming more ordered. The latter conclusion contrasts with that deduced by NMR methods.

Again, the problem of interpreting the meaning of the data derived using probe molecules arises. This is another example of a probe molecule which can perturb its environment. Furthermore, it has to be remembered that although the fluorescent probe molecule is supposed to mimic a lipid molecule or chain, the DPH molecule is rigid and much larger than the C-C bonds of a lipid chain and may not be sensitive to any small disorder of the CH_2 groups[14]. A possible interpretation of the data is that when proteins are present in the lipid bilayer, the motion of the rigid probe is affected mainly by collisions with these intrinsic proteins rather

126

than providing a measure of the 'microviscosity' of the lipid matrix or the static order of the lipid chains.

PROTEIN-PROTEIN CONTACTS

In contrast with the usual assumption that each and every protein has its own shell of lipids separating it from its neighbouring proteins, recent fluorescence polarization and EPR data indicate that protein-protein contacts may occur within the lipid bilayer[14]. This interpretation is consistent with the deuterium NMR results.

The steady state fluorescent probe studies which show that the steady state polarization increases until, at high protein concentrations, a saturation effect occurs with little further change in polarization are interesting. The shape of the curves of steady state DPH polarization versus protein and large for a large protein. This behaviour can be understood when we consider that as the protein concentration increases in the lipid bilayer matrix, protein-protein contact becomes increasingly probable. Therefore, the maximum allowable contact of probe molecule with the protein decreases. These fluorescent probe studies thus indicate that the number of lipid chains influenced by an intrinsic protein is not constant but depends upon the protein concentration in the lipid matrix. Some spin label data on protein-lipid systems, e.g. cytochrome oxidase reconstituted systems, can also be fitted to the form of an exponential equation where the exponent corresponds to the number of lipid chains which at high dilution can contact the protein[14].

MEMBRANES IN A GEL STATE

So far we have discussed intrinsic protein perturbation and organization within a fluid bilayer. Many biological membranes are in the fluid state under physiological conditions. However, in some cell membranes, especially those without cholesterol, the phospholipids may crystallize when they are cooled below a critical temperature, Tc. This phase transition temperature separates a fluid from a crystalline (or gel) state. Below Tc, i.e. in the gel state, proteins can be forced by the crystallizing lipid into patches of high protein:lipid ratio.

This crystalline state is easily produced in reconstituted systems containing phospholipids with fully saturated fatty acids, and so these systems are useful in the study of intrinsic proteins in crystalline bilayers. The 'melting' of the crystalline lipid is a co-operative endothermic process, and consequently it can be conveniently monitored by calorimetric techniques. Of these, the most widely used is differential scanning calorimetry (DSC). Synthetic phospholipids containing a single type of fatty acid display,

127

when fully hydrated, highly co-operative endotherms appearing as narrow peaks upon heating across their main Tc transition temperature. DSC studies of intrinsic proteins reconstituted with defined phospholipids at various protein:lipid ratios show that, as the protein concentration is increased, the main transition peak broadens and decreases in height[15]. The area under the main peak corresponds to the enthalpy of the remaining transition.

To interpret these enthalpy data in terms of protein-lipid perturbation effects, it is necessary to ask: do the proteins remain in a similar disposition in the plane of the lipid bilayer in the fluid and crystalline states? The answer, in many cases if not all, is no. Freeze fracture electron microscopy studies show that upon crystallization of the lipid chains, a phase separation occurs leading to the production of high protein:lipid patches separate from areas corresponding mainly to pure lipid. Fig. 1. In some instances, the proteins within these patches may even adopt a crystalline arrangement. This occurs with reconstituted systems of bacteriorhodopsin[16]. In other cases, the lipids within the high protein:lipid patches continue to adopt a crystalline lipid arrangement (as occurs with the Ca-ATPase protein reconstituted into lipid-water systems[15]). Of course, as more protein is included into the lipid system, the aggregated patch increases in size at the expense of the remaining crystalline region, and the average transition enthalpy is reduced.

THE MODULATION OF PROTEIN STRUCTURE AND FUNCTION BY THE PHOSPHOLIPIDS

Apart from the effect of proteins on lipids, lipids also affect proteins. Although it is not clear whether protein conformation is appreciably affected, it has been found in many cases that protein activity, e.g. enzymatic activity, can be modulated by the lipid. An example of this is the Ca-ATPase in reconstituted DPPC lipid-water systems[17]. X-ray diffraction experiments show that, upon heating, the lipid within the patches melts at 30°C; this is in contrast to the pure lipid which undergoes its phase transition at 41°C. The patch melts at a lower temperature because of its high protein content. This has been confirmed by fluorescence depolarization studies carried out on the same system[15], where changes in polarization occur at both 30°C and 41°C. See Fig. 2. This melting at 30°C has a dramatic effect upon the protein as marked increase in rotational motion of this protein occurs and the enzyme activity sharply increases at this temperature. See Fig. 3 showing the marked change in rotation at 30°C.

It seems that the effects caused by membrane lipids on intrinsic proteins are mainly non-specific. Although some polar-head specificity has been claimed for certain enzymes this is still uncertain[18].

Fig. 1.   Freeze-fracture micrograph of DMPC-ATPase reconstituted systems.

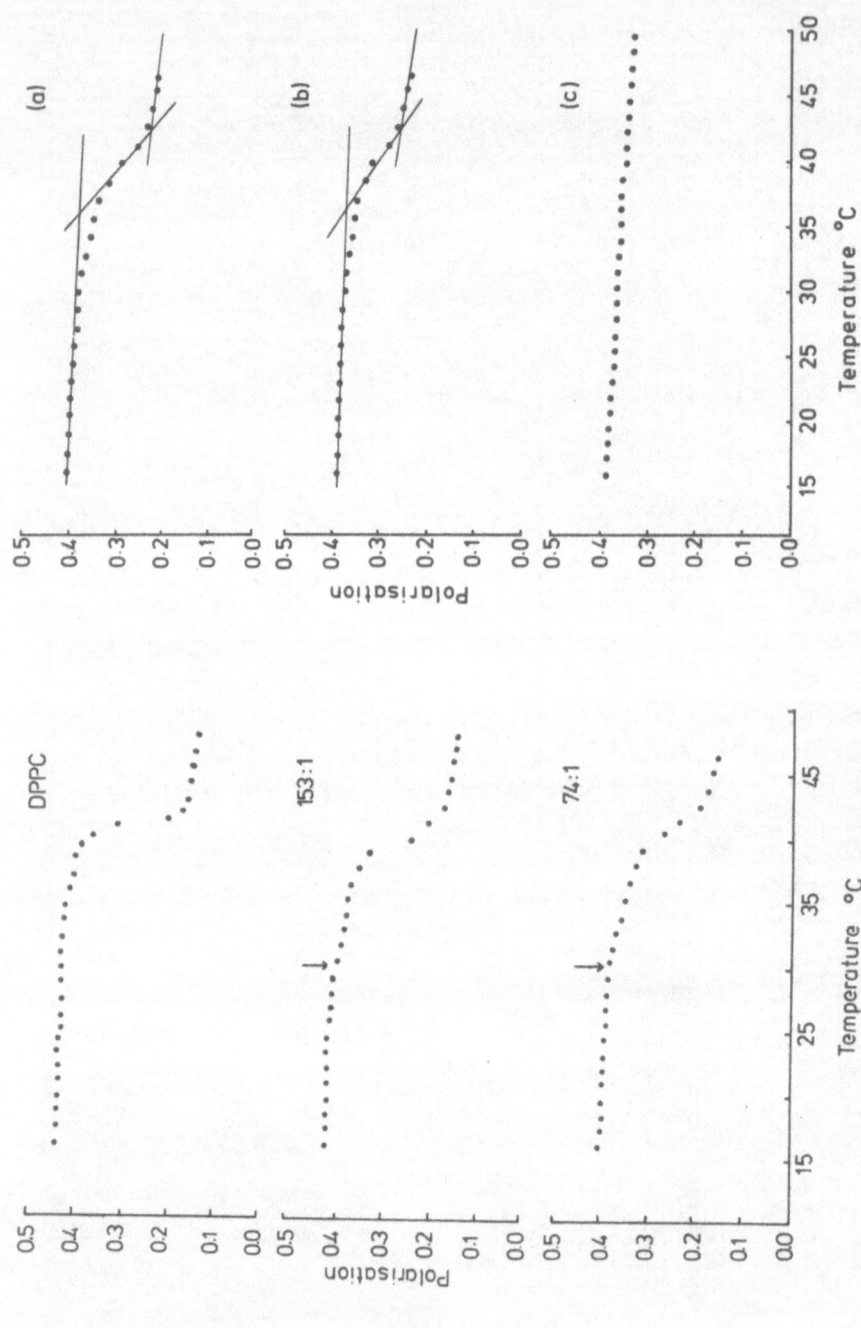

Fig. 2. (i) Temperature dependence of DPH polarisation (P) in DPPC-ATPase recombinants. The lipid protein molar rates are as indicated (the arrows mark 30°C). (ii) The temperature dependence of P for (a) 37:1 DPPC-ATPase recombinant, (b) 19:1 DPPC-ATPase recombinant and (c) 1:1 DPPC-cholesterol mixture.

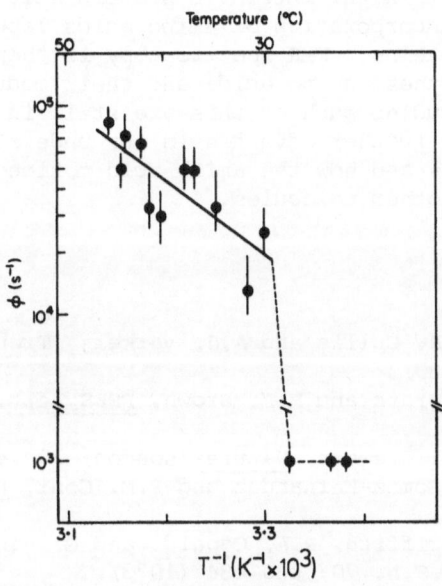

Fig. 3. Arrhenius plot of rotational motion parameter for a Ca^{2+}-ATPase recombitant having a lipid:protein molar ratio of 86:1. The line is drawn by linear regression; from the slope an activation energy of 67kJ mol^{-1} (16 kcal mol^{-1}) is calculated.

Non-specific factors governing the influence of lipids on intrinsic proteins arise from the fact that these proteins have to be at least partially surrounded by the hydrophobic part of an amphiphilic environment. The natural amphiphiles for membrane proteins are the phospholipids; however, phospholipids are not sp specifically required since they can be substituted by externally added amphiphiles (detergents)[19].

NEW METHODS

New methods are becoming available for studying how lipids (and other molecules) may perturb the dynamics of the proteins and also of the amino acids within the intrinsic protein structure. One new method involves the incorporation of amino acids labelled with ^{2}H and ^{15}N into membrane proteins. NMR spectroscopy is then used to determine the dynamics of these amino acids and their modulation within the biomembranes. Studies such as this are still in progress but the future should see further advances in our understanding of membrane protein dynamics and how the amino acid motion can be modulated by interactions with other molecules.

REFERENCES

1. B. de Kruijff, P.R. Cullis and A.J. Verkeij, Trends in Biochem. Sci. 5, 79-81 (1980).
2. A.J. Deese, E.A. Dratz and M.F. Brown, FEBS Lett. 124, 92-99 (1981).
3. S.W. Hui, and T.P. Stewart, Nature (London) 290,427 (1981).
4. D. Chapman, J.C. Gomez-Fernandez and F.M. Goni, FEBS Lett. 98, 211-223 (1979).
5. P.C. Jost, O.H. Griffith, R.A. Capaldi and G. Vanderkooi, Proc. Nat. Acad. Sci. U.S.A. 70, 480-484 (1973).
6. T.R. Hesketh, G.A. Smith, M.D. Houslay, K.A. McGill, N.J.M. Birdsall, J.C. Metcalfe and G.B. Warren, Biochemistry 15, 4145-4151 (1976).
7. M.F. Brown, G.P. Miljanich and E.A. Dratz, Biochemistry 16, 2640-2648 (1977).
8. E. Oldfield, R. Gilmore, M. Glaser, H.S. Gutowski, J.C. Hschung, S. Kang, M. Meadows and D. Rice, Proc. Nat. Acad. Sci. U.S.A. 75, 4657-4660 (1978).
9. A. Seelig and J. Seelig, Hoppe Seyler's Z. Physiol. Chem. 359, 1747-1756 (1978).
10. D. Rice, M.D. Meadows, A.O. Scheinman, F.M. Goni, J.C. Gomez-Fernandez, M.A. Moscarello, D. Chapman and E. Oldfield, Biochemistry, 18, 5893-5903 (1979).
11. D. Rice and E. Oldfield, Biochemistry 18, 3272-3279 (1979).

12. H. Traube and E. Sackmann, <u>J. Am. Chem. Soc.</u> 94, 4499–4510 (1972).
13. D.A. Pink, A. Georgallas and D. Chapman, <u>Biochemistry</u> 20, 7152–7157 (1981).
14. W. Hoffmann, D.A. Pink, C.J. Restall and D. Chapman, <u>Eur. J. Biochem.</u> 114, 585–589 (1981).
15. J.C. Gomez-Fernandez, F.M. Goni, D. Bach, C.J. Restall and D. Chapman, <u>Biochim. Biophys Acta</u> 598, 502–516 (1980).
16. A. Alonso, C.J. Restall, M. Turner, J.C. Gomez-Fernandez, F.M. Goni and D. Chapman, <u>Biochim. Biophys. Acta</u> 689, 283–289 (1982).
17. W. Hoffmann, M.G. Sarzala, J.C. Gomez-Fernandez, F.M. Goni, C.J. Restall, D. Chapman, G. Heppeler and W. Kreutz, <u>J. Mol. Biol.</u> 141, 119–132 (1980).
18. H. Sandermann Jr, <u>Biochim. Biophys. Acta</u> 515, 209–237 (1978).
19. W.L. Dean and C. Tanford, J. Biol. Chem. 252, 3351–3353 (1977).

CONFORMATIONAL STUDIES OF MEMBRANE PROTEINS BY HIGH-RESOLUTION NMR

Larry R. Brown

Michigan Molecular Institute

Midland, Michigan 48640

The lack of conformational information for membrane proteins is a great impediment to understanding the functional mechanisms of biological membranes. The problem of associated lipids has prevented extensive application of X-ray diffraction or high-resolution NMR to conformational characterization of membrane proteins. Difficulties in obtaining crystals suggests that high-resolution NMR might be a more promising approach for such proteins. With currently available methods, high-resolution ^{1}H NMR spectra suitable for detailed conformational analyses are not obtained from proteins bound to native membranes[1,2] or from reconstituted systems consisting of proteins bound to phospholipid vesicles.[3,4] We have been able to obtain appropriate NMR spectra for membrane proteins and polypeptides incorporated into fully deuterated micelles. Proteins which have been tested include melittin,[5] glucagon,[6] phospholipase A_2, glycophorin A, snake venom cardiotoxin and myelin basic protein (unpublished results). For these proteins, the conformation of the membrane protein is preserved in the micelle and the micellar protein-lipid complex is homogeneous and stoichiometrically well-defined[5-7] (unpublished results).

Although it has long been recognized that high-resolution NMR has the potential to provide many-parameter characterizations of protein conformation and dynamics, practical applications have been limited by difficulty in resolving and assigning resonances corresponding to individual nuclei within a protein molecule. This situation has recently been dramatically changed by the advent of two-dimensional NMR methods.[8,9] For conformational studies of proteins, particularly useful methods are 2D correlated experiments,[8,10] which allow spin systems from many amino acids to be identified in a single experiment;[11,12] 2D J-resolved experiments,[13,14] which allow large numbers of scalar coupling constants to be measured accurately;[11] and 2D cross-relaxation experiments,[15,16] which allow pairwise spatial proximity between large numbers of pairs of nuclei to be simultaneously identified.[17,18] The large numbers of spectral parameters which can be obtained with these new spectroscopic methods has led to the development of a new computational procedure, i.e. a distance-geometry algorithm, for using NMR data to calculate three-dimensional conformations.[19,20] With the new NMR and computational methods, it is now possible in principle to determine three-dimensional conformations for small proteins from NMR data. These new methods are illustrated with a study of the conformation of melittin bound to dodecylphosphocholine micelles.

Obtaining a Suitable High-Resolution ^{1}H NMR Spectrum

From Figure 1, it is apparent that the ^{1}H NMR spectra of monomeric melittin free in solution and of melittin bound to deuterated micelles show comparable spectral resolution. This means that any high-resolution NMR experiments appropriate for conformational studies of the water-soluble form of the polypeptide will also be applicable to the micelle-bound form of melittin.

Assigning ^{1}H NMR Resonances to Amino Acid Types

Melittin is a medium sized polypeptide. Nonetheless, the

Gly-Ile-Gly-Ala-Val-Leu-Lys-Val-Leu-Thr-Thr-Gly-Leu-Pro-Ala-
1   5   10   15

Leu-Ile-Ser-Trp-Ile-Lys-Arg-Lys-Arg-Gln-Gln-NH$_2$
16   20   25

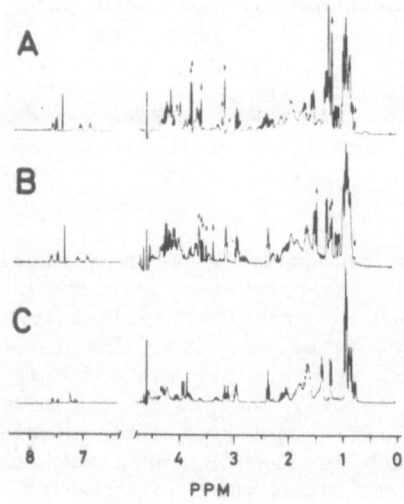

Figure 1. 360 MHz ^{1}H NMR spectra of:

(A) melittin bound to deuterated dodecylphosphocholine micelles
(B) melittin bound to deuterated dodecyl-(oxyethylene)$_8$-glycol micelles
(C) monomeric melittin free in solution
(Taken from Ref. 5 with permission)

one-dimensional ^{1}H NMR spectrum is complicated and not easily analyzed (Fig. 1). Figure 2 shows a 360 MHz 2D spin-echo correlated ^{1}H NMR spectrum of micelle-bound melittin. In experiments of this type, the structural features of the different types of amino acid residues are directly reflected in the connectivity patterns observed for the NMR resonances. For example, the connectivities shown for alanine, threonine and valine in Figure 2 allowed assignment of the respective resonances to specific hydrogen atoms in these amino acids. Of the 179 non-labile hydrogen atoms in melittin, resonances corresponding to 165 hydrogens were identified and resonances corresponding to 150 hydrogen atoms were assigned to specific amino acid types.[12]

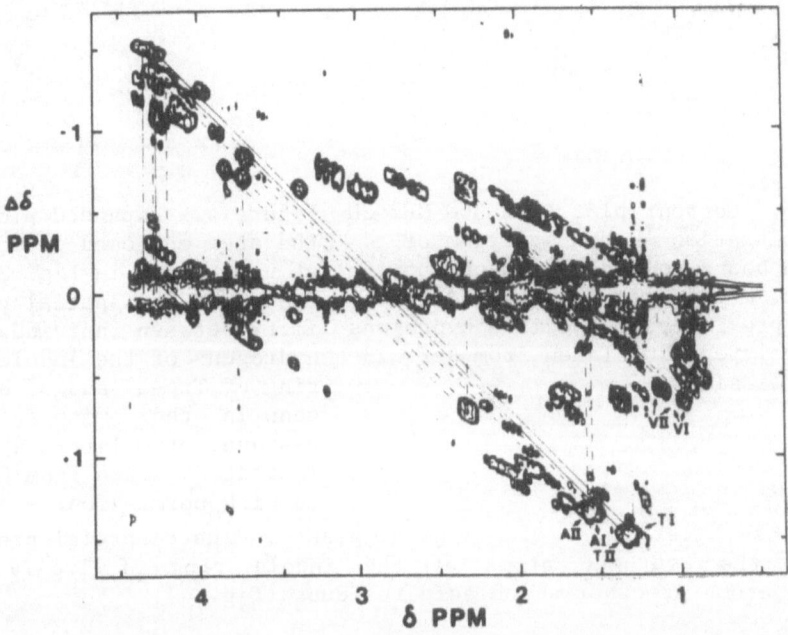

Figure 2. Contour plot of a 360 MHz spin-echo correlated ^{1}H NMR spectrum of melittin bound to deuterated dodecylphosphocholine micelles. Connectivities between the individual components of the following types of spin systems are indicated: ——— Thr; ----- Ala; —•—• Val (Taken from Ref. 12 with permission).

Detection of Nearest Neighbors in Space

In 2D cross-relaxation ^{1}H NMR spectra, the off-diagonal peaks correspond to pairs of hydrogen atoms which are close together in space. The folding of the polypeptide chain may bring hydrogen atoms into proximity even though there are many intervening bonds along the covalent structure. The off-diagonal peaks enclosed in the dotted area at upper left in Fig. 3 correspond to

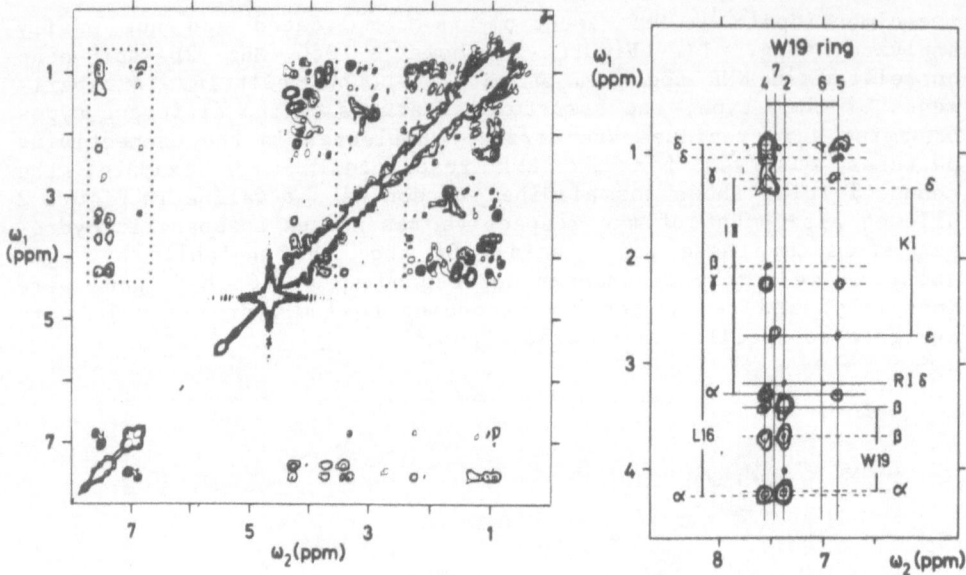

Figure 3. Contour plot of a 360 MHz 2D cross-relaxation ^{1}H NMR spectrum of melittin bound to dodecylphosphocholine micelles. Off-diagonal peaks indicate spatial proximity of specific hydrogens in the polypeptide. Taken from Ref. 18 with permission.

Figure 4. Expanded plot of the area enclosed in dots at upper left in Fig. 3. Peaks showing spatial proximity between individual hydrogens of the indole ring of Trp-19 and of hydrogens in other amino acid residues of melittin are indicated. Taken from Ref. 18 with permission.

such a case, i.e. these off-diagonal peaks indicate spatial proximity of the hydrogen atoms of the indole ring of Trp-19 to hydrogen atoms of other amino acid residues (Fig. 4).

Assignments to Specific Positions in the Amino Acid Sequence

2D correlated and 2D cross-relaxation ^{1}H NMR experiments can be performed under conditions where the labile amide NH hydrogens can be observed in the NMR spectra. In favorable cases, these experiments allow the NMR resonances to be assigned to specific residues in the amino acid sequence.[21] Experiments of this type are in progress. So far, specific assignments have been obtained for residues 10–24, e.g. isoleucine resonances detected in Figure 2 have been specifically assigned to Ile-17 and Ile-20.

Calculation of Three-Dimensional Structures

Based on the experimental parameters used to obtain the 2D cross-relaxation spectrum in Figure 3, it was assumed that the

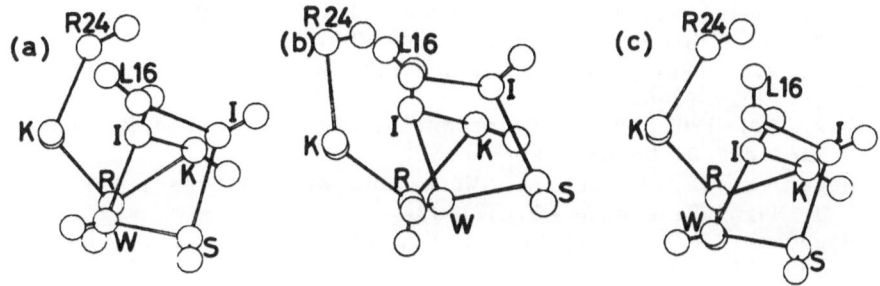

Figure 5. Computer drawings of three different structures for residues 16–24 of micelle-bound melittin. Only the positions of the α– and β–carbon atoms are shown. Taken from Ref. 18 with permission.

existence of an off-diagonal peak indicated that the corresponding hydrogen atoms were separated by at most 4Å.[18] These upper bounds were then used as input to a distance-geometry algorithm in order to calculate a three-dimensional conformation for residues 16–24 of micelle-bound melittin. Because the experimental data was used only semi-quantitatively to give upper bound constraints on inter-atomic distances, more than one structure will be consistent with the constraints. A group of structures which satisfy the constraints for residues 16–24 of micelle-bound melittin is shown in Fig. 5. Although the structures are not identical, in each case the polypeptide backbone shows two irregular right-handed turns. Because it is observed in all calculated structures, this feature, which resembles a right-handed α–helix, is considered to correspond to the real structure of micelle-bound melittin.[18]

ACKNOWLEDGMENTS

Acknowledgements. I acknowledge the contributions to this work made by many colleagues at the E.T.H., Zurich.[3,6,7,12,18,20]

REFERENCES

1. Brown, L.R., Bradbury, J.H., Austin, K. and Stewart, P.R., 1975, J. Membrane Biol. 24, 35–54.
2. Austin, K. Brown, L.R. and Stewart, P.R., 1975, J. Membrane Biol. 24, 55–69.
3. Brown, L.R. and Wüthrich, K., 1977, Biochim. Biophys. Acta 468, 389–410.
4. Feigenson, G.W., Meers, P.R. and Kingsley, P.B., 1977, Biochim. Biophys. Acta 471, 487–491.
5. Brown, L.R., 1979, Biochim. Biophys. Acta 557, 135–148.
6. Bösch, C., Brown, L.R. and Wüthrich, K., 1980, Biochim. Biophys. Acta 603, 298–312.

7. Lauterwein, J., Bösch, C., Brown, L.R. and Wüthrich, K., 1979, Biochim. Biophys. Acta 556, 244-264.
8. Aue, W.P., Bertholdi, E. and Ernst, R.R., 1976, J. Chem. Phys. 64, 2229-2246.
9. Bax, A., 1982, "Two-Dimensional Nuclear Magnetic Resonance in Liquids" D. Reidel, Dordrecht.
10. Nagayama, K., Kumar, A., Wüthrich, K. and Ernst, R.R., 1980, J. Magn. Resonance 40, 321-334.
11. Nagayama, K. and Wüthrich, K., 1981, Eur. J. Biochem. 114, 365-374.
12. Brown, L.R. and Wüthrich, K., 1981, Biochim. Biophys. Acta 647, 95-111.
13. Aue, W.P., Karhan, J. and Ernst, R.R., 1976, J. Chem. Phys. 64, 4226-4237.
14. Nagayama, K., Bachmann, P., Wüthrich, K. and Ernst, R.R., 1978, J. Magn. Resonance 31, 133-148.
15. Macura, S. and Ernst, R.R., 1980, Mol. Phys. 41, 95-117.
16. Kumar, A., Ernst, R.R. and Wüthrich, K., 1980, Biochem. Biophys. Res. Commun. 95, 1-6.
17. Wagner, G. and Wüthrich, K., 1982, J. Mol. Biol. 155, 347-366.
18. Brown, L.R., Braun, W., Kumar, A. and Wüthrich, K., 1982, Biophys. J. 37, 319-328.
19. Crippen, G.M. and Havel, T.F., 1978, Acta Cryst. A34, 282-284.
20. Braun, W., Bösch, C., Brown, L.R., Gō, N. and Wüthrich, K., 1980, Biochim. Biophys. Acta 667, 377-396.
21. Billeter, M., Braun, W. and Wüthrich, K., 1982, J. Mol. Biol. 155, 321-346.

WATER ORGANIZATION IN REVERSED MICELLES

C. A. Boicelli, F. Conti, M. Giomini
and A. M. Giuliani

CNR, University of Bologna, Italy; University of
Rome, Italy; CNR, Rome, Italy

1. INTRODUCTION

Biological membranes control cell life, modulating transport and exchange processes between the endocellular and the extracellular compartments. In other words, the metabolism of the cells and their relationships with the environment are strongly dependent on membrane function. Membrane activity, in turn, is influenced by the physical and chemical properties of the inner and outer compartments.

It is difficult to study the physical chemistry of the cellular interior, because it cannot be obtained as an isolated system and the exchange phenomena with the external medium always give an averaged picture of the two environments. Besides, biological membranes are a quite complex system and it is necessary to use models, like liposomes and single-bilayer vesicles, to get some insight into their properties.

Moreover, the models cannot give reliable information about the internal compartment since they suffer from the same drawback, i.e. exchange phenomena, as the real system. The use of a more simplified model thus becomes necessary.

A suitable experimental model to gain access to the properties of the internal region of the liposomes (and cells) seems to be reversed micelles, where the inside-outside exchange processes are reduced to a minimum and are, in any case, very slow.

In particular, the microdynamics of the micellar water pool becomes accessible. Nuclear magnetic resonance (NMR) and infrared spectroscopy (IR) provide suitable tools for this study, since their

141

time windows match precisely the time scale of the various aspects of the water dynamics.

2. REVERSED MICELLES AS MODEL

Reversed micelles (RM) seem to be a suitable working model to obtain some information about the internal compartment of liposomes. The micelles are obtained by dissolving a phospholipid, namely L-α-phosphatidylcholine from egg yolk (EPC), in a solvent sparingly miscible with water (i.e. benzene), adding the required amount of water and mildly sonicating until a translucent suspension is obtained. The RM which are prepared in this way contain water on the inside.

In the experimental system which we have used as a model, ca. 30 water molecules per phosphocholine head were present and the average micellar diameter (as determined by electron microscopy) was 130 ± 20 Å; each micelle contained from 700 to 800 EPC molecules. The use of RM to describe the interior of liposomes is supported by the proton relaxation data of the system under discussion and of some related systems[1], which are reported in Table 1.

The data reported in Table 1 clearly show that the dynamics of the choline polar group $-\overset{+}{N}(CH_3)_3$ are the same in our system as in liposomes, where the water-lipid interface is found at the level of the ester groups[3], while the fatty acid residues behave as in organic solvents (benzene or chloroform), and the phospholipid is present in the form of micelles. The relaxation behavior of the glycerol and choline $-CH_2$-groups, which are located at the organic solvent-water interface (the polar heads have been determined to be parallel to the micellar wall[4] is different from both the liposomal water dispersion and the micellar suspensions. This and other features of the data in Table 1 are discussed elsewhere[1]. In our systems the EPC used contained between 3 and 4 water molecules per polar head, which have not been removed. It may be concluded that phospholipids in RM behave as in liposomes as far as the moiety in the water pool is concerned, and their use as models for the liposomal interior is therefore justified.

3. MAGNETIC RELAXATION AND MICRODYNAMICS

The most abundant chemical species in RM is certainly water and water can provide a great deal of information about the internal region, by measuring its relaxation times T_1 and T_2[5].

The longitudinal or spin-lattice relaxation time, T_1, characterizes the exchange of energy with the environment (i.e. lattice), which is an enthalpic process. The transversal or spin-spin relax-

Table 1. Proton Longitudinal Relaxation Times T_1 of EPC in Different Media

| Medium | T_1 (ms) | | | |
	$\overset{(+)}{-\text{N}}(CH_3)_3$	$-CH_2-$ (glyc. + chol.)	$-CH_2-$ (acyl chains)	$-CH_3$
$C_6{}^2H_6$ a)	218	488	901	1518
C^2HCl_3 b)	150		890	3290
$C_6{}^2H_6/{}^2H_2O$ a)	384	885 c)	1118	2852
${}^2H_2O$ b)	390		470	540

a) Measured at 20 MHz; accuracy of the measurements is ±20 ms and reproducibility ±10 ms. The system contains 3-4 moles H_2O/mole EPC

b) Measured at 100 MHz; contains 1 mole H_2O/mole EPC[2].

c) Determined in reversed micelles of dimyristoil-L-α-phosphatidyl-choline-d$_{54}$, deuterated at the fatty-acid chains.

ation is an entropic process and the time T_2 characterizes the exchange of energy within the spin system. Both T_1 and T_2 are sensitive to several factors; the concentration and the mobility of the involved nuclei, the medium, the presence of solutes, the viscosity, geometrical constraints from the environment, temperature, the presence of different phases, are some examples. All these factors modify the motional properties of the nuclear system under study in different ways, since the correlation time τ_c changes and the relaxation times are modified accordingly.

The correlation time τ_c gives an indication of how fast a certain type of motion occurs; the translational correlation time τ_t specifies how fast a system bearing the observed nucleus moves through a distance comparable with its molecular dimensions, the rotational one τ_r indicates the time necessary for a system to rotate through a radian and so on. The dependence of the relaxation times on the correlation time is well known[6-9] and is shown in Figure 1. The pertinent theory applies to homogeneous ideal fluids but it can be extended, even if only qualitatively, to heterogeneous systems. In the case of water, which is our main interest in this context, it can be said that the shorter the relaxation times, the longer are the correlation times and the more organized is the water in the system under investigation. For instance, the longitudinal relaxation time of water changes from 12 s at 95°C, near the boiling point, to 1.7 s at 0°C (when ice is formed).

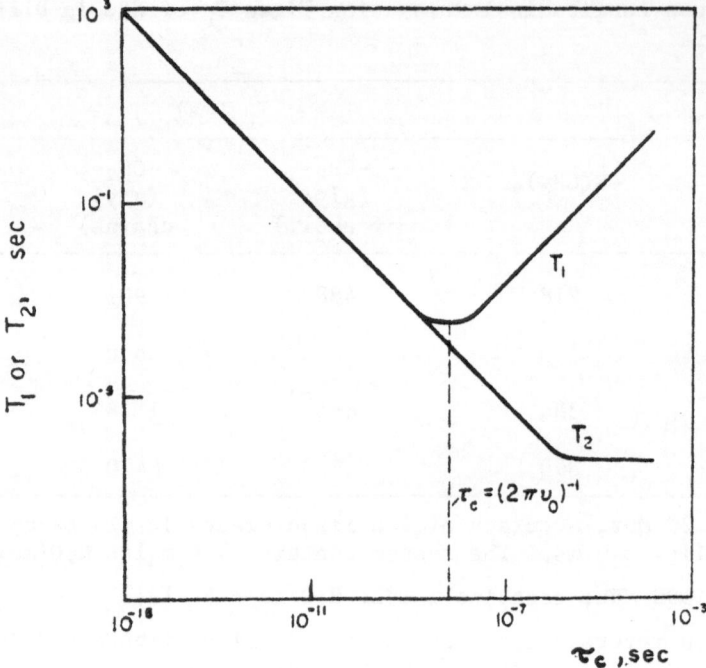

Fig. 1. The behavior of the relaxation times T_1 and T_2 as a function of the correlation time τ_c.

In biological samples and in model systems, however, the water longitudinal relaxation time is even shorter than in the ice and in some cases more than one value for T_1 and/or T_2 has been measured[10]. Several hypotheses have been proposed to explain these experimental results, for example, the existence of different types of water in exchange with one another or a multiple sites exchange model. In these suggested models, if the exchange rate is slow, different relaxation times can, under certain conditions, be measured[10] and the relative populations of the different types of water or the percentage of time spent in the various sites can also be evaluated. On the other hand, for exchange processes fast on the nuclear magnetic resonance (NMR) time scale only an average value of T_1 and/or T_2 can be determined[11]. Alternative models involve a discontinuous distribution of correlation times with fast exchange of the water molecules from one situation to the other[12] or a continuous distribution of correlation times with the water molecules slowly exchanging among different conditions[13].

Another approach to the problem of water relaxation in biological specimen is known as "cage-model": it invokes low-dimensional effects, whereby the mere presence of geometrical constraints

144

modifies the water microdynamics even in the absence of direct interactions of the water molecules with the "cage" boundary[14].

Another phenomenon which modifies the relaxation times of water
in biological systems is cross-relaxation by spin-diffusion[15].
This mechanism markedly affects the relaxation of water through the
dipolar interactions between the protons of the macromolecules and
those of water in the hydration layer[16]; as a result of such interactions, the relaxation times of the protons of the water and of the
macromolecule tend to become equal.

4. THE MEASUREMENTS OF T_1.

The longitudinal relaxation time T_1 is the time constant characteristic of the relaxation process along the $\vec{B}_0$ field direction. All
the methods for its determination entail tipping the macroscopic
magnetization vector $\vec{M}_0$ away from the static field direction and
observing how the net magnetization recovers with time along the same
axis. Among the various procedures for the determination of
T_1[17-19], the most convenient in the case of biological samples is
the so called "inversion-recovery"[17], because it is well suited for
the short T_1 values involved and it allows an easy detection of
different relaxation rates, when they exist.

The inversion-recovery method rests on the application of the
pulse sequence $(\pi-\tau-\pi/2-T)$ where π and $\pi/2$ (flip angles) indicate
the pulses which tip the magnetization of π and $\pi/2$ radians, respectively. During the variable time interval τ between the two pulses
the magnetization recovers to a value M_τ. T is the delay which must
be introduced before repeating the sequence, to restore the magnetization to its equilibrium value.

The plot $\ln(M_0 - M_\tau)$ vs τ is linear and its slope $-1/T_1$ yields
the T_1 value, while the intercept with the vertical axis gives the
value of the equilibrium magnetization. If more than one relaxation
time characterizes the system under study, the plot will appear as a
sequence of linear segments, provided the T_1 values are not too
similar[10]. The slope of each segment yields a T_1 value, while the
intercepts on the y axis allow the percentage of the different types
of protons to be determined.

Also T_2 has been measured for a few reverse micellar systems and
the correlation time hence derived with standard calculations. It
has been found that for our model system it falls in the region of
the times of Figure 1, where $T_1 \simeq T_2$.

Therefore we have measured only T_1 and the methods for determing
T_2 and its significance will not be discussed in this paper.

5. INFRARED SPECTROSCOPY AND WATER

The application of vibrational spectroscopy to the problem of structure determination of systems of biological interest requires studies in aqueous solution, since water forms the natural environment for the majority of them. However, because of the complexity of the systems considered, vibrational spectra can be explained only if an adequate understanding of the spectroscopic behavior of water itself has been attained.

Infrared spectroscopy is a useful technique for the study of the structure of water, since the periods of vibration for both the intramolecular and the intermolecular modes are short compared with the time of diffusional motions of the molecules (in the order of 10^- s and 10^{-11} s, respectively). The analysis of the infrared spectrum gives information on the local environment of the molecules and on their relative positions and offers the opportunity to learn details both of the dynamic behavior of the components of complex systems, like solutions, and of the interactions among them[20,21].

The isolated water molecule has three modes of vibration, namely the symmetric and the antisymmetric stretching and the bending. In the liquid phase, such a simple situation is complicated by the "polymeric" nature of water, due to H-bonds which give rise to changes in the effective local symmetry of the system.

In the 4000-1000 cm^{-1} wavenumber range, the infrared spectrum of water is characterized by the presence of three separate bands.

1) A most prominent, broad, irregularly shaped band with the maximum located near 3490 cm^{-1}. The O-H stretching vibrations are responsible for this absorption, with a probable contribution of the first overtone of the H-O-H bending mode: the presence of an extended network of H-bonds broadens this band and increases its integrated intensity. This spectral region of water is the most thoroughly studied and gives the most interesting information.
2) A very broad but very weak band centered near 2125 cm^{-1}, which may be due to overtones of intermolecular modes and/or to the combination of the bending vibration with such modes.
3) A band located near 1650 cm^{-1}, which arises from the H-O-H bending mode of the molecules[22].

6. ORGANIZATION OF THE WATER IN THE MICELLAR CAVITY

The NMR relaxation parameters of RM change with the amount of water present inside the micelle.

RM prepared by adding increasing amounts of water to EPC in benzene show increasing values of the proton[23] and ^{31}P[1] relaxation times, until the molar ratio of water to EPC (w_0) becomes larger than ca. 25, when these parameters become constant (Figure 2).

All the data discussed in this paper were obtained for systems with a w_0 value of ca. 30, that is in conditions where the measured parameters were independent of the amount of water.

The proton relaxation times have been measured with a low resolution NMR spectrometer (operating at 20 MHz) since with a high resolution instrument a considerable part of the magnetization is lost before sampling, because of the instrumental dead time. Moreover, at 20 MHz, T_1 is still sensitive to dynamic phenomena occurring with a correlation time of the order of nanoseconds or longer[24]. The logarithmic plots of the proton magnetization decay, obtained by application of the inversion-recovery pulse sequence, show that the protons in the sample relax at distinctly different rates. In Figure 3 such plots are reported for RM containing only water in the internal compartment: in one case (a) deuterated water has been used to observe the relaxation rates of EPC protons; in the other (b), also, the protons of the added water H_2O contribute to the relaxation.

From the linear segments of such plots, several T_1 values can be extracted. The fraction of protons contributing to each segment can be calculated from the intercepts on the vertical axis and the known total number of protons of the systems. The presence of two types of water with two different time constants[1] can be easily deduced comparing the two plots. The relative amounts and dynamic characteristics of these two types of water are readily obtained from the experimental data.

One type of water is more "structured", has a shorter relaxation time (305 ms) and its relation population is 35±5%, while the other has a longer relaxation time (597 ms) and a lesser degree of "organization". We will indicate, from now on, the former population as type "A" water and its relaxation time at T_1^A, and the other as type "B" water with a relaxation time T_1^B.

The infrared spectrum of our RM containing only water, recorded against a micellar solution of EPC in benzene, shows in the O-H stretching region a broad absorption, which may be separated in three components by instrumental deconvolution[25]. These components have maxima centered at ca. 3650 cm^{-1} (due to free water molecules not involved in the H-bond network), ca. 3500 cm^{-1} (associated with dimeric water) and ca. 3300 cm^{-1} (originated from "polymeric" water with an extended H-bond network). Infrared spectroscopy (IR) thus provides evidence for the existence within the RM of several types of water, which are in slow exchange on the IR time scale. NMR spectroscopy, which is a slower technique, sees only two water populations and the third one must be in fast exchange (on the NMR time scale)

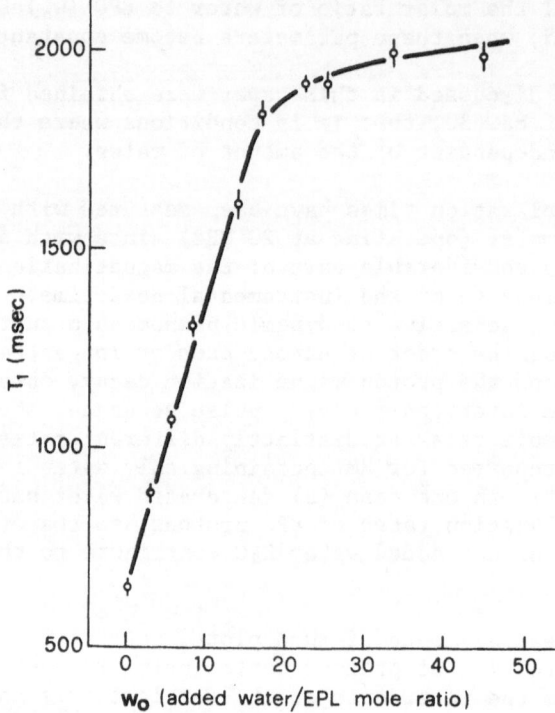

Fig. 2. Dependence of the ^{31}P longitudinal relaxation time on the added water/EPC mole ratio in reversed micelles.

with one of them. It may be supposed that some other technique, faster than infrared, could detect other types of water.

The description of the water organization given by the infrared data is unfortunately only qualitative since the extinction coefficients of the three components of the O-H stretching absorption are at present unknown. Only the area under the deconvoluted bands can therefore be evaluated, but not the three relative water populations.

7. THE EFFECT OF HYDROPHILIC PERTURBATORS

Many important biological processes take place in the region near cellular membranes, where the water properties are distinctly different from those of bulk water. It is of critical interest to understand the nature of the interactions of this interfacial water with membrane proteins or at least with the solvent exposed amino-acidic residues.

Since the water inside RM seems to have properties quite similar to the water adjacent to cell membranes, it is interesting to invest-

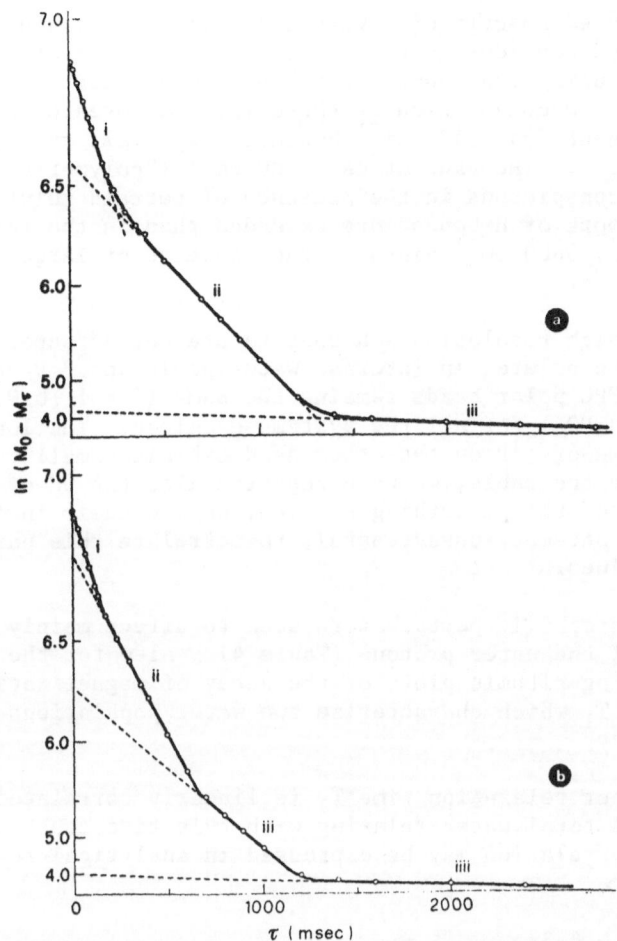

Fig. 3. Logarithmic plots of the ^1H magnetization decay for reversed micelles of EPC. (a) EPC solution in $C_6{}^2H_6$ (50 mg/ml) + ^2H_2O (30 µl/ml): (i) $\overset{+}{N}(CH_3)_3$ + choline + glycerol protons; (ii) fatty acyl chains protons; (iii) terminal methyl protons. (b) EPC solution in $C_6{}^2H_6$ (50 mg/ml) + H_2O (30 µl/ml): (i) $\overset{+}{N}(CH_3)_3$ + H_2O^A; (ii) choline + glycerol protons + H_2O^B; (iii) fatty acyl chains protons; (iiii) terminal methyl protons.

igate how the presence of amino-acids and small peptides (the "solutes") in the micellar cavity modifies the physico-chemical characteristics of the water pools. The perturbations induced by these species can arise either from direct chemical interactions or from physical effects and may afford information on the dynamic properties of the system.

The infrared spectra of reversed micelles containing selected amino-acids and peptides in the internal aqueous compartment have been recorded using the "unperturbed" system as reference (Table 2 and Figure 4). Deconvolution of these spectra reveals that the central component (ca. 3490 cm^{-1} becomes very small and cancels out in several cases. The band at ca. 3280 cm^{-1} ("polymeric" water) is instead very conspicuous in the presence of certain solutes, suggesting a network of H-bonds more extended than in the reference and the band at ca. 3600 cm^{-1} hints at the presence of larger amounts of "monomeric" water.

The ^{13}C high resolution NMR spectra are not affected by the presence of the solutes in internal water pools and ^{31}P chemical shift of the EPC polar heads remains the same ($\delta = 1.16\pm0.02$ ppm from external 85% H_3PO_4) for all the systems examined. The longitudinal relaxation time of ^{31}P on the other hand exhibits small variations (Table 3). In the table, we have reported also the pH of the starting solutions of the perturbing species, before their inclusion in the RM, in an attempt, unsuccessful, to correlate this parameter with the ^{31}P T_1 values.

These hydrophilic perturbators seem to affect mainly the relaxation times of the water protons (Table 4). Also for the perturbed systems, the logarithmic plots of the decay of magnetization yields two values of T_1 which characterize two water populations of different mobility.

The shorter relaxation time T_1 is linearly correlated to the fraction x_A of total water relaxing with this time constant (Figure 5) and this correlation may be expressed in analytical form by the equation

$$T_1^A = 532 - 6.3x_A \qquad (1)$$

A maximum allowed value for x may be calculated for $T_1^A = 0$; this corresponds to a w_θ value of $\simeq25$.

It has been reported[26] that only when the number of water molecules per polar head exceeds 23, a "bulk" (or "free") water population in slow exchange with the water tightly associated with the phospholipids may be observed. This result is in excellent agreement with our findings.

8. THE pH IN REVERSED MICELLES

The presence of solutes in the water pools of RM seems to affect mainly the relative amounts of the different water populations. It is, however, possible that the pH of the aqueous solutions used to prepare the micelles plays an important role in determining their physico-chemical properties.

Table 2. O-H Stretching Band (cm^{-1}) for the Reversed Micelles[a)]

Solute	Wave Number		
TRP	3610 (10)	3470 (77)	3260 (13)
CYS	3570 (14)	3420 (73)	3260 (13)
THR	3600 (7)	3480 (69)	3260 (24)
ALA	3600 (13)	3460 (50)	3230 (37)
TYR	3640 (30)	3430 (31)	3280 (39)
LYS.2HCl	3580 (29)	3440 (28)	3300 (43)
GLU	3610 (28)	3470 (17)	3300 (55)
PRO	3560 (39)	3420 (13)	3280 (47)
MIF [b)]	3630 (25)	3480 (9)	3280 (66)
PHEVAL	3620 (35)	3490 (6)	3300 (59)
LEU	3615 (25)	3470 (5)	3320 (70)
GLY	3625 (22)	3500 (4)	3310 (74)
VAL	3605 (54)	----------	3280 (46)
PHE	3600 (30)	----------	3280 (70)
HIS	3600 (42)	----------	3300 (58)
PHELEU	3630 (43)	----------	3260 (57)
LUE^{5}ENK [b)]	3630 (33)	----------	3250 (67)

[a)] Deconvoluted bands; in parentheses are the relative areas in %.
[b)] MIF = PROLEUGLY   LEU^{5}ENK = TYRGLYGLYPHELEU.

This issue immediately brings into discussion two open problems, namely the real nature of the water in the water pools, which is a different solvent from bulk water, and the closely related questions of the determination in such solvent of the pH (and pK's) which may be different in RM and in the isotropic aqueous solution[27].

The ^{31}P chemical shift of inorganic phosphates is an excellent intrinsic pH indicator in cellular systems, since, as a function of pH, it follows a titration curve[28]. It would seem simple to use the ^{31}P chemical shift of phosphate buffers to define an acidity scale in RM. It must, however, be kept in mind that the phosphorus chemical shift in our systems remains constant in spite of the largely different pH values of the solutions secluded (par. 7). Besides, the longitudinal relaxation times of the water protons do not show any correlation with the variations of the starting pH (Table 4) and the water chemical shift, measured by high-resolution proton NMR, remains unchanged. The presence of an extensive network of H-bonds in the water pools may account for the observed constancy of chemical shifts.

To clarify this point, however, we have prepared RM containing phosphate buffers in the internal aqueous compartment and studied their NMR behavior. The pertinent proton T_1 values are reported in

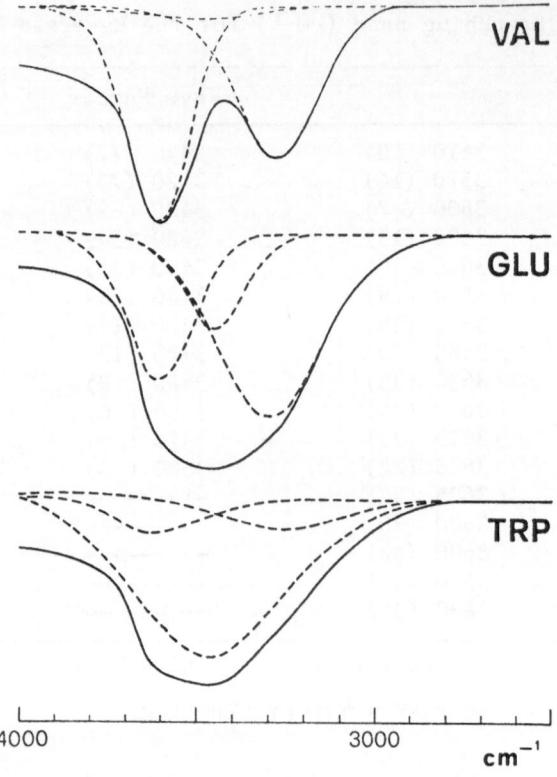

Fig. 4. Infrared spectral pattern of reversed micelles containing
different solutes.

Table 5: while the relaxation of the more structured water is unaffected, the value of T_1^B increases, but not regularly, with the pH.
On the other hand, a titration curve is obtained for the ^{31}P longitudinal relaxation time as a function of pH (Figure 6) and the apparent pK value obtained from the plot is 6.85±0.05, in agreement
with the value reported[33].

Clearly, the results obtained for the NMR parameters of RM
containing amino-acids or small peptides in the water pool are not
related to the behavior just described, and thus the effect of the
starting pH can be regarded as negligible for the systems under
discussion.

9. THE EFFECT OF LIPOPHILIC PERTURBATORS

The "solutes" so far considered were all solubilized in the
micellar water pool and different effects may be expected when the
added species is entrapped in the micellar wall.

Table 3. ^{31}P NMR Longitudinal Relaxation Times of EPC in Reversed Micelles

Solute	pH	T_1 (ms) [a]
None	6.50	2050
CLY	7.53	2113
VAL	7.07	2059
LEU	7.06	2113
PRO	6.16	2070
PHE	5.56	2234
TYR	5.74	2007
PHEVAL	5.79	2110
PHELEU	6.03	2064
PHEPHE	5.86	2054
MIF	9.08	2145
LEU^{5}ENK	3.64	2053

[a] Accuracy: ±20 ms; reproducibility: ±10 ms.

A suitable molecule to use for such studies is the coenzyme Q (Q_{10}), and its shorter homologues, since its quinonoid moiety is hydrophilic enough to lay at the phospholipid-water interface while the prenyl side-chain is arranged parallel to the fatty acid chains.

The three water populations observed by infrared spectroscopy are modified also in this case, and the effect is different for the short and the long side-chain coenzyme Q homologues (Figure 7) (C. A. Boicelli, M. Giomini and L. Masotti, unpublished data). The modifications observed for the O-H stretching band are large, but not as dramatic as observed for hydrophilic solutes. This is consistent with the fact that Q_{10} and its homologues are entrapped in the micellar wall and not in the internal compartment, as are the amino acids and the small peptides considered before.

On the other hand, the ^{31}P relaxation time is differently changed by the presence of the different coenzyme Q homologues (Table 6). The observed variation is small for the short chain Q's (Q_0 and Q_3), but quite dramatic for the Q_{10}. A T_1 value similar to that measured in this case is found for RM in which $w_0 \simeq 17$, suggesting that a rather large amount of water might have been removed from the polar head of EPC. The hypothesis of a large variation in the water population at the lipid-water interface is supported by the proton T_1 data (Table 7). It is evident that a major change in the "A" water population occurs for Q_{10}, which, because of its length, can span completely the lipid layer.

Table 4. ^{1}H Relaxation Times of Water in Reversed Micelles
(measured at 20 MHz)

Solute	pH	T_1^A (ms) [a]	T_1^B (ms) [a]
None	6.50	305	537
GLY	7.53	307	697
VAL	7.07	472	787
LEU	7.06	396	512
PRO	6.16	256	475
PHE	5.56	247	631
TYR	5.74	438	971
PHEVAL	5.79	335	538
PHELEU	6.03	318	400
PHEPHE	5.86	303	469
MIF	9.08	340	439
LEU^{5}ENK	3.64	197	508

[a] Accuracy: ±20 ms; reproducibility: ±10 ms.

A possible explanation, consistent with the experimental data, is that Q_{10}, because of its molecular dimensions, creates a discontinuity in the micellar wall architecture and the water molecules, as a consequence, can move more freely within the internal compartment.

To ascertain this point, we have measured the ^{2}H longitudinal relaxation time of ^2H_2O in liposomes of EPC. The proton relaxation data of water in liposomes cannot give information on its mobility in the internal compartment not only because of the exchange phenomena with bulk water (as mentioned in the Introduction) but also because of the strong dipolar intermolecular contributions to the relaxation. The relaxation mechanism of deuterium, instead, is essentially quadrupolar, and intramolecular, and the intermolecular dipolar contributions, very important in the case of protons, are negligible. Thus, the deuterium relaxation time of heavy water in liposomes can give information on the exchange processes between sites[30]. In the case of RM, the longitudinal relaxation time measured for ^2H_2O (170 ms) is the same in the unperturbed system and when hydrophilic or lipophilic species are present. Only one relaxation time is obtained in this case, because a high resolution spectrometer has been used for the measurements and the shorter T_1 is lost because of the too long dead time of the instrument.

The deuterium T_1's of ^2H_2O in liposomes of EPC, where different coenzyme Q homologues have been added, are reported in Table 8.

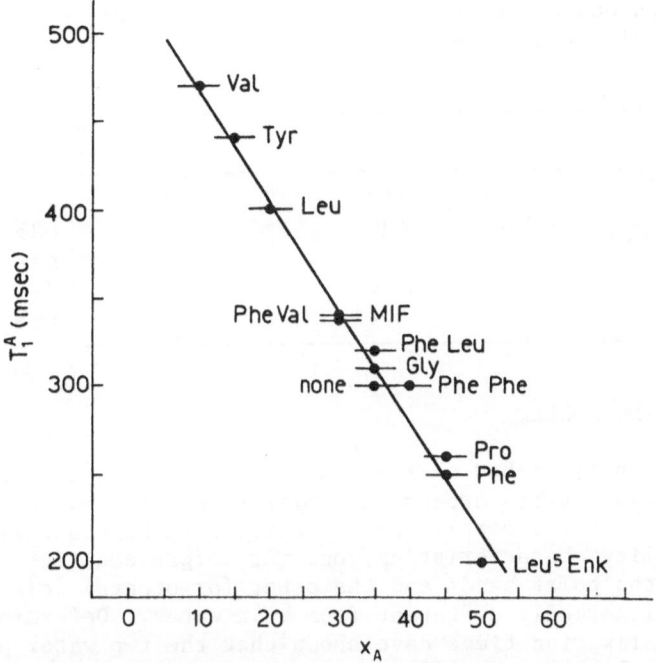

Fig. 5. Dependence of T_1^A on the fraction x_A of type "A" water
(measured at 20 MHz).

If fast exchange of ^2H_2O between the internal compartment and
bulk water is assumed, the observed relaxation behavior may be described by the relationship

$$\frac{1}{T_1^{obs}} = \frac{a}{T_1^{int}} + \frac{(1-a)}{T_1^{ext}} \qquad (2)$$

where T_1^{obs} is the experimental T_1 value, T_1^{int} is the ^2H_2O relaxation
time inside the liposome and the value of 170 ms measured in RM is
used, $T_1^{ext} = 480$ ms is the value for bulk, free ^2H_2O and "a" is the
probability for a water molecule to be in an environment where its
relaxation time is T_1^{int}.

It is readily calculated that, when Q_{10} is entrapped in the
phospholipidic bilayer, the probability "a" increases by a factor 2.6
indicating a more efficient exchange of the water through the bilayer. Consequently, the number of water molecules that are effectively found around the phospholipid polar heads per unit time markedly decreases. This result can explain the sharp drop in the ^{31}P
relaxation time observed in RM when Q_{10} is present and discontinuities in the micellar wall are formed.

Table 5. pH Dependence of the Water Protons Longitudinal
Relaxation Times in Reversed Micelles (measured
at 20 MHz)

Buffer pH	T_1^A (ms)	T_1^B (ms)
5.96	447	752
6.51	430	823
7.02	448	831
7.56	458	847
7.97	457	945

10. THE EMERGING MODEL

It is well established that in RM and liposomes water is distributed in hydration layers around the polar heads of phospholipids[26,31]. Two (or more) types of water have been observed with
different mobility characteristics, one more organized tightly associated with the polar heads and the other (or others) relatively
free, but still markedly different from bulk water. Determinations
of deuterium relaxation times have shown that the two water populations are in slow exchange[26].

Our measurements of proton relaxation times have confirmed this
picture in the case of RM: two water populations have been identified, one with a short T_1 (type "A") and the other (type "B") characterized by a greater mobility, with a longer T_1. On the other hand,
three types of water have been observed by infrared spectroscopy.
These results suggest that the number of water types which can be
detected depend on the time window of the experimental technique used
for the measurements. Thus, the water pool in RM might be thought of
as a distribution of sites characterized by individual correlation
times. The τ_c values depend on the geometrical constraints to which
the water molecules are exposed (cage effect) and on the dipolar
interactions of the water molecules with each other and with the
phosphocholine groups. A whole range of motional properties may be
envisaged; however, only a limited number of populations, each
characterized by a single common correlation time can be experimentally observed, since the individual motional properties may be
averaged by fast exchange within groups of molecules. Only when the
exchange between a certain group of molecules (a "class" of correlation times) and other groups is slow on the time scale of the
detection system used, may a definite type of water be identified and
characterized.

The experimental results obtained for RM, when various perturbing species, hydrophilic or lipophilic, are contained in the water
pool, support this description of the water microdynamics inside the

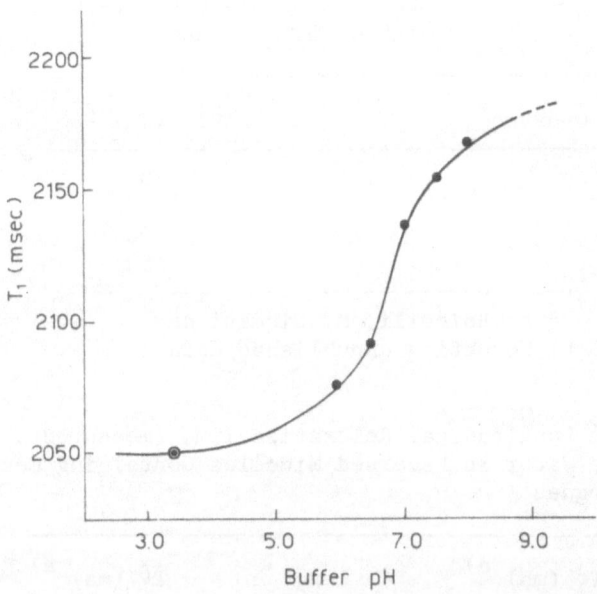

Fig. 6. Effect of the pH of phosphate buffers on the ^{31}P longitudinal relaxation time in reversed micelles.

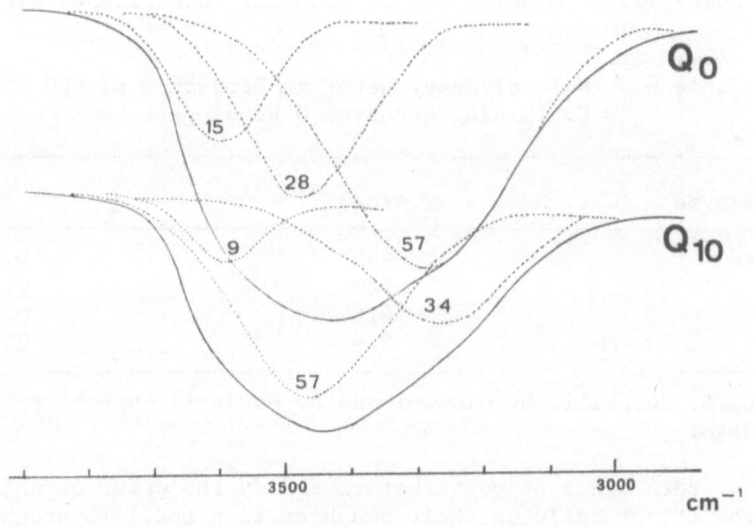

Fig. 7. Infrared spectral pattern of reversed micelles containing Q homologues (relative areas of the individual components of the O–H stretching band are indicated).

157

Table 6. ^{31}P Longitudinal Relaxation
Time of the Phosphorus of
EPC in Reversed Micelles
with Coenzyme Q Homologues

Coenzyme	T_1 (ms) [a]
None	2058
Q_0	2076
Q_3	2005
Q_{10}	1503

[a] C. A. Boicelli, M. Giomini and
L. Masotti: unpublished data.

Table 7. Proton Longitudinal Relaxation Time (measured at 20 MHz)
of the Water in Reversed Micelles Containing Coenzyme Q
Homologues

Coenzyme	T_1^A (ms) [a]	x_A	T_1^B (ms) [a]	x_B
None	305	30±5	537	70±5
Q_0	307	30±5	518	70±5
Q_3	299	30±5	512	70±5
Q_{10}	395	10±5	561	90±5

[a] C. A. Boicelli, M. Giomini and L. Masotti: unpublished data.

Table 8. ^2H T_1 of Heavy Water in Liposomes of EPC
Containing Coenzyme Q Homologues

Coenzyme	T_1 (ms) [a]	"a"
None	435	0.057
Q_0	420	0.078
Q_3	415	0.087
Q_{10}	379	0.187

[a] C. A. Boicelli, M. Giomini and L. Masotti: unpublished
data.

micelles. Both kinds of perturbators modify the water organization
within the RM, or build up their own hydration shell (hydrophilic
species) or disrupt the continuity of the lipidic matrix by inter-

calation and thus changing its permeability (lipophilic species). It has already been pointed out that both types of water ("A" and "B") revealed by NMR and RM, are rather structured and their relaxation times much shorter than the time measured for pure water.

The proton relaxation data reflect this high degree of organization, since water relaxes together with the most rigid parts of the phospholipid molecule (Figure 3). This behavior can be ascribed to cross-relaxation, a mechanism which is effective only in systems with an high degree of organization[15,16,33-35]. The water molecules in RM, when they are tightly associated with the phospholipid polar heads ("A" water), can transfer magnetization to the EPC matrix, across the phase boundary. The transfer occurs by mutual spin flips between bound water protons and protons in the internal micellar surface. The magnetization is then transferred from the surface to the bulk of the micellar wall via spin-diffusion and cross-relaxation through the phospholipidic matrix proton network.

This implies that the tightly packed polar heads which constitute the interface behave as a macromolecule.

More details about the organization of the water in the RM can be deduced from the proton relaxation data. It may be reasonable to suppose that the group of water molecules characterized by the shorter value of T_1 (type "A") are confined in a layer of limited thickness at the interface. If this is true any change of the "A" water population should modify the mean free path of the molecules and consequently their motional properties and the associated correlation times. If x_A increases, the packing of the water molecules in this limited layer should become tighter and the intermolecular dipolar interactions more efficient. Such sequence of events should lead to shorter relaxation times, and this is experimentally observed as described by the empirical relationship[1].

In conclusion, we describe the "A" water as a class of molecules confined in a layer of definite thickness with fast exchange (on the NMR time scale) among the sites within the layer, and characterized by a single average correlation time, and longitudinal relaxation time. The type "B" water, in our view, represents another class of water molecules, which exchanges slowly with the "A" class, and is identified by a single relaxation time because of the fast exchanges among different situations within the population itself.

The hydrophilic species in the internal compartment of RM change x_B by building up their own hydration sphere and "A" molecules move consequently to restore the $A \leftrightarrow B$ equilibrium and both T_1^A and T_1^B are modified accordingly. The lipophilic species, on the other hand, which are intercalated in the lipidic matrix, displace water from the "A" layer loosening its packing and T_1^A increases as a consequence.

REFERENCES

1. C. A. Boicelli, F. Conti, M. Giomini, and A. M. Giuliani, Interactions of small molecules with phospholipids in inverted micelles, Chem.Phys.Lett., 89:490 (1982).
2. A. G. Lee, N. J. M. Birdsall, Y. K. Levine, and J. C. Metcalfe, High-resolution proton relaxation studies of lecithins, Biochim.Biophys.Acta, 255:43 (1975).
3. G. Govil and R. V. Hosur, Organization of phospholipids in biological membranes, Int.J.Quant.Chem., 16:19 (1979).
4. D. L. Yeagle, Phospholipid headgroup behavior in biological assemblies, Accounts Chem.Res., 11:321 (1978).
5. T. Axenrod, in "Physical Methods on Biological Membranes and their Model Systems", Plenum Publ. Co., in press (1984).
6. A. Abragam, "Principles of Nuclear Magnetism", Clarendon Press, Oxford, chapter VIII (1961).
7a. K. J. Packer, Nuclear spin relaxation studies of molecules adsorbed on surfaces,
 b. E. L. Mackor and C. MacLean, Relaxation processes in systems of two non-identical spins,
 c. H. G. Hertz, Microdynamic behavior of liquids as studied by NMR relaxation times, in: "Progress in Nuclear Magnetic Resonance Spectroscopy", Pergamon Press, Oxford (1967).
8. F. Noak, Nuclear magnetic relaxation spectroscopy, in: "NMR Basic Principles and Progress", vol.3, Springer-Verlag, Berlin (1971).
9. T. L. James, "Nuclear Magnetic Resonance in Biochemistry", Academic Press, New York (1975).
10. NMR basic principles and progress, vol.19, "NMR in Medicine", R. Damadian, ed., Springer-Verlag, Berlin (1981).
11. J. R. Zimmermann and W. E. Brittin, Nuclear magnetic resonance studies in multiple-phase systems: lifetime of a water molecule in an adsorbing phase on silica gel, J.Phys.Chem., 61:1328 (1957).
12. E. D. Finch and L. D. Homer, Proton NMR relaxation measurements in frog muscle, Biophys.J., 14:907 (1974).
13. K. Hallenga and S. H. Koenig, Protein rotational relaxation as studied by solvent proton and deuteron magnetic relaxation, Biochem., 15:4255 (1976).
14. A. De Ambrosis, S. Aldrovandi, G. Bonera, and M. Villa, The NMR response of heterogeneous systems: anisotropy and low-dimensionality effects, in: "Magnetic Resonance in Colloid Interface Science", NATO ASI, 61 (1980).
15. A. Kalk and H. J. C. Berendsen, Proton magnetic relaxation and spin-diffusion in proteins, J.Magn.Res., 24:343 (1976).
16. B. M. Fung and T. W. McGaughy, Cross-relaxation in hydrated collagen, J.Magn.Res., 39:413 (1980).
17. R. Freeman and H. D. W. Hill, High-resolution studies of nuclear spin-lattice relaxation, J.Chem.Phys., 51:3140 (1969).

18. R. Freeman and H. D. W. Hill, Fourier transform study of NMR spin-lattice relaxation by "Progressive Saturation", J.Chem. Phys., 54:3367 (1971).

19a. J. L. Markley, W. J. Horsley, and M. P. Klein, Spin-lattice relaxation measurements in slowly relaxing complex spectra, J.Chem.Phys., 55:3604 (1971).

  b. G. G. McDonald and J. S. Leigh, Jr., New method for measuring longitudinal relaxation times, J.Magn.Res., 9:358 (1973).

20. U. P. Fringeli and Hs. H. Gunthard, Infrared membrane spectroscopy, in: "Membrane Spectroscopy", E. Grel, ed., Springer-Verlag, Berlin (1981).

21. B. E. Conway, "Ionic Hydration in Chemistry and Biophysics", Elsevier Publ. Co., Amsterdam, chapter VII (1981).

22. D. Eisenberg and W. Kauzmann, "The Structure and Properties of Water", Clarendon Press, Oxford, chapter IV (1969).

23. B. M. Fung and J. L. McAdams, The interaction between water and the polar head in inverted phosphatidylcholine micelles. A ^{2}H and ^{31}P relaxation study, Biochim.Biophys.Acta, 451:313 (1976).

24. G. J. Béné, Foundations and preliminary results on medical diagnosis by nuclear magnetism, Adv.Electron.Electron.Phys., 49:85 (1979).

25. C. A. Boicelli, M. Giomini, and A. M. Giuliani, Effect of small solutes on the water populations in inverted micelles, Spectrochim.Acta, 37A:559 (1981).

26. E. G. Finer, Interpretation of deuteron magnetic resonance spectroscopic studies of the hydration of macromolecules, J.Chem.Soc.,Faraday Trans.II 69:1590 (1973).

27. IV ESF Workshop on polymer sciences, "Biological and Technological Relevance of Reverse Micelles and other Amphiphilic Structures in Apolar Media", Rigi–Kaltbad, Sept.29–Oct.2 (1982).

28. D. P. Hollis, Phosphorus NMR of cells, tissues and organelles, in: "Biological Magnetic Resonance", vol.2, L. J. Berliner and J. Reuben, eds., Plenum Press, New York (1980).

29   D. G. Gadian, G. K. Radda, R. E. Richards, and P. J. Seeley, ^{31}P NMR in living tissue: the road from a promising to an important tool in biology, in: "Biological Applications of Magnetic Resonance", R. Shulman, ed., Academic Press, New York (1979).

30. H. H. Mantsch, H. Saito, and I. C. P. Smith, Deuterium magnetic resonance, applications in chemistry, physics and biology, in: "Progress in NMR Spectroscopy", J. W. Emsley, J. Feeney, and L. H. Sutcliffe, eds., Pergamon Press, Oxford, vol.11 (1977).

31. E. G. Finer and A. Darke, Phospholipid hydration studied by deuteron magnetic resonance spectroscopy, Chem.Phys.Lipids, 12:1 (1974).

32. G. Klose and F. Stelzner, NMR investigations of the interaction of water with lecithin in benzene solutions, Biochim.Biophys. Acta, 363:1 (1974).

33. P. J. Andree, The effect of cross-relaxation on the longitudinal relaxation times of small ligands binding to macromolecules, <u>J.Magn.Res.</u>, 29:419 (1978).
34. H. T. Edzes and E. T. Samulski, The measurement of cross-relaxation effects in the proton NMR spin-lattice relaxation of water in biological systems: hydrated collagen and muscle, <u>J.Magn.Res.</u>, 31:207 (1978).
35. B. M. Fong, Proton and deuteron relaxation of muscle water over wide ranges of resonance frequencies, <u>Biophys.J.</u>, 18:235 (1977).

LIPID-PROTEIN INTERACTIONS: FROM CRYSTALLINE LIPO-
PROTEINS TO INTACT MEMBRANES

Joachim Seelig

Biocenter of the University of Basel
4056 Basel
Switzerland

POSSIBLE MODES OF PHOSPHOLIPID INTERACTION

The molecular structure of phospholipid molecules suggests at least three types of interactions: (1) ionic interactions of the polar lipid headgroups with metal ions and other charged molecules dissolved in the aqueous phase, (2) lipid-water interactions (hydration) at the level of the polar groups and the glycerol backbone region, and (3) hydrophobic interactions of the fatty acyl chains with non-polar molecules. ^{2}H-NMR is a particularly sensitive method to detect such interactions as is illustrated in figure 1 for the binding of metal ions to bilayers of 1,2-dipalmitoyl-sn-glycero-3-phosphocholine (DPPC). In these experiments the choline head group was selectively deuterated at the α-CH$_2$ segment and the ^{2}H-NMR spectra were recorded as a function of metal ion concentration at 59°C [1]. In the absence of ions the quadrupole splitting of the α-segment was 6 kHz, but decreased considerably upon addition of ions. Monovalent ions like Na$^+$ and K$^+$ had only small effects (of about a few hundred Hz) on the quadrupole splitting, whereas multivalent ions induced changes up to the order of 10 kHz. ^{2}H-NMR may thus be used as a convenient means to monitor ion binding to bilayer membranes [1]. If the ion-binding studies are repeated with the deuterium label in the fatty acyl chains, no change in the quadrupole splitting is

163

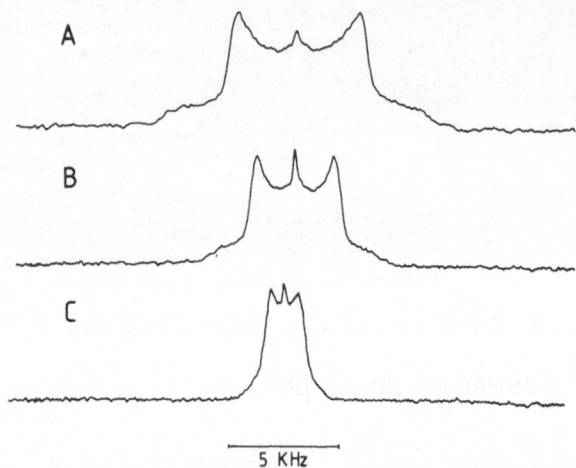

Fig. 1 Metal ion binding to phospholipid bilayers ^{2}H-NMR spectra of [+]NCH$_2$CD$_2$O-DPPC at 59°C. (A) no ions; (B) 0.1 M CaCl$_2$; (C) 0.05 M LaCl$_3$ (from [1]).

observed. The effect of ion-binding is thus limited to a conformational change in the head group region of the phosphatidylcholine bilayer.

In contrast, the addition of cholesterol has only little or no effect on the polar head groups [2] but has quite a strong effect on the hydrocarbon chains as illustrated in figure 2 [3]. In this experiment the [2]H-label is attached at the C-5 segment of both fatty acyl chains of DPPC. In the pure lipid bilayer above the phase transition the quadrupole splitting is ∿ 25 kHz but increases up to about 55 kHz in the presence of cholesterol at a 1:1 molar ratio.

The gylcerol backbone, finally, is remarkably insensitive to the addition of cholesterol or external agents [4]. These examples may suffice to demonstrate that phospholipids are indeed sensitive to different types of molecular interactions, be it at the level of the polar groups or be it in the hydrophobic core of the bilayer membrane.

Most membrane proteins are spanning the lipid bilayer and from simple geometric arguments one may therefore expect interactions at the level of both the

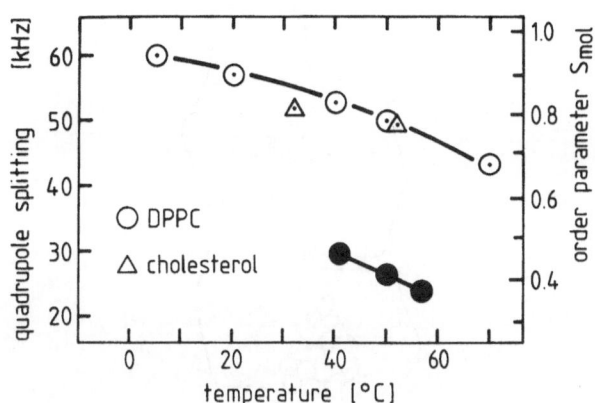

Fig. 2   Quadrupole aplittings of deuterated C - 5
         DPPC and cholesterol in a fluid bilayer. (●) DPPC
         without cholesterol; (○) DPPC with an equimolar
         amount of cholesterol; (△)[3 - ^{2}H] cholesterol
         (from [3]).

polar groups and the hydrocarbon chains. This shall be
investigated in the following for three typical cases:
(1) A crystalline lipoprotein isolated from frog-yolk,
(2) reconstituted cytochrome c oxidase membranes, and
(3) intact E. coli membranes.

LIPOVITELLIN-PHOSVITIN. A CRYSTALLINE LIPOPROTEIN

     The structural properties of the Lipovitellin-Phos-
vitin (LV:PV) complex have been summarized recently [5].
In brief, the complex has a molecular weight of 456 000
and is composed of two identical subunits. Each subunit
consists of one lipovitellin chain (made up from A, B,
and C chains) and one phosvitin chain. About 35 - 40
phospholipid molecules (mainly phosphatidylcholine
and phosphatidylethanolamine) are associated with the
lipovitellin chain. The phosvitin chain, on the other
hand, is also a very unusual protein in that 66 out
of 155 residues are serine of which about more than
90% are phosphorylated. The phosvitin chain may thus
be likened to a negatively charged polyelectrolyte, explaining the good solubility of the LV:PV complex in

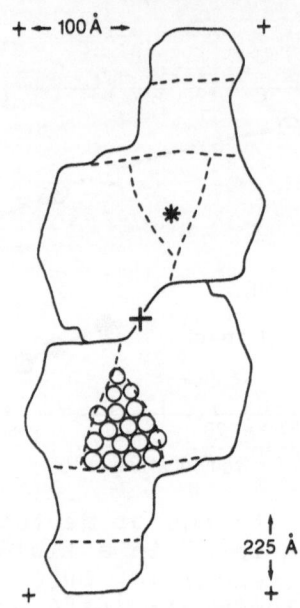

Fig. 3   Low-resolution model of the lipovitellin:
         phosvitin complex (from [6]).

monovalent salt solutions.

     The structure of the crystalline LV:PV complex
at about 20 Å resolution was derived by Banaszak and
co-workers using a combination of X-ray and electron
diffraction techniques and is reproduced in figure 3
[6]. The dimer is roughly 250 Å long, 115 Å wide, and
55 Å thick. The diffraction data further reveal a region of low electron density as indicated by the das-
hed lines surrounding the asterix. The interesting
question then is of how the lipids are accommodated
in the protein structure. Are the lipid molecules
tightly bound to individual sites much like fatty
acids are bound to bovine serum albumin? Should
the protein be pictured more like a sponge in which
droplets of micellar phospholipids are dissolved?
Or is there perhaps some other structure?

     These questions have been studied in some de-
tail with ^{31}P- and ^{2}H-NMR and the main conclusions
will be summarized briefly (an extensive report of
this work has been published [7, 8]).Since the
LV:PV complex can be dissolved in 0.5 M NaCl it is
possible to record high-resolution ^{31}P-NMR spectra, a
typical example of which is displayed in figure 4. The

166

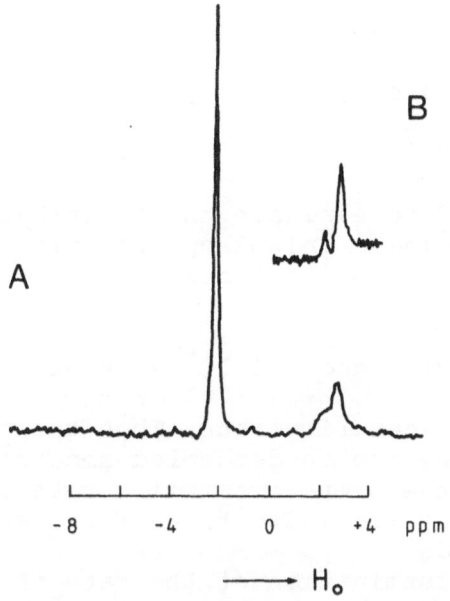

Fig. 4   High-resolution ^{31}P-NMR studies of the LV:PV
         complex. (A) non-decoupled spectrum; (B) pro-
         ton-decoupled (from [7]).

spectrum is characterized by an intense resonance which
must be assigned to the phosphorylated serines and two
somewhat weaker resonances at higher field which arise
from phosphatidylcholine and phosphatidylethanolamine.
The linewidth of the latter two resonances is further
reduced upon proton-decoupling as indicated in the insert of figure 4. A linewidth of 20 - 30 Hz as observed
for the two lipid resonances already rules out a complete immobilization of the lipids on the protein. If
the lipoprotein is approximated by a rotational ellip-
soid of length 2a = 250 Å and thickness 2b = 85 Å the
calculation of the rotational diffusion constant yields
$R_\| = 10^6 \ s^{-1}$ and $R_\perp = 4.5 \times 10^5 \ s^{-1}$ for the rotation
around the long and the short axis, respectively, assu-
ming a viscosity of η = 0.01 Poise. These rotational
rates are large compared to the anisotropies of the
phosphorus chemical shielding tensor ($\Delta\sigma_{max} \simeq$ 230 ppm,
corresponding to 2×10^4 at 7 Tesla field strength [9])
and are thus sufficiently fast to average completely
the ^{31}P chemical shielding anisotropies. However, the
chemical shielding anisotropies also enter into the
linewidth. The rotational diffusion coefficients are
related to rotational correlation times according to

167

$$\tau_{2m}^{-1} = 6R_\perp + m^2 (R_{\parallel} - R_\perp)$$

Insertion of the above numbers yields

$$\tau_{20} = 4 \times 10^{-7} \text{ s} \qquad \tau_{22} = 2 \times 10^{-7} \text{ s}$$

which can be used to evaluate the contributions of the chemical shielding anisotropy and also of dipole-dipole interactions to the linewidth. Conservative estimates lead to

$$\Delta\nu_{1/2}^{CSA} \simeq 70 \text{ Hz} \quad \text{and} \quad \Delta\nu_{1/2}^{DIP} \simeq 40 \text{ Hz}$$

for the linewidth contributions. Thus the relative sharp lines in the proton-decoupled spectra can not be explained by the overall tumbling rate of the whole LV:PV complex, but are indicative of rather freely moving phospholipids.

A direct determination of the rate of phosphate group motion is possible via ^{31}P spin lattice relaxation time measurements as reproduced in figure 5. The T_1 relaxation time of the phosphoserine residue is about 2s and exhibits only a small temperature dependence. Quite a different behavior is observed for the phospholipid head groups which exhibit a well-defined relaxation minimum around $23°C$. At the minimum the molecular correlation time τ_c is the reciprocal of the resonance frequency ω_o yielding $\tau_c \sim 1$ ns at $23°C$. For the pure phospholipid extract the relaxation minimum is found at a somewhat lower temperature, namely $5°C$. It can be concluded from these measurements that the correlation time of the phosphate group is only slightly increased in the presence of protein.

We may mow proceed to the solid state ^{31}P-NMR spectra of wet crystals of the LV:PV complex. The spectra consist again of a sharp central resonance which can be assigned to isotropically moving phosphorylserines and a broader signal due to the phospholipids. Closer inspection of the phospholipid "powder" pattern reveals that it is virtually identical with a fluid phospholipid bilayer spectrum. This is further supported by ^{2}H-NMR studies in which the natural phosphatidylcholine of the LV:PV complex is partially exchanged against deuterated phosphatidylcholine. The ^{2}H-NMR spectrum of the deuterated LV:PV complex also shows the characteristic signature of a fluid lipid bilayer. Hence the NMR results lead to the conclusion that the phospho-

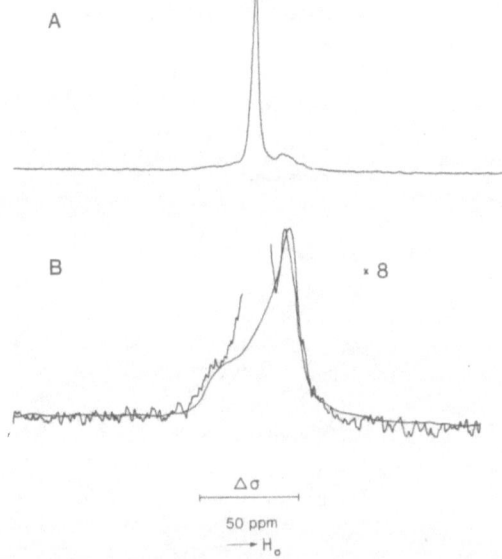

Fig. 5 ^{31}P NMR spectra of the crystalline LV:PV
complex. (A) Suspension of LV:PV in buffer
(B) 8 fold magnification of B. Smooth trace
is spectrum of lipid extract (from [7]).

lipids in the LV:PV complex were arranged in a bilayer-
like structure. Since each subunit of lipovitellin contains a microdomain of about 35 - 40 lipids, at least
50% of these lipids must be in direct contact with the
protein surface. Nevertheless, there is no appreciable
effect of the protein on the phospholipid conformation
or the dynamics of the segmental motions. This can be
understood by assuming a liquid-like match between the
protein surface due to flexible protein side-chains and
the fluid bilayer interface. The bilayer domain is probably identical with the area of low-electron density
detected by diffraction techniques. In support of the
bilayer structure it should be noted that the thickness
of the protein crystal (55 Å) corresponds roughly to that
of a fluid bilayer (45 Å).

CYTOCHROME C OXIDASE CONTAINING MEMBRANES

Cytochrome c oxidase is an intrinsic membrane protein which catalyzes the transfer of electrons from
cytochrome c to molecular oxygen. The monomer is roughly 110 Å long whereas the thickness of the lipid bi-

169

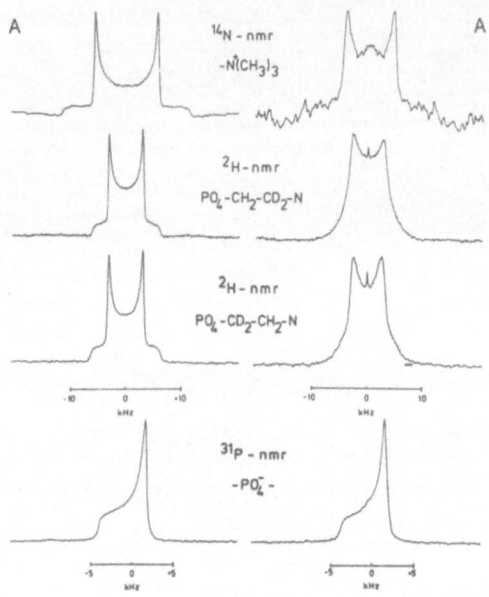

Fig. 6 NMR spectra of lipid bilayers with (A') and
without (A) cytochrome c oxidase (from [10]).

layer is only about 45 Å. Thus from the dimensions
and the location of the protein one may expect hydrophobic interactions with the fatty acyl chains as well
as polar interactions with the phosphocholine groups.
In order to study the interaction of cytochrome c oxidase with the surrounding lipids the enzyme was reconstituted into bilayers of 1-palmitoyl-2-oleoyl-sn-gly-
cero-3-phosphocholine (POPC) selectively deuterated at
either methylene segment of the choline residue [10].
Figure 6 compares the ^{2}H-, ^{14}N- and ^{31}P-NMR spectra
of pure POPC bilayers with those containing more than
50 wt.% protein. Because of the high protein content
at least 50% of the phospholipid molecules should be
in direct contact with the protein surface, yet the
corresponding NMR spectra of the pure lipid bilayers
and the reconstituted cytochrome c oxidase membranes
are remarkably similar. This is even more surprising
since it has been demonstrated earlier that the polar
head groups are quite sensitive to the addition of
external agents such as metal ions.

Qualitatively similar results are obtained when
the enzyme is incorporated into unsaturated phosphatidylcholine bilayers with the deuterium label in the

hydrophobic core of the bilayer. Even though the protein spans the lipid bilayer, the motional freedom
of the fatty acyl chains is practically identical with
and without protein. The principal conclusion that must
be drawn from these NMR studies is that neither the
conformation of the phosphocholine head group nor the
average order of the hydrocarbon chains [10, 11] are
influenced to any significant extent by cytochrome c
oxidase.

STUDIES OF ESCHERICHIA COLI CELLS

        For the bacterium Escherichia coli a promising
approach to achieve selective deuteration is the
use of mutants defective in the synthesis of precursors of phospholipid synthesis. The precursors are
then supplied in deuterated form to the growth medium and are incorporated into the membrane phospholipids. E. coli mutants defective in fatty acids synthesis have been employed successfully to incorporate deuterated fatty acids into E. coli inner and outer
membranes [12, 13] and figure 7 compares the order pro-

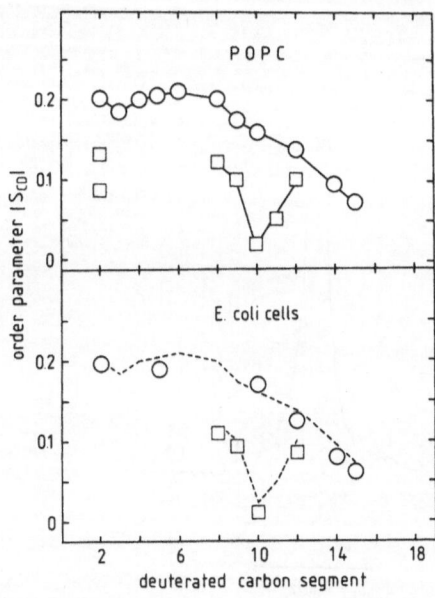

Fig. 7   Comparison of order profiles of intact
         E. coli cells and phospholipid model membra-
         nes (from [12]).

file of a synthetic, unsaturated phospholipid bilayer
(POPC) with that of intact E. coli cells [12]. The order profile is a measure of the flexibility and average orientation of the individual chain segments and
the figure reveals a remarkable similarity between
the pure lipid bilayer and intact E. coli cells, the
latter containing an inner and outer membrane with more than 50 wt.% protein. The conspicuous dip in the
order profile of the oleic acyl chain is indicative of
a slight tilting of the cis-double bond with respect
to the bilayer normal and is very sensitive to a change
in the tilt angle. However, it is obvious from the figure that the membane proteins in the bacterial cell
membrane do not even affect the tilt angle of the cis-
double bond.

By the same method it has also become possible to
study the polar head groups region of E. coli membranes.
Mutants defective in both the biosynthesis and degradation of glycerol incorporate externally added glycerol specifically into membrane phospholipids. Figure
8 shows the outcome of an experiment in which E. coli
glycerol auxotrophs were grown on perdeuterated glycerol [14]. The two main phospholipids in the membrane are phosphatidylethanolamine ($\sim$ 80 wt.%) and phos-

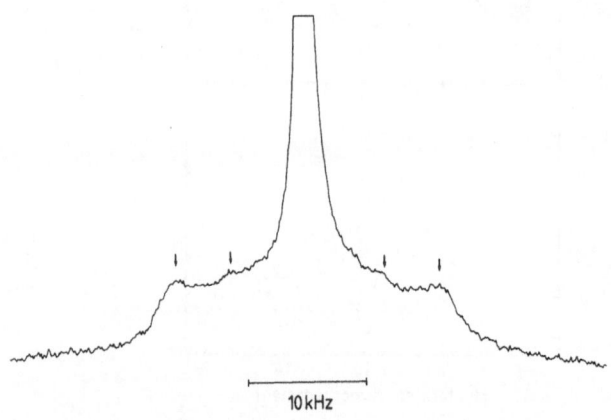

10 kHz

Fig. 8 ^{2}H NMR spectrum of E. coli cells grown
on perdeuterated glycerol (from [14]).

172

phatidylglycerol (∿ 20 wt.%). The latter compound is
labeled in the glycerol backbone and the glycerol head
group while phosphatidylethanolamine is labeled in the
glycerol backbone only. The ^{2}H-NMR spectrum of the intact membranes is not particularly revealing, since
it consists of rather broad, overlapping resonances.
However, after extraction of the lipids from the membrane and dispersing them in buffer much better resolved
spectra are obtained as shown in figure 9. Since the
total lipid extract contains only ∿ 20% phosphatidylglycerol the weak inner quadrupole splittings must be
assigned to the head group resonances of phosphatidylglycerol whereas the more intense outer quadrupole
splittings arise from the glycerol backbone deuterons
of both phosphatidylethanolamine and phosphatidylglycerol. An unambiguous assignment became possible when
these experiments were repeated with selectively deuterated glycerols. Figure 10 shows a comparison of
the ^{2}H-NMR spectra of the total phospholipid extracts
for the complete set of selectively deuterated glycerols. The final step is then to compare the ^{2}H-NMR
spectra of a selectively labeled head groups in a
pure bilayer with that of the same head group in an

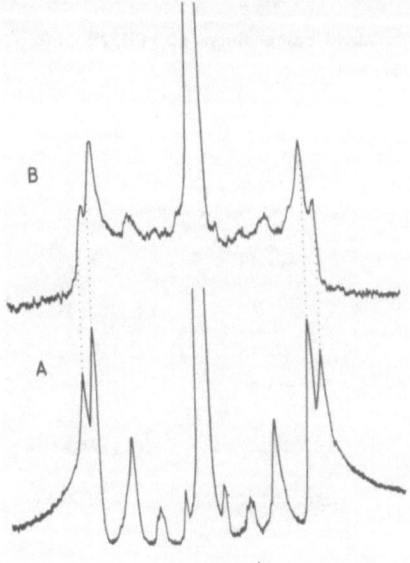

⊢ 10 kHz ⊣

Fig. 9 ^{2}H NMR spectra of multilamellar liposomes
        derived from E. coli phopsholipids. (A) sing-
        le pulse mode; (B) quadrupole echo mode
        (from [14]).

173

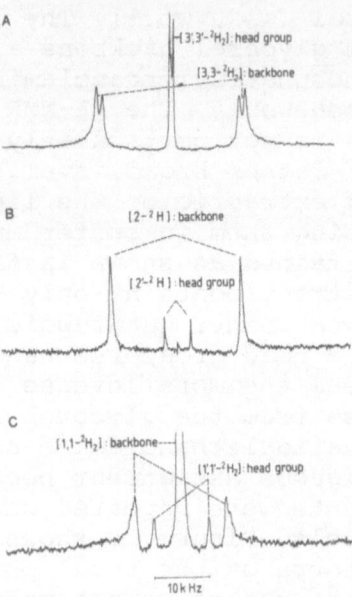

Fig. 10 ^{2}H-NMR spectra of multilamellar liposomes derived from total phospholipid extract (from [14]).

lipid composition : 80 % PE, 20 % PG

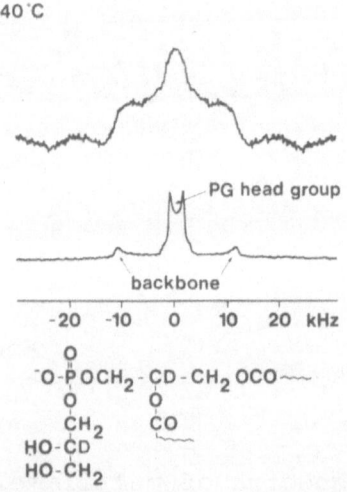

Fig. 11 ^{2}H-NMR spectra of E. coli cells (upper trace) and extracted phosphatidylglycerol (lower trace) (from [15]).

intact $\underline{E}$. $\underline{coli}$ membrane. The result of such a study is shown in figure 11 [15]. The cells were grown on [2-^{2}H]-glycerol and the figure displays the ^{2}H-NMR spectra of intact cells and of purified phosphatidylglycerol. The membrane spectrum clearly consists of two components with quite different quadrupole splittings. Evaluation of the areas underneath the broad and the narrow component yields a ratio of about 4:1 (broad component: narrow component). Based on this intensity ratio as well as on previous assignments with deuterated 1,2-dipalmitoyl-$\underline{sn}$-glycero-3-phosphoglycerol [16] the narrow component must be assigned to the 2' segment fo the phosphatidylglycerol head group while the broad component originates from the 2-segment of the glycerol backbone of both phosphatidyl-glycerol and -ethanolamine. This is further supported by comparison with the ^{2}H-NMR spectrum of the purified phosphatidylglycerol alone. The lineshapes of the aqueous dispersions of the pure phosphatidylglycerol at temperatures above the phase transition are sharp, well-defined "powder" patterns, charactersitic of random dispersions of fluid lipid bilayers [cf. 17]. For the $\underline{E}$. $\underline{coli}$ cells a definite broadening of the lineshapes is observed. However, the quadrupole splittings, i.e. the separation of the peaks in the spectra, shows little or no change at all. Similar results are obtained if [1,1-^{2}H$_2$]-glycerol is employed. This allows the immediate conclusion that the conformation of the glycerol head group as well as that of the glycerol backbone are practically identical in intact $\underline{E}$. $\underline{coli}$ membranes and in pure lipid bilayers. The differences in intrinsic linewidth are however not understood at present and must await further experimentation.

REFERENCES

[1] H. Akutsu and J. Seelig, Interaction of metal ions with phosphatidylcholine bilayer membranes, Biochemistry 20:7366 (1981).
[2] M.F. Brown and J. Seelig, Influence of cholesterol on the polar region of phosphatidylcholine and phosphatidylethanolamine bilayers, Biochemistry 17:381 (1978).
[3] H. Gally, A. Seelig and J. Seelig, Cholesterol in-

duced rod-like motion of fatty acyl chains in lipid
bilayers. A deuterium magnetic resonance study, Hoppe
Seyler's Z. Physiol. Chem. 357:1447 (1976).

[4] R. Ghosh and J. Seelig, On the interaction of cholesterol with bilayers of phosphatidylethanolamine,
Biochim. Biophys. Acta 691:151 (1982).

[5] L.J. Banaszak, J.M. Ross and R.F. Wrenn, Lipovitellin and the yolk lipoprotein complex, in "Lipid-
Protein Interactions",Wiley, N.Y. 1:223 (1982).

[6] D.H. Ohlendorf, R.F. Wrenn and L.J. Banaszak,
Three-dimensional structure of the lipovitellin-phosvitin complex from amphibian oocytes, Nature 272:28
(1978).

[7] L. Banaszak and J. Seelig, The lipid domains in
the crystalline lipovitellin:phosvitin complex. A ^{31}P
NMR study, Biochemistry 21:2436 (1982).

[8] G.B. Birrell, P.B. Anderson, P.C. Jost, O.H.
Griffith, L.J. Banaszak and J. Seelig, Lipid environments in the yolk lipoprotein system. A spin labeling
study of the lipovitellin:phosvitin complex from Xe-
nopus laevis, Biochemistry 21:2444 (1982).

[9] R.G. Griffin, Observation of the effect of water
on the ^{31}P nuclear magnetic resonance spectra of dipalmitoyllecithin, J. Amer. Chem. Soc. 98:851 (1976).

[10] L.K. Tamm and J. Seelig, Lipid solvation of cytochrome c oxidase. ^{2}H-, ^{14}N- and ^{31}P-NMR studies on
the phosphocholine headgroup and on cis-unsaturated
fatty acyl chains, Biochemistry, in press, (1983).

[11] A. Seelig and J. Seelig, Lipid-protein interaction in reconstituted cytochrome c oxidase/phospho-
lipid membranes, Hoppe Seyler's Z. Physiol. Chem.
359:1747 (1978).

[12] H.U. Gally, G. Pluschke, P. Overath and J. Seelig,
Structure of Escherichia coli membranes. Phospholipid
conformation in model membranes and cells as studied
by deuterium magnetic resonance, Biochemistry 18:5605
(1979).

[13] H.U. Gally, G. Pluschke, P. Overath and J. Seelig,
Structure of Escherichia coli membranes. Fatty acyl
chain order parameters of inner and outer membrane and
derived liposomes, Biochemistry 19:1638 (1980).

[14] H.U. Gally, G. Pluschke, P. Overath and J. Seelig,
Structure of Escherichia coli membranes. Glycerol
auxotrophs as a tool for the analysis of the phospholipid head group region by deuterium magnetic resonance, Biochemistry 20:1826 (1981).

[15] F. Borle and J. Seelig, unpublished results.
[16] R. Wohlgemuth, N. Waespe-Sarcevic and J. Seelig, Bilayers of phosphatidylglycerol. A deuterium and phophorus nmr study of the head group region, Biochemistry 19:3315 (1980).
[17] J. Seelig, Deuterium magnetic resonance: theory and application to lipid membranes, Quart. Rev. Biophys. 10:353 (1977).

MEMBRANE FUSION AND LIPID POLYMORPHISM

A.J. Verkleij[1], R. Van Venetië[2], J. Leunissen-Bijvelt[2],
B. de Kruijff[1], M. Hope[3] and P.R. Cullis[3]

[1]Institute of Molecular Biology and [2]Department of Molecular Cell Biology, State University of Utrecht, Utrecht The Netherlands; [3]Department of Biochemistry, University of British Columbia, Vancouver, Canada

INTRODUCTION

Membrane fusion is an extremely important phenomenon in biology. During this process two membranes, which can be two different membranes or two sites of one membrane, come in close contact, join and subsequently fuse, resulting in an intermixing of membrane lipids and proteins of the two membranes. Moreover, aqueous compartments, which were separated before the fusion, will intermix (see Fig. 1). If fusion is stopped at the stage of joining and the two membranes stay connected one may call it arrested fusion.

Among the important biological phenomena where membrane fusion is involved are: (i) fusion of the sperm and the egg membrane which leads to fertilization; (ii) secretion of neurotransmitters, insulin and other hormones plus digestive enzymes of the respective storage vesicles inside the gland cells called exocytosis, and (iii) the uptake of particles, bacteria (phagocytosis) and the uptake of viruses and the removal of receptors from the surface (receptor-mediated endocytosis).

In fact, membrane fusion is the potency of any membrane and this potency may be revealed more in one membrane than in the other. In most intracellular membrane types like endoplasmic reticulum, lysozymes, Golgi system, fusion events take place continuously.

Many studies have been undertaken to understand the fusion process itself and internal parameters which are involved in and/or activity modulate the fusion process. In recent years this basic interest in the membrane fusion process has been further stimulated.

179

One realized that the application of artificially-induced fusion is a powerful tool for hybridization of cells in order to produce monoclonal antibodies, for introducing membrane components in a cell membrane, and also in relation with the potential to introduce drugs in a cell (targeting).

At present, many effectors are known to trigger membrane fusion (Schramm et al., 1982), like Ca^{2+} in exocytosis, antibodies and hormones in receptor-mediated endocytosis, pH during the fusion of endocytotic vesicles and the lysosomes. Moreover, the involvement of many other substances like ATP, cAMP, GTP, drugs and Ca^{2+} binding proteins, including calmodulin and synexin, has been reported. Also membrane proteins appear to be involved in the fusion (Schramm et al. 1982), including the state of the cytoskeleton, the extent of glycosylation and the distribution of membrane spanning proteins. With respect to the latter aspect, it was assumed that the membrane proteins are cleared from the fusion site before actual fusion can start. This was based on freeze-fracture experiments which exhibited smooth bilayer patches without intramembraneous particles (IMP). However, later on it has been demonstrated that this phenomenon can be attributed to the use of cryoprotectants and/or chemical fixation. No particle clearance has been found using fast-freezing methods (Chandler and Heuser, 1979). So there is no requirement for a visibled lateral reorganization of intrinsic proteins before membrane fusion. In fact, only a small area of lipid may be enough for fusion, which is consistent with the local point fusion hypothesis.

As discussed above, many factors are thought to play a role in membrane fusion, but lipids actually fuse. The lipids of the bilayer have to become in close opposition, which requires a reduction in electrostatic repulsion and in the hydration forces of the lipids.

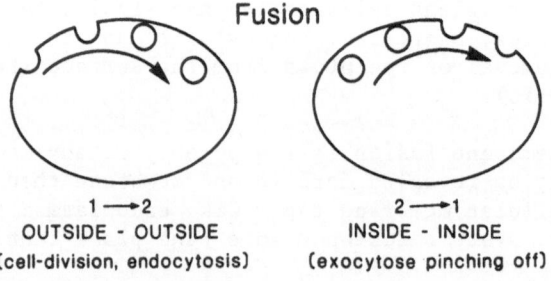

Fig. 1.

Subsequently, the lipids of both bilayers have to join, which implies that they temporarily will leave the bilayer configuration at the fusion point. Finally, both bilayers fuse with each other, which includes bilayer restabilization.

The question is: "Is there one general mechanism involved in the fusion with respect to the lipids?" From model studies a variety of possible factors have been proposed, including the presence of lysolecithin (Lucy, 1979), the presence of solid and lipid domains which can be triggered isothermally in mixtures of phosphatidylserine and neutral phospholipid (Papahadjopoulos et al., 1978). Although these principles may play a role in membrane fusion they only fit for some distinct fusion processes. In this respect it may be mentioned that one can divide biological fusion in two types. Firstly, there is fusion in which the outer (exoplasmic) monolayer of the membrane is involved, like in endocytosis and cell fusion. Secondly, the opposite situation is encountered in exocytosis, fusion between intracellular organelles and cell division in which the inner (cytoplasmic) monolayer is primarily involved (Fig. 1).

So, the involvement of for instance phosphatidylserine can only be relevant for exocytosis of eukaryotic cells, since this phospholipid is present in sufficient quantity only in the cytoplasmic monolayer of the plasma membrane and exocytotic vesicles, phosphatidylserine as a result of phospholipid asymmetry in these membranes (Verkleij et al., 1973; Rothman and Lenard, 1977). Furthermore, the fact that phospholipid compositions of biomembranes in nature may vary widely – compare the lipid composition of the membranes of gram-negative bacteria, the plasma membrane of eukaryotic cells and the chloroplast membranes – it is quite clear that membrane fusion does not require the presence of a special lipid.

HEXAGONAL II LIPIDS

Recently it has been proposed that lipids which prefer the hexagonal II phase upon isolation from the membrane are involved in membrane fusion (Cullis and Hope, 1978; Verkleij et al., 1979; Cullis and De Kruijff, 1979). Such a type of lipid is present in almost every membrane. The organization of lipids in the hexagonal II phase can be detected with X-ray diffraction (Luzzati and Husson, 1962; Luzzati et al., 1968, 1974), freeze fracturing (Deamer et al., 1970; Verkleij and De Gier, 1981) and [31]P Nuclear magnetic resonance (^{31}P NMR; Cullis and De Kruijff, 1978). Figure 2 shows the characteristic ^{31}P NMR spectra and freeze-fracture morphology of the bilayer and hexagonal II phase.

Examples of hexagonal II phase-forming lipids are unsaturated phosphatidylethanolamines (Reis-Husson, 1967; Rand et al., 1971; Cullis et al., 1978a), monogalactosyldiglycerides (Shipley, 1973) and monoglucosyldiglycerides (Wieslander et al., 1978; De Kruijff

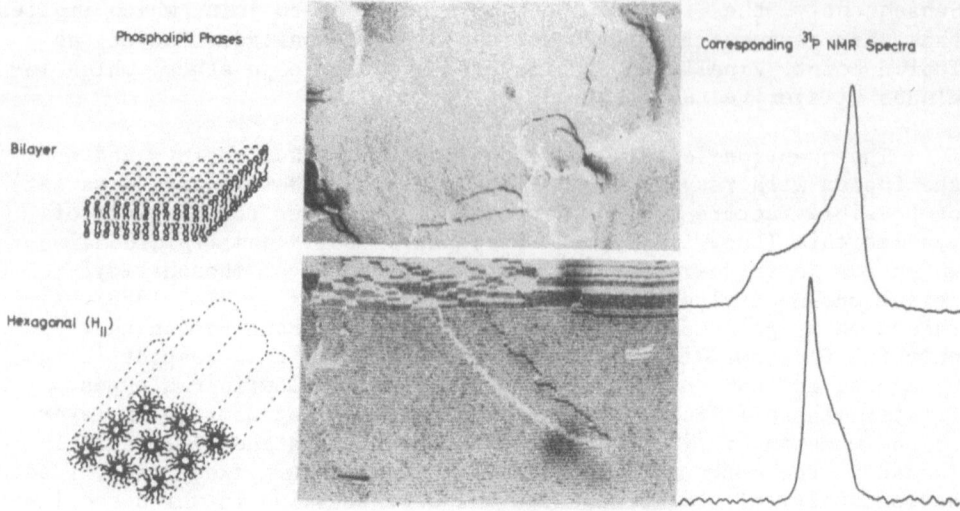

Fig. 2. Molecular arrangements of phospholipids in the bilayer and hexagonal II phases, with their characteristic freeze-fracture morphology and ^{31}P NMR. Micrographs x 50,000.

et al., 1979). The lipid cylinders in the hexagonal II structures have diameters of about 50-70 Å (Luzzati et al., 1968; Rand and Sengupta, 1972). Other hexagonal II phase-forming lipids are the negatively charged phospholipids cardiolipin (Deamer et al., 1970; Rand and Sengupta, 1972; Cullis et al., 1978b) and phosphatidic acid at neutral pH (Papahadjopoulos et al., 1976; Verkleij et al., 1981), which can be converted isothermally from the bilayer structure to the hexagonal II phase by the addition of Ca^{2+} and Mg^{2+} ions, respectively. In the case of cardiolipin and phosphatidic acid it has been shown that the local anaesthetics dibucaine and chlorpromazine can produce a similar phase change (Cullis et al., 1978a; Verkleij et al., 1981). Recently it has been found that hydrophobic peptides like gramicidin can induce the hexagonal II phase in phosphatidylcholine bilayers (Van Echteld et al., 1981).

The bilayer-hexagonal II transition in systems of purified unsaturated phosphatidylethanolamines is remarkably abrupt and occurs within a temperature range of only a few degrees. Differential scanning calorimetric studies show that the enthalpy change involved in this structural rearrangement is very small compared to the amount of heat taken up needed to melt the solid bilayer phase. Since the polymorphic transition occurs above the gel-liquid-crystalline transition of the bilayers, it is likely that the hexagonal II phase is in the liquid-crystalline state.

MIXTURES OF BILAYER AND HEXAGONAL II FORMING LIPIDS

The presence of hexagonal II forming lipids in mixtures with bilayer forming lipids will tend to destabilize the bilayer structure and possibly allows the occurrence of non-bilayer phases in membranes. With ^{31}P NMR an unexpected behavior in these lipid mixtures was found revealing the presence of hexagonal II phase, the lamellar phase and of a narrow spectral component, indicating isotropic motional averaging. With freeze fracturing a variety of structural features have been encountered which in principle can be found in a variety of lipids of which one of the lipids prefers the hexagonal II phase.

First of all, one can find a conglomerate of particles like in phosphatidylcholine/cardiolipin in excess of Ca^{2+} (Verkleij et al., 1979; Van Venetië and Verkleij, 1981), cardiolipin and chlorpromazine (Verkleij et al., 1981), monogalactosyldiglyceride/digalactosyldiglyceride (Sen et al., 1981), phosphatidylcholine/phosphatidylethanolamine (Hui et al., 1982) and monoglucosyldiglyceride/diglucosyldiglyceride (Fig. 3). These particles vary from 100 to 150 Å in diameter in dependency of the system. The particles are packed in a hexagonal or rectangular lattice, which is consistent with inverted micelles in a close packing. Alternatively, this structure may reflect a cubic phase in which two separated aqueous compartments are present, a model which has been put forward by Fontell (1981).

A second feature found in these lipid mixtures is the presence of cylinders with different diameters. Figure 4 shows two examples of this feature. Table 1 shows the diameter of the hexagonal II cy-

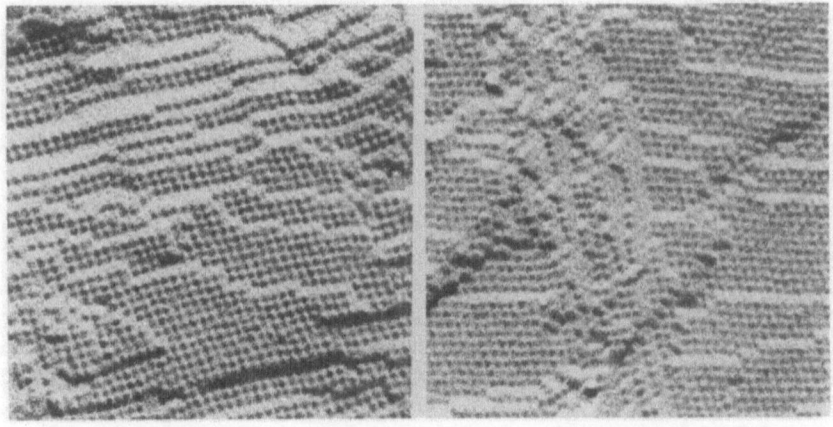

Fig. 3. Monoglucosyldiglyceride/diglucosyldiglyceride in a molar ratio of 3:1 in excess of water. x 100,000 (Wieslander and Verkleij, to be published), frozen from 50°C with jet-freezing.

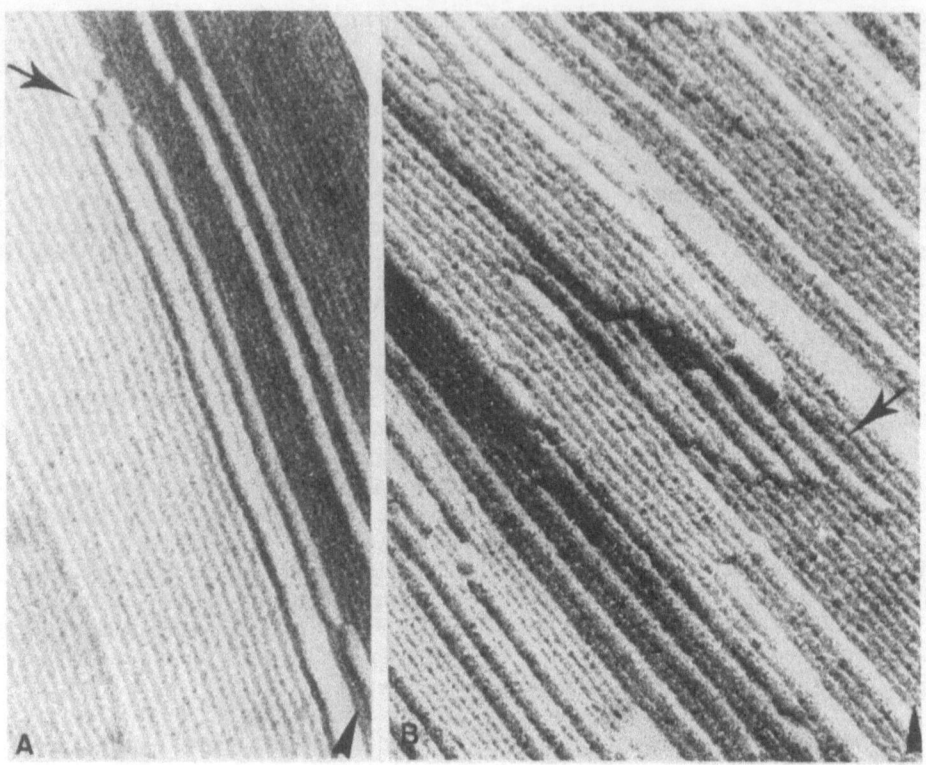

Fig. 4. Two types of hexagonal II tubes with distinct diameters exist in phosphatidylcholine/cardiolipin mixtures, in the presence of Ca^{2+} (A) and Mg^{2+} (B). The arrows indicate thick type tubes. x 200,000.

linders in pure hexagonal II phases and those in mixed lipid systems (Van Venetië and Verkleij, 1981).

A third feature is the presence of smooth fracture faces being continuous with respect to their fracture plane with the hexagonal II phase tubes, i.e., the tubes divergate into smooth fracture faces as for instance in Fig. 5 (Van Venetië and Verkleij, 1981). A model for this transition can be seen in Fig. 6.

Finally, one finds "lipidic particles" and their complementary pits on smooth fracture faces. The particles may be arranged in three dimensional agglomerates as mentioned above or lined up, frequently changing into hexagonal II phase tubes (Verkleij et al., 1980; Van Venetië and Verkleij, 1981) or they can be present solely on fracture faces of smooth fracture faces (De Kruijff et al., 1979; Fig. 7).

Other types of particles (cusp-like) have been found as well in addition to the more well-defined particles (Hui and Stewart, 1981).

Table 1. Periodicity measurements of the hexagonal II phase tubes.

	Diameter (nm)			
	H_{II}	$H_{II}*$	$H_{II}**$	Lipidic particle
DOPE	7.4			
DOPE/DOPC/cholesterol	8.6	–	–	9.5
Ca/DPG	5.2			
Ca/DPG/PC	–	7.3	14.0	8.5
Mg/DPC	6.5			
Mg/DPG/PC	6.4	11.5	–	13.0
Mn/DPG	7.5			
Mn/DPG/PC	7.3	9.7	–	13.0

$H_{II}*$ indicates the thick type tubes. $H_{II}**$ represents the extra thick tube type, as found in the Ca/DPG/PC system only. The diameter value of the lipidic particles is represented as the mean of the particle and pit diameter.

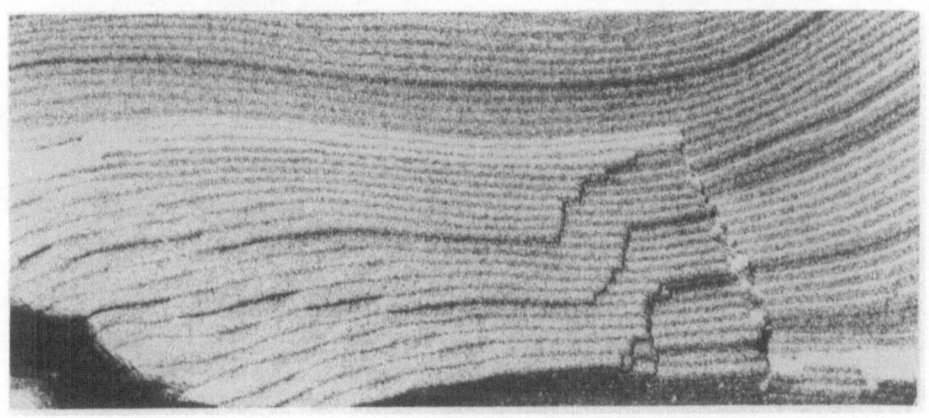

Fig. 5. Hexagonal II tubes, gradually divergating into stacked layers in dioleoylphosphatidylethanolamine/dioleoylphosphatidylcholine/cholesterol mixture (molar ratio: 3:1:2). x 100,000.

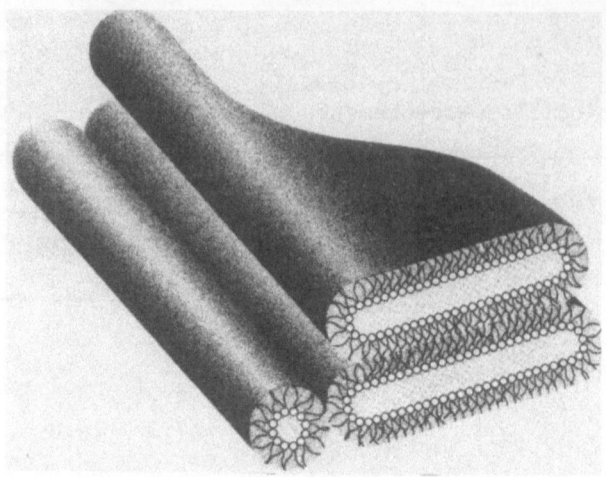

Fig. 6. The hexagonal II to lamellar phase transition.

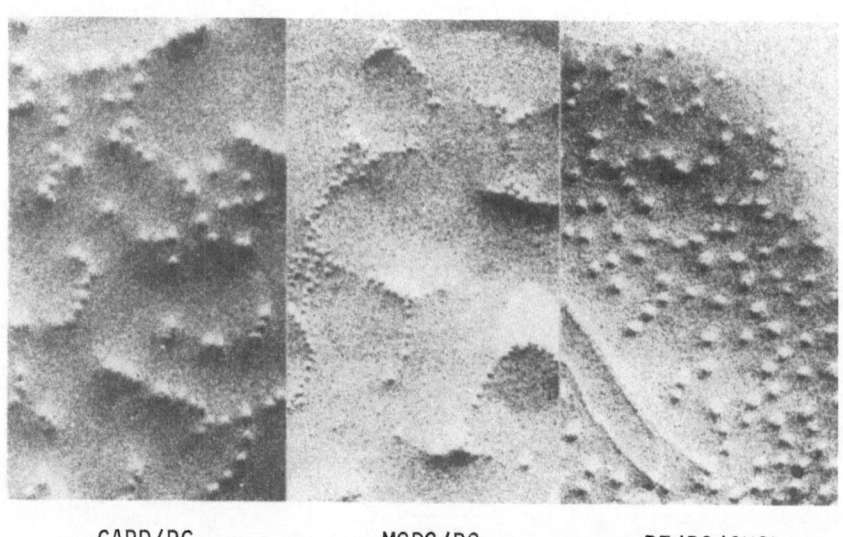

CARD/PC   MGDG/PC   PE/PC/CHOL

Fig. 7. Examples of phospholipid systems showing intramembraneous
particles; cardiolipin and egg phosphatidylcholine (CARD/PC;
molar ratio 1:1) in the presence of Ca^{2+}; monoglucosyldigly-
ceride and egg phosphatidylcholine (MGDG/PC; molar ratio
1:1) at $10^{\circ}C$, after being heated to $60^{\circ}C$; dioleoylphosphati-
dylethanolamine/dioleoylphosphatidylcholine/cholesterol
(PE/PC/CHOL; molar ratio 3:1:2) at $10^{\circ}C$, after being heated
to $60^{\circ}C$. x 200,000.

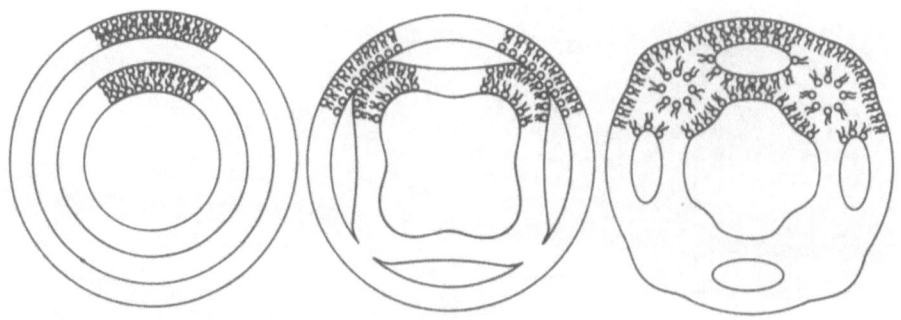

Fig. 8. Model for fusion between bilayers via inverted micelles.

At present the discussion with respect to the nature of the particles is not completed but it is generally accepted that these lipidic particles represent structures at the fusing site or nexus of two or more membranes. The cusp-shaped particles could be prefusing contact structures whereas the well-defined lipidic particles are inverted micelles at the fusing site as depicted in Fig. 8.

MEMBRANE FUSION

As visualized in Fig. 8, non-bilayer structures like inverted micelles and small tubes of lipids, organized in a similar way as the hexagonal II phase, could be intermediary stages in membrane fusion. Such an intermediary structure has been proposed earlier (Lau and Chan, 1975; Pinto da Silva and Noqueira, 1977). This hypothesis has been precised by Cullis and Hope (1978) who suggested that the fusogenic capacity, which proceeds via a non-bilayer "inverted micelle", is derived from lipids which prefer the hexagonal II phase.

The involvement of hexagonal II type lipids in fusion has been confirmed by experiments with model membrane systems. It has been shown that vesicles composed of an equimolar mixture of bovine heart cardiolipin and egg-lecithin fuse upon the addition of Ca^{2+} (Fig. 9; Verkleij et al., 1979).

Similar phenomena have been observed with vesicles of dioleoylphosphatidylethanolamine/dioleoylphosphatidylcholine/cholesterol at a molar ratio of 3:1:2 (Verkleij et al., 1980). These vesicles, obtained by sonication, fuse upon increasing the temperature. This behavior has been found for many mixtures of lipids of which one of the phospholipids prefers the hexagonal II phase (Nayar et al., 1982).

It has been found that in all these systems the fusion is associated with the formation of lipidic particles, frequently found at the fusion interface (Fig. 10). This phenomenon has led to the

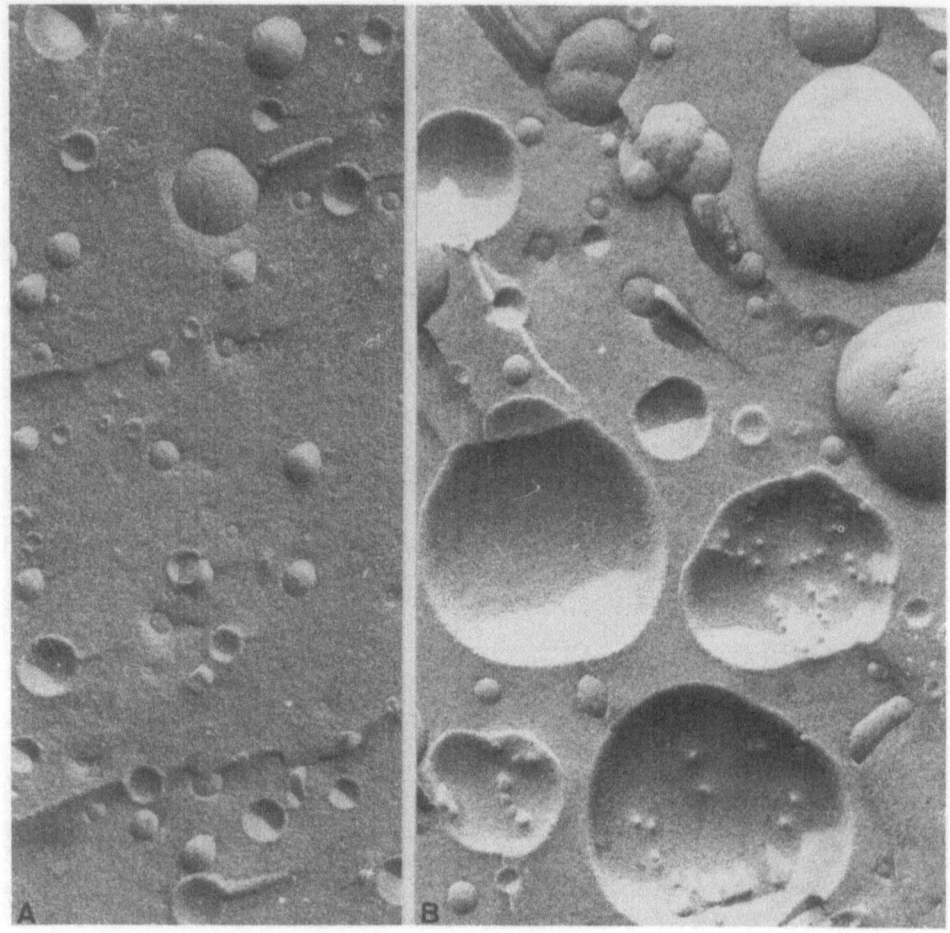

Fig. 9. Fusion of lipid vesicles composed of an equimolar mixture of cardiolipin and lecithin before (A) and after (B) addition of Ca^{2+}.

hypothesis that these lipidic particles represent inverted micelles, which are intermediary structures in fusion. Such micrographs were obtained after incubation times longer than 10 min. Moreover, glycerol as cryoprotectant has been used.

Recent kinetic experiments (Wilschut et al., 1982) have demonstrated that fusion of vesicles prepared from an equimolar mixture of cardiolipin and lecithin is extremely fast (on the time scale of sec) and that this fusion is non-leaky. Therefore we (Verkleij and Wilschut, unpublished results) have repeated our initial fusion experiments using the fast-freezing method of spray freezing. Vesicles

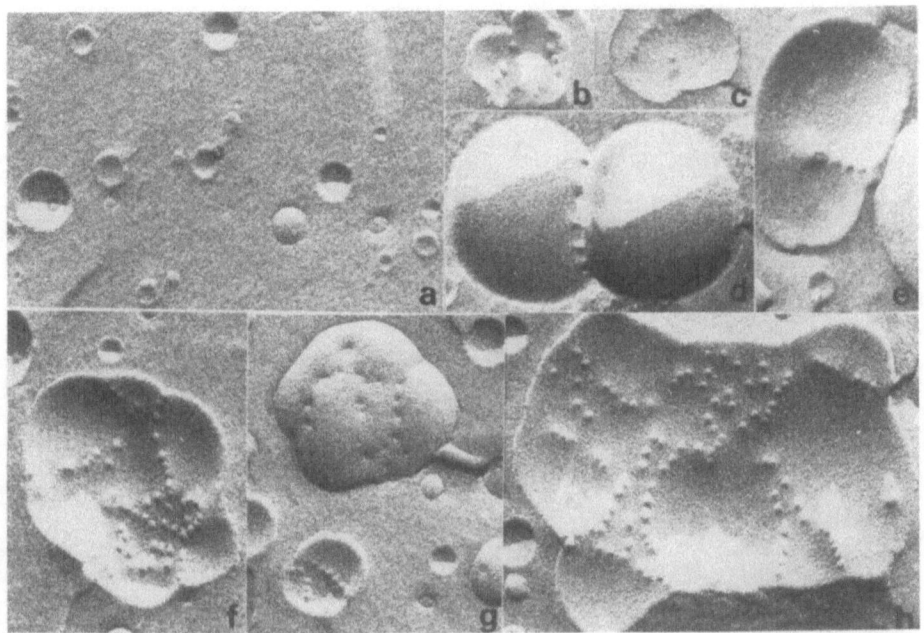

Fig. 10. Fusion of lipid vesicles composed of an equimolar mixture
of cardiolipin and egg-lecithin. (a) Without Ca^{2+}, (b-h)
representative for Ca^{2+} concentrations between 2-10 mM
$CaCl_2$. x 100,000.

composed of cardiolipin and lecithin were quenched within seconds
upon addition of Ca^{2+}. Figure 11 shows that larger vesicles are
formed within this time scale. However, lipidic particles at the
points of fusion between vesicles have not been identified. Similar
results have been obtained in other mixtures like phosphatidyletha-
nolamine/phosphatidylcholine/cholesterol/phosphatidylserine upon
addition of Ca^{2+}. These mixtures show lipidic particles upon longer
incubation times. So at present it is not clear whether lipidic particles are really intermediary in membrane fusion or only equilibrium structures after fusion. On the other hand, it is not excluded
that the life time of initial intermediate fusion structures is too
short (less than a thousand of a second) to be visualized with freeze
fracture electron microscopy.

In summary, it is clear that vesicles composed of a lipid
mixture, of which one of the lipids prefers the hexagonal II phase,
fuse. This fusion is extremely fast and, as has been shown for one
mixture, non-leaky. On the other hand, the nature of the intermediate structure during fusion is not revealed.

189

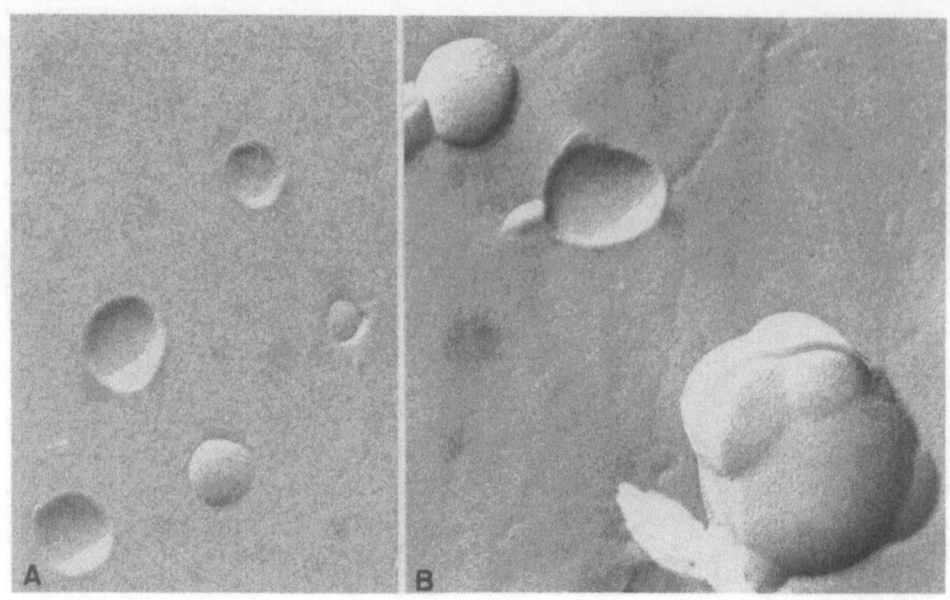

Fig. 11. Fusion of lipid vesicles composed of an equimolar mixture of cardiolipin and egg-lecithin made by reverse phase evaporation (Wilschut et al., 1982), quenched by spray freezing. (A) Without Ca^{2+}, (B) 2 sec after addition of Ca^{2+} in concentrated form (20 mM) resulting in a final Ca^{2+} concentration of 5 mM.

REFERENCES

Chandler, D.E. and Heuser, J.E., 1980, Arrest of membrane fusion events in mast cells by quick freezing, J. Cell Biol., 86: 666.

Cullis, P.R. and Hope, M.M., 1978, Effects of fusogenic agent on membrane structure of erythrocyte ghosts and the mechanism of membrane fusion, Nature, 271: 672.

Cullis, P.R. and De Kruijff, B., 1978, Lipid polymorphism and the functional role of lipids in biological membranes, Biochim. Biophys. Acta, 559: 399.

Cullis, P.R., Verkleij, A.J. and Ververgaert, P.H.J.Th., 1978a, Polymorphic phase behavior of cardiolipin as detected by ^{31}P NMR and freeze-fracture techniques. Effects of calcium, dibucaine and chlorpromazine, Biochim. Biophys. Acta, 513: 11.

Cullis, P.R., Van Dijck, P.W.M., De Kruijff, B. and De Gier, J., 1978b, Effects of cholesterol on the properties of equimolar mixture of synthetic phosphatidylethanolamine and phosphatidylcholine. A ^{31}P NMR and differential scanning calorimetry study, Biochim. Biophys. Acta, 513: 21.

Deamer, D.W., Leonard, R., Tardieu, A. and Branton, D., 1970, Lamellar and hexagonal lipid phases visualized by freeze etching, Biochim. Biophys. Acta, 219: 47.

Echteld, C.J.A. van, Van Stigt, R., De Kruijff, B., Leunissen-Bij-
    velt, J., Verkleij, A.J. and De Gier, J., 1981, Gramicidin pro-
    motes formation of the hexagonal II phase in aqueous dispersions
    of PE and PC, Biochim. Biophys. Acta, 648: 287.
Fontell, K., 1981, Liquid crystallinity in lipid water systems, Mol.
    Cryst. Liq. Cryst., 63: 59.
Hui, S. and Boni, L.L., 1982, Lipidic particles and cubic phase
    lipids, Nature, 296: 175.
Hui, S. and Stewart, T., 1981, Lipidic particles are intermembrane
    attachment sites, Nature, 290: 427.
Kruijff, B. de, Verkleij, A.J., Van Echteld, C.J.A., Gerritsen, W.J.,
    Mombers, C., Noordam, P.C. and De Gier, J., 1979, The occurrence
    of lipidic particles in lipid bilayers as seen by ^{31}P NMR and
    freeze-fracture electron microscopy, Biochim. Biophys. Acta,
    555: 200.
Lau, A.L.Y. and Chan, S.J., 1975, Alamethicin-mediated fusion of
    lecithin vesicles, Proc. Natl. Acad. Sci. USA, 72: 2170.
Lucy, L.A., 1970, The fusion of biological membranes, Nature, 227:
    814.
Luzzati, V. and Reiss-Husson, F., 1962, The structure of the liquid
    crystalline phases of lipid-water systems, J. Cell Biol., 12:
    207.
Luzzati, V. and Tardieu, A., 1974, Lipid phases: structure and
    structural transitions, Ann. Rev. Phys. Chem., 25: 79.
Luzzati, V., Gulik-Krzywicki, T. and Tardieu, A., 1968, Polymorphism
    of lecithins, Nature, 218: 1031.
Nayar, R., Hope, M.J. and Cullis, P.R., 1982, Phospholipids as ad-
    juncts for calcium ion stimulated release of chromaffin granules
    contents: implications for mechanism of exocytosis, Biochemistry,
    21: 4583.
Papahadjopoulos, D., Portis, A. and Pangborn, W., 1978, Calcium-
    induced lipid phase transitions and membrane fusion, Ann. New
    York Acad. Sci., 308: 50.
Papahadjopoulos, D., Vail, W.J., Pangborn, W. and Poste, G., 1976,
    Studies on membrane fusion. -II. Induction of fusion in pure
    phospholipid membranes by calcium ions and other divalent me-
    tals, Biochim. Biophys. Acta, 448: 265.
Pinto da Silva, P. and Nogueira, M.L.J., 1977, Membrane fusion during
    secretion. A hypothesis based on EM observation of Phytophora
    palmivora zoospores during encystment, J. Cell Biol., 73: 166.
Rand, R.P. and Sengupta, S., 1972, Cardiolipin forms hexagonal struc-
    tures with divalent cations, Biochim. Biophys. Acta, 255: 484.
Rand, R.P., Tinker, D.O. and Fast, F.G., 1971, Polymorphism of phos-
    phatidylethanolamines from two natural sources, Chem. Phys.
    Lipids, 6: 333.
Reiss-Husson, F., 1967, Structures des phases liquide-cristallines
    de differents phospholipides, monoglycerides, sphingolipides,
    anhydres ou en presence d'eau, J. Mol. Biol., 25: 363.
Rothman, J.E. and Lenard, J., 1977, Membrane asymmetry: the nature
    of membrane asymmetry provides clues to the puzzle of how mem-

branes are assembled, Science, 195: 743.

Schramm, M., Oates, J., Papahadjopoulos, D. and Loyter, A., 1982, Fusion and implantation in biological membranes, Trends in Pharmacol. Sci., 3: 221.

Sen, A., Williams, W.P., Brain, A.P.R., Dickens, M.J. and Quinn, P.J., 1981, Formation of inverted micelles in dispersions of mixed galactolipids, Nature, 293: 488.

Shipley, G.G., 1973, Recent X-ray diffraction studies of biological membranes and membrane components, in: "Biological membranes," Chapman, D. and Wallace, D.F.H., eds., Academic Press, London and New York, Vol. 2, pp. 1-89.

Venetië, R. van and Verkleij, A.J., 1981, Analysis of the hexagonal II phase and its relation to lipidic particles and the lamellar phase. A freeze-fracture study, Biochim. Biophys. Acta, 645: 262.

Verkleij, A.J. and De Gier, J., 1981, Freeze-fracture studies on aqueous dispersions of membrane lipids, in: "Liposomes: From Physical Structure to Therapeutic Applications", Elsevier, Amsterdam, Vol. 4, pp. 83-103.

Verkleij, A.J., Van Echteld, C.J.A., Gerritsen, W.J., Cullis, P.R. and De Kruijff, B., 1980, The lipidic particle as an intermediate structure in membrane fusion processes and bilayer to hexagonal HII transitions, Biochim. Biophys. Acta, 600: 620.

Verkleij, A.J., De Maagd, R., Leunissen-Bijvelt, J. and De Kruijff, B., 1982, Divalent cations and chlorpromazine can induce nonbilayer structures in phosphatidic acid containing model membranes, Biochim. Biophys. Acta, 684: 255.

Verkleij, A.J., Mombers, C., Gerritsen, W.J., Leunissen-Bijvelt, J. and Cullis, P.R., 1979, Fusion of phospholipid vesicles in association with the appearance of lipidic particles as visualized by freeze fracturing, Biochim. Biophys. Acta, 555: 358.

Verkleij, A.J., Zwaal, R.F.A., Roelofsen, B., Comfurius, P., Kastelijn, P. and Van Deenen, L.L.M., 1973, The asymmetric distribution of phospholipids in the human red cell membrane. A combined study using phospholipase and freeze-etch electron microscopy, Biochim. Biophys. Acta, 323: 178.

Wieslander, A., Ulmius, J., Lindblom, G. and Fontell, K., 1978, Water binding and phase structures for different Acholeplasma laidlawii membrane lipids studied by deuteron NMR and X-ray diffraction, Biochim. Biophys. Acta, 512: 241.

Wilschut, J., Holsappel, M. and Jansen, R., 1982, Ca^{2+}-induced fusion of cardiolipin/phosphatidylcholine vesicles monitored by mixing of aqueous contents, Biochim. Biophys. Acta, 690: 297.

CONTROL OF MEMBRANE FUSION BY DIVALENT CATIONS,

PHOSPHOLIPID HEAD-GROUPS AND PROTEINS

Nejat Düzgüneş[a,b,d], Jan Wilschut[e] and
Demetrios Papahadjopoulos[a,c]

[a*]Cancer Research Institute and Departments of
[b]Anesthesia and [c]Pharmacology, University of California,
San Francisco, CA 94143; [d]Bruce Lyon Memorial Research
Laboratory, Children's Hospital Medical Center, Oakland,
CA 94609, U.S.A. and [e]Laboratory of Physiological
Chemistry, University of Groningen, Bloemsingel 10,
9712 KZ Groningen, The Netherlands

INTRODUCTION

Membrane fusion is a vital process for many cellular activities
such as exocytosis, endocytosis, formation of secondary lysosomes,
membrane biosynthesis and cell division. During exocytosis secretory
vesicles containing proteins or neurotransmitters fuse with the
plasma membrane and release their contents into the extracellular
space. Such fusion events have been observed by electron-microscopy
in many cell types, for example during histamine release in mast
cells (Lagunoff, 1973; Lawson et al., 1977), insulin secretion in
pancreatic B-cells (Orci et al., 1977), mucocyst secretion in
Tetrahymena (Satir et al., 1973), chromaffin granule extrusion in
the adrenal medulla (Douglas, 1968) and neurotransmitter release at
the neuromuscular junction (Ceccarelli et al., 1972; Heuser et al.,
1979). The involvement of Ca^{2+} in exocytosis has been shown by
microinjection of Ca^{2+} into cells (Miledi, 1973; Kanno et al., 1973),
or by using Ca^{2+}-ionophores (Foreman et al., 1973; Cochrane and
Douglas, 1974). Llinas and Nicholson (1975) have shown the increase
in intracellular Ca^{2+} concentration following electrical stimulation
of the squid giant synapse, and Baker and Knight (1978) have
demonstrated the dependence of catecholamine release from adrenal
medullary cells on the free Ca^{2+}-concentration. The site of action

---

* Address for correspondence.

of Ca^{2+} or the mechanisms which control membrane fusion during excytosis are not known, however.

Several model systems have been utilized to study the molecular basis of the Ca^{2+} requirement for exocytosis and of the membrane fusion reaction which accompanies it. Studies on isolated secretory vesicles have suggested that membrane glycoproteins may be involved in the sensitivity and selectivity for Ca^{2+} in the fusion of the vesicles (Gratzl and Dahl, 1978; Gratzl et al., 1980). Liposomes derived from the phospholipids and glycolipids of these membranes do not exhibit this property and require higher concentrations of Ca^{2+} for fusion compared to the intact secretory vesicle membranes (Gratzl et al., 1980; Dahl et al., 1979; Ekerdt et al., 1981). In these liposomes, however, the lateral and transbilayer organization of the lipids which may exist in the intact membranes is lost. Domains of certain phospholipids may be necessary for a membrane to be susceptible to fusion under certain ionic conditions. Membrane proteins may be involved in creating these domains, as observed in a number of model systems (Boggs et al., 1977a,b). It is important to understand the properties of individual phospholipid species with respect to their ability to undergo fusion in order to infer the fusion susceptibility of lipid domains which might be present in a biological membrane.

METAL ION SPECIFICITY IN THE AGGREGATION AND FUSION OF PHOSPHOLIPID VESICLES

The role of individual phospholipids in divalent cation-induced membrane fusion has been studied extensively by means of liposomes made from purified phospholipids. Membrane fusion has been demonstrat॑ by a number of assays ranging from monitoring the increase in vesicle size (Papahadjopoulos et al., 1975; Kantor and Prestegard, 1975; Liao and Prestegard, 1979; Koter et al., 1978; Day et al., 1977) to following the mixing of membrane components (Maeda and Ohnishi, 1974; Papahadjopoulos et al., 1974, 1976; Miller and Racker, 1976; Ohki and Düzgüneş, 1979; Vanderwerf and Ullman, 1980; Struck et al., 1981; Hoekstra, 1982a). We define membrane fusion as the coalescence of the internal aqueous contents of membrane vesicles with concomitant intermixing of the membrane components. In this sense, our definition corresponds to the step in exocytosis called "fission" by Palade (1975). The formation of a pentalaminar structure between two membranes which has been termed "fusion" (Palade, 1975) is the initia॑ step in the overall process of membrane fusion and would correspond to aggregation in our description. Hence, following Nir et al. (1982a,b), we separate the fusion process into two distinct but kinetically coupled stages: (i) Vesicle aggregation which brings the membranes of two vesicles into close apposition, and (ii) membrane fusion, which involves the local and momentary destabilization and loss in structural integrity of the phospholipid bilayer, resulting in the intermixing of membrane components and of the aqueous compartments of the two vesicles. Aggregation of vesicles has been detected by light absorbance (Düzgüneş and Ohki, 1977; Nir et al., 1981; Ohki et al., 1982) or 90° light scattering (Portis et al., 1979; Wilschut and Papahadjopoulos, 1979; Wilschut et al., 1980; Düzgüneş et al., 1981a). Several assays have been developed for monitoring the intermixing of aqueous contents of fusing vesicles (Ingolia and Koshland, 1978; Holz and Stratford, 1979; Hoekstra et al., 1979; Wilschut and Papahadjopoulos, 1979; Wilschut et al., 1980). We have used an assay based on the formation of a chelation complex between dipicolinic acid (DPA) and Tb which results in an increase of Tb fluorescence intensity by four orders of magniture. This reaction is extremely fast, enabling us to analyze the rapid kinetics of membrane fusion (Wilschut and Papahadjopoulos, 1979; Wilschut et al., 1980). Tb/citrate or Tb/nitrilotriacetate is encapsulated in one population of vesicles and dipicolinate in another. Vesicle fusion results in the formation of the fluorescent Tb/DPA complex. The reaction is inhibited by EDTA and divalent cations, which are present in the external medium; thus any contents which may be released into the medium do not contribute to the fluorescence (Wilschut et al., 1980). The leakage of contents is determined independently by using vesicles containing carboxy-fluorescein (CF) at self-quenched concentrations and following the relief of self-quenching of the fluorescence when the contents are diluted into the medium (Weinstein et al., 1977; Portis et al., 1979) or by using Tb vesicles and monitoring the appearance of Tb fluorescence in a medium containing DPA in the absence of EDTA (Düzgüneş et al., 1982a; Hoekstra, 1982b).

Liposomes composed of negatively charged phospholipids suspended in physiological salt solutions experience mutual electrostatic repulsion which prevents the formation of stable aggregates. As the concentration of cations in the medium is increased the surface charge density is reduced due to ion binding to the membrane surface. Within the framework of the Derjaguin-Landau-Verwey-Overbeek (DLVO) theory the free energy of interaction between colloidal particles is expressed as the sum of the attractive Van der Waals interactions and repulsive electrostatic interactions (Verwey and Overbeek, 1948; Nir, 1977). In general, vesicles will aggregate (dimerize) only if there is a local minimum in the free energy of interaction, and the average time which the vesicles can spend in this configuration depends on the depth of the minimum. If the depth of the potential well is less than kT, where k is Boltzmann's constant and T is the absolute temperature, the vesicles will diffuse apart. There are two types of local potential minimum between vesicles, one at surface separations of < 10 Å, called the primary minimum, and the other at 20-100 Å, called the secondary minimum with a potential barrier in between (Nir et al., 1982a). Cations in solution are concentrated near the negative surface in the double layer region according to a Boltzmann distribution. The screening of the negative charge causes a significant reduction in the repulsive barrier to close approach. This barrier can be reduced further by binding of cations which decreases the magnitude of the surface charge density. For small unilamellar vesicles (SUV) of 250 Å diameter composed of phosphatidyl serine (PS) significant aggregation occurs in the presence of 1 mM Ca^{2+} or 4 mM Mg^{2+} in 100 mM NaCl, pH 7.4 (Papahadjopoulos et al., 1977; Düzgüneş and Ohki, 1977; Portis et al., 1979), a result which can be predicted by the DLVO theory, accounting explicity for ion binding (Nir and Bentz, 1978). The difference between Ca^{2+} and Mg^{2+} can be explained by the difference in their affinity for PS (Newton et al., 1978; Ohki and Sauvé, 1978; Portis et al., 1979). Binding of monovalent and divalent cations to PS has been determined by various techniques such as atomic absorption spectroscopy (Portis et al., 1979; Düzgüneş et al., 1981a), micro-electrophoresis (Eisenberg et al., 1979; McLaughlin et al., 1981), ^{23}Na- and ^{31}P-NMR (Kurland et al., 1979a,b), surface potential measurements (Ohki and Sauvé, 1978; Ohki and Kurland, 1981) and ion-selective electrodes (Rehfeld et al., 1981; Ekerdt and Papahadjopoulos, 1982). The sequence of effectiveness of divalent cations for inducing aggregation of PS vesicles is $Mn^{2+} > Ba^{2+} > Ca^{2+} > Sr^{2+} > Mg^{2+}$ and that for monovalent ions is $H^+ > Na^+ > Li^+ > K^+ > TMA^+$ (tetramethylammonium) (Ohki et al., 1982). Aggregation is a dynamical process in which vesicles aggregate and deaggregate in primary or secondary minima with characteristic forward and backward rates. The secondary minimum for SUV is too shallow to support measurable aggregation and hence they can aggregate only in the primary minimum (Nir and Bentz, 1978). Large unilamellar vesicles (LUV) of > 1000 Å diameter, however, can be expected to aggregate in the secondary minimum under certain conditions. This mode of aggregation has been demonstrated for PS vesicles in the presence of a mixture of Na^+ and Mg^{2+}, which does not induce any aggregation of SUV in the primary minimum (Nir et al., 1981).

The next step in the overall fusion reaction outlined above is the destabilization and merging of bilayers resulting in communicatic between the internal aqueous compartments. In the presence of > 1 mM Ca^{2+}, PS SUV undergo fusion (Papahadjopoulos et al., 1974) and eventually form cochleate lipid cylinders, which can be converted into large unilamellar vesicles upon addition of an excess of EDTA (Papahadjopoulos et al., 1975). The final cochleate product of Ca^{2+} induced fusion of PS vesicles is a highly condensed structure with essentially no water between the bilayers (Papahadjopoulos et al., 1978); it is therefore obvious that the contents of the vesicles are released during this reorganization, consistent with early observations on the increase in permeability of PS bilayers in the presence of Ca^{2+} (Papahadjopoulos and Bangham, 1966; Papahadjopoulos and Ohki, 1969). The kinetics of this release has been found to be slower than that of the aggregation of the vesicles (Portis et al., 1979; Wilschut and Papahadjopoulos, 1979).

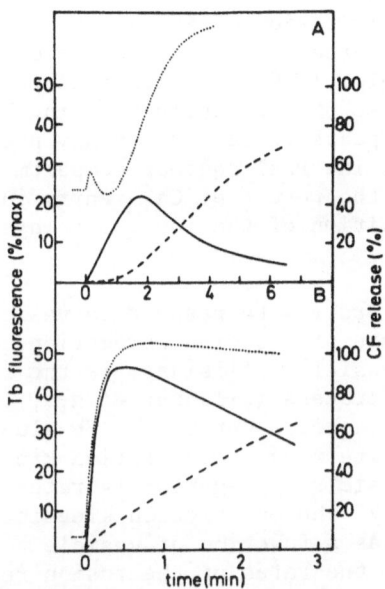

Fig. 1. Ca^{2+}-induced aggregation and fusion of phosphatidylserine
vesicles. (A) LUV. (B) SUV. Ca^{2+} was added at t = 0 to a
final concentration of 3 mM (LUV), or 1.5 mM (SUV). Lipid
concentration, 50 µM. T = 25°C. Tb fluorescence (——) is
given as the percentage of maximal fluorescence attained
when all the Tb in vesicles reacts with excess DPA.
Excitation was at 276 nm, emission at 545 nm with a Corning
3-68 cut-off filter (> 530 nm) between sample and monochromator. 90° light scattering at 276 nm (....) was followed
simultaneously with Tb fluorescence. CF release (---) is
given as the percentage of fluorescence obtained when the
vesicles are lysed with detergent. Excitation was at 430 nm
for LUV and 493 nm for SUV, emission at > 530 with a cut-off
filter (From Wilschut *et al.*, 1980).

The relative rates of aggregation, fusion and release of contents
of PS vesicles in the presence of Ca^{2+} are shown in Fig. 1. The
threshold Ca^{2+} concentration, $[Ca^{2+}]_t$, for the fast aggregation and
fusion of LUV is higher (2.4 mM) than that of SUV (1.2 mM). In either
case aggregation and fusion proceed on the same time scale, whereas
leakage is a slower process. The initial stages of LUV fusion are
essentially non-leaky, the release of contents occurring after a
significant delay. This delay in the release has been shown not
only by carboxyfluorescein fluorescence, but also by the release of
Tb into the medium (Düzgüneş *et al.*, 1982a). Ca^{2+}-induced fusion of
PS SUV is more leaky than the fusion of LUV (Fig. 1). It has been

estimated that initially fusion of SUV is accompanied by the release
of about 10% of the internal contents per fusion event (Wilschut
et al., 1980; Nir *et al.*, 1980). Also in this case Tb release is
identical to the release of carboxyfluorescein. The eventual complete
release of contents appears to be a secondary phenomenon, presumably
due to collapse of the internal aqueous compartment of the fused
products. This may be the result of Ca^{2+} entry into the vesicles,
inducing a close apposition of the inner monolayer on opposite sides
of the vesicle.

Fusion is second-order with respect to vesicle concentration
indicating that the rate of vesicle aggregation, determined by the
frequency of vesicle-vesicle collisions, is the rate-limiting step
of the overall fusion process (Wilschut *et al.*, 1980). Furthermore,
the time course of the development of the Tb fluorescence intensity
can be described adequately by a mass action kinetic model which
also indicates that vesicle aggregation is rate-limiting (Nir
et al., 1980). Recently, the mass action kinetic analysis of the
fusion of PS vesicles as a function of vesicle concentration has
yielded information on the rates of the fusion reaction per se in
addition to rates of the aggregation step (Nir *et al.*, 1982a,b;
Bentz *et al.*, 1982a). The analysis indicates that the rate of fusion
of SUV is about two orders of magnitude faster than that of LUV. On
the basis of the estimated rate constants fusion is expected to occur
within several msec of dimerization, depending of the $Ca2+$ concentration. Indeed, experiments utilizing fast-freezing freeze-fracture
electron microscopy have shown that fusion of PS SUV occurs within
10 msec of Ca^{2+} addition (Miller and Dahl, 1982).

Mixing of membrane components during Ca^{2+}-induced fusion of PS
vesicles can be demonstrated by resonance energy transfer between
fluorescently labelled phospholipids incorporated into two populatior
of vesicles (Vanderwerf and Ullman, 1980; Hoekstra, 1982a). The Ca^{2+}
concentration dependence of the initial rate of fusion detected by
lipid mixing and by coalescence of aqueous contents is very similar
(Düzgüneş *et al.*, 1982). Followed by either method the fusion reactic
is arrested immediately upon addition of excess EDTA at different
times after initiation (Wilschut *et al.*, 1980; Hoekstra, 1982a).

Ca^{2+} has a marked specificity over Mg^{2+} in inducing the fusion
of PS vesicles. At $25^{\circ}C$ LUV aggregate when the Mg^{2+} concentration
exceeds a threshold of 5 mM, but neither fusion nor release of
contents is observed (Wilschut *et al.*, 1981; Fig. 2). Aggregation
is completely reversible when EDTA is added. In contrast, Mg^{2+} can
induce the fusion of SUV, but only to a limited extent. The process
ceases after a few rounds of fusion, presumably because the strain
of the highly curved bilayer is relieved. The Mg^{2+}-induced fusion
of SUV is delayed with respect to aggregation and is initially
significantly more leaky than Ca^{2+}-induced fusion. However, with
Mg^{2+} fusion is not followed by collapse of the internal aqueous space

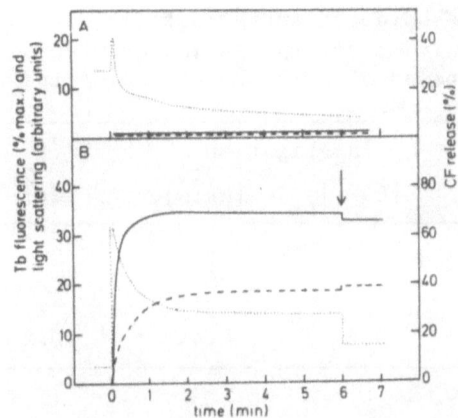

Fig. 2. Mg^{2+}-induced aggregation and fusion of phosphatidylserine
vesicles. (A) LUV. (B) SUV. Mg^{2+} was added at t=0 to a final
concentration of 20 mM (LUV) or 8 mM (SUV). (——) Tb
fluorescence; (...) light scattering at 276 nm; (---) CF
release. The arrow in (B) indicates the addition of excess
EDTA. Other conditions as in Fig. 1. (From Wilschut *et al.*,
1981).

of the vesicles and consequently a significant fraction of the
contents is eventually retained.

The Ca^{2+}/Mg^{2+} specificity must arise from the differences in
the physico-chemical properties of the metal ion/PS complex (Newton
et al., 1978; Papahadjopoulos *et al.*, 1978; Portis *et al.*, 1979;
Düzgüneş and Papahadjopoulos, 1982). In addition to any specific
effects of an isothermal phase transition (Papahadjopoulos *et al.*,
1977), the dehydration of the membrane surface may be crucial to
the induction of bilayer fusion (Portis *et al.*, 1979). The nature
of the membrane surface and interfacial water have been proposed
to be important parameters in membrane fusion (McIver, 1979; Ohki
and Düzgüneş, 1979). Binding of cations to the membrane will not
only dehydrate the ions, at least partially, but also alter the
hydration of the lipid/water interface, depending on the chemical
nature of the ion and the lipid. The hydration layer adjacent to
the phospholipid head-group results in strong repulsive forces
between lipid bilayers at short distances of separation (Le Neveu
et al., 1976; Cowley *et al.*, 1978; Rand, 1981). Ca^{2+} is more readily
dehydrated than Mg^{2+} (Gresh, 1980) and it binds more strongly to
PS (Newton *et al.*, 1978; Nir *et al.*, 1978). That the Ca^{2+}/PS complex
is essentially anhydrous is indicated by the very short lamellar
repeat distance in low angle X-ray diffraction (Portis *et al.*, 1979)
or by ^{1}H-NMR (Hauser *et al.*, 1975). The Mg^{2+} complex retains a layer

Table 1. Threshold concentrations (mM) of Ca^{2+} and Mg^{2+} for inducing aggregation and fusion of LUV composed of pure phospholipids (pH 7.4; T=25°C).

Lipid	Aggregation		Fusion	
	$[Ca^{2+}]_t$	$[Mg^{2+}]_t$	$[Ca^{2+}]_t$	$[Mg^{2+}]_t$
PC	–	–	–	–
PI	3	6	–	–
PS	2	5	2	–
PA	0.2	0.4	0.2	0.4

of water between the lipid bilayers. The specificity of Ca^{2+} over Mg^{2+} is likely to arise during interbilayer contact, since the anhydrous complex of Ca^{2+} with PS is formed only if two bilayers are allowed to come into contact (Portis *et al.*, 1979). Thus, when PS vesicles aggregate a new *trans* binding mode of Ca^{2+} becomes possible and the apparent affinity of Ca^{2+} for PS increases abruptly (Portis *et al.*, 1979; Rehfeld *et al.*, 1981; Ekerdt and Papahadjopoulos, 1982).

The destabilization and fusion of SUV made of PS (or its mixtures with phosphatidylcholine) requires a critical ratio of bound divalent cations/PS (Düzgüneş *et al.*, 1981a), the magnitude of the ratio depending on whether the kinetics of fusion is rate-limited by the aggregation step or the fusion reaction (Bentz *et al.*, 1982b). The sequence of effectiveness of divalent cations in inducing fusion of PS SUV is, in terms of the bulk concentration in the presence of 100 mM Na^+, the same as the sequence for aggregation (Düzgüneş *et al.*, 1981b; Ohki, 1982). In the presence of higher Na^+ concentrations where the fusion rate becomes rate limiting Ca^{2+} appears to be more effective than Ba^{2+} (Bentz *et al.*, 1982b). Then the resulting sequence is $Ca^{2+} > Ba^{2+} > Sr^{2+} > Mg^{2+}$. The Ca^{2+} or Mg^{2+} concentration required to induce fusion is increased when the Na^+ concentration is increased, in accordance with the ability of Na^+ to displace bound divalent cations (Düzgüneş *et al.*, 1981a). At high monovalent cation concentrations the sequence of effectiveness of the ions in inhibiting divalent cation-induced fusion of PS is $Li^+ > Na^+ > K^+ > TMA^+$ (Nir *et al.*, 1982c), in accordance with the sequence of binding constants determined by microelectrophoresis (Eisenberg *et al.*, 1979).

LIPID SPECIFICITY

Other acidic phospholipids than PS also undergo fusion in the presence of divalent cations whose threshold concentrations depend on the phospholipid species (Table 1). $[Ca^{2+}]_t$ varies from about 10^{-4} M for phosphatidate (PA, phosphatidic acid) (Papahadjopoulos

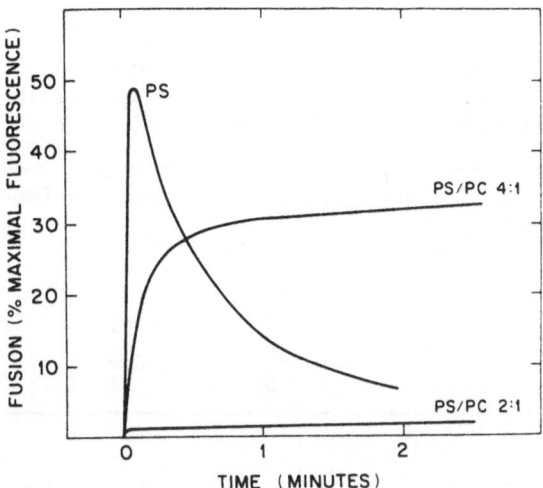

Fig. 3.   The effect of membrane composition on the kinetics of
          vesicle fusion induced by 3 mM Ca^{2+}. SUV were composed of
          pure phosphatidylserine (PS) or mixtures of PS with
          phosphatidylcholine (PC). Details of the experiments are
          given in Fig. 1. (From Düzgüneş *et al.*, 1981a).

et al., 1976; Sundler and Papahadjopoulos, 1981) to 10^{-2} M for
phosphatidylglycerol (PG) (Papahadjopoulos *et al.*, 1976). Pure
phosphatidylcholine (PC) vesicles above their T_c do not fuse
(Papahadjopoulos *et al.*, 1974; Schullery *et al.*, 1980). LUV made of
phosphatidylinositol (PI) aggregate in the presence of divalent
cations but do not fuse (Sundler and Papahadjopoulos, 1981).

Membranes composed of mixtures of PS and PC require higher
concentrations of divalent cations to initiate fusion and vesicles
containing less than 50% PS are resistant to fusion (Papahadjopoulos
et al., 1974). The inhibitory effect of PC on the fusion
susceptibility of the membrane is demonstrated in Fig. 3, where the
time course of fusion induced by 3 mM Ca^{2+} in vesicles (SUV) composed
of pure PS or mixed PS/PC is shown. In the case of pure PS the
fluorescence intensity initially increases after initiation of rapid
fusion and drops subsequently due to leakage of vesicle contents into
the medium. The PS/PC (4:1) vesicles do not collapse after fusion
and the Tb/DPA complex is sequestered from the external medium inside
intact vesicles after about 20% of the vesicle contents have leaked
during initial vesicle fusion (Düzgüneş *et al.*, 1981a). The inclusion
of PC in PS SUV also raises the threshold concentration of divalent
cations required for their fusion with PS monolayers, planar bilayers
of LUV (Ohki and Düzgüneş, 1979; Düzgüneş and Ohki, 1981).

Table 2. Threshold concentrations (mM) of Ca^{2+} and Mg^{2+} for inducing aggregation and fusion of LUV composed of equimolar mixtures of acidic and zwitterionic phospholipids (pH 7.4, T=25°C).

Lipid	Aggregation		Fusion	
	$[Ca^{2+}]_t$	$[Mg^{2+}]_t$	$[Ca^{2+}]_t$	$[Mg^{2+}]_t$
PS/PC	7	15	–	–
PS/PE	3	5	3	5
PI/PE	5	30	10	50
PA/PC	3	5	3	–

LUV composed of an equimolar mixture of PS and PC aggregate above 7 mM Ca^{2+}, but do not fuse or release their contents even in the presence of very high (50 mM) Ca^{2+} concentrations (Düzgüneş et al., 1981c; Table 2). A different result is obtained, however, with an equimolar mixture of PS and phosphatidylethanolamine (PE), another zwitterionic phospholipid. These vesicles do undergo fusion at Ca^{2+} concentrations slightly above the threshold for pure PS. Mixtures containing as low as 25% PS in PE can fuse rapidly in the presence of Ca^{2+} (Fig. 4.). Similar observations on the difference between PC and PE in modulating Ca^{2+}-induced fusion have been made by Miller and Racker (1976) and Vanderwerf and Ullman (1980). Another striking feature of PS/PE membranes is that they may be induced to fuse by Mg^{2+}. Since this ion does not induce fusion of PS LUV, PE appears to play a particularly important role in Mg^{2+}-induced fusion. Preincubation of these vesicles with near-threshold levels of Mg^{2+} increases the rate at which they fuse when sub-threshold Ca^{2+} is introduced (Düzgüneş et al., 1981c), suggesting that the presence of Mg^{2+} facilitates the close approach of the vesicles allowing Ca^{2+} to induce fusion at a rate comparable to that obtained with pure PS vesicles in the presence of considerably higher Ca^{2+} concentrations. In PS/PE/PC mixtures, in which the PC content is maintained at 25% the initial rate of fusion decreases, with decreasing PS content, indicating that PE cannot substitute for PS. The fusogenic capacity of Mg^{2+} is abolished when 10% of the PE in PS/PE (1:1) membranes is replaced by PC.

The propensity of PE to facilitate fusion and of PC to inhibit it may be related to various physico-chemical differences between the two molecules. The head-groups of PE form a compact lattice via electrostatic interactions and hydrogen bonding between the phosphate oxygens and ammonium nitrogens, whereas the quaternary ammonium groups in PC do not form these bonds. The phosphate groups of PC are intercalated by water molecules (Hauser et al., 1981). PC appears to have a higher affinity for water than PE (Jendrasiak

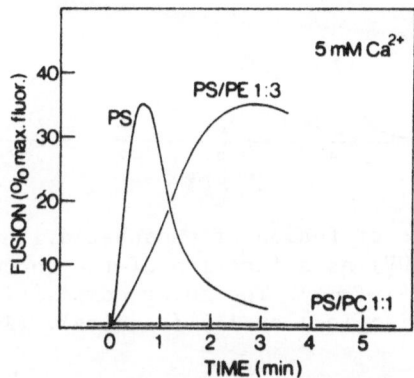

Fig. 4.   Ca^{2+}-induced fusion of LUV composed of pure phosphatidylserine (PS) and its mixtures with phosphatidylcholine (PC)
or phosphatidylethanolamine (PE). (From Düzgüneş et al.,
1980).

and Hasty, 1974). Fully hydrated PC always maintains a lamellar
structure, whereas PE transforms into the hexagonal (H$_{II}$) phase at
a characteristic temperature which depends on the nature of the
hydrocarbon chains (Reiss-Husson, 1967; Cullis and De Kruijff,
1978). Repulsive hydration forces between PE membranes are effective
at shorter separations than between PC membranes allowing the former
to establish closer contact under appropriate ionic conditions
(Rand, 1981).

PE, when included with PI in LUV, renders these membranes
susceptible to fusion by Ca^{2+} or Mg^{2+}, whereas pure PI vesicles
are resistant to fusion. As the PI content is decreased the
threshold concentration of either ion decreases (Sundler et al.,
1981; Fig. 5), indicating a direct involvement of PE in membrane
fusion. The role of PE acyl chains, and hence the possible
involvement of the bilayer-hexagonal (H$_{II}$) transition in membrane
fusion, has been examined by the temperature dependence of Ca^{2+}-
induced fusion in PI/dimyristoyl-PE or PI/egg-PE vesicles. Fusion

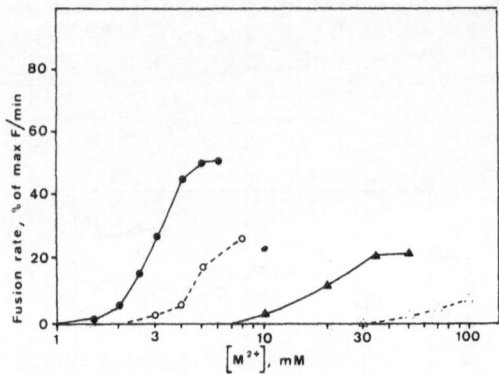

Fig. 5. Initial rate of fusion of phosphatidylinositol/egg-PE vesicles (LUV) as a function of the concentration of Ca^{2+} (——) or Mg^{2+} (---). The molar composition of the vesicles was 1:1 (triangles) or 1:4 (circles). (From Sundler *et al.*, 1981).

is dependent on the gel to liquid-crystalline transition of the mixture, but not on the phase behavior of the PE species. Since dimyristoyl-PE does not transform to a non-bilayer phase up to $50^{\circ}C$ the bilayer-hexagonal transition does not appear to be necessary for fusion in this system (Sundler *et al.*, 1981). A similar argument can be made for the fusion of PS/PE vesicles in the presence of Mg^{2+}, which, unlike Ca^{2+} does not induce the formation of the hexagonal phase even at temperatures where the PE itself would be expected to be in the hexagonal phase (Düzgüneş *et al.*, 1981c). It appears that the aggregation of PI/PE vesicles by Ca^{2+} or Mg^{2+}, and of PS/PE vesicles by Mg^{2+}, leads to close molecular contact between the PE molecules due to their low level of hydration and the partial dehydration of the lipid/water interface by the divalent cations. The different packing arrangements of the PE and the acidic phospholipid molecules under these conditions may create local defects which expose the hydrophobic interior of the bilayers at apposed regions of the two membranes. The lipids may intermix and from a continuous bilayer.

The specificity of PA over PI in membrane fusion raises the possibility that the conversion of PI to PA in several cell types following secretory stimuli (Michell, 1975) may be involved in the fusion of secretory vesicles with the plasma membrane, particularly since this phenomenon is associated with mobilization of cellular Ca^{2+}. The effect of substitution of PA for PI in a complex lipid mixture (which may be found in cellular membranes) on the Ca^{2+}

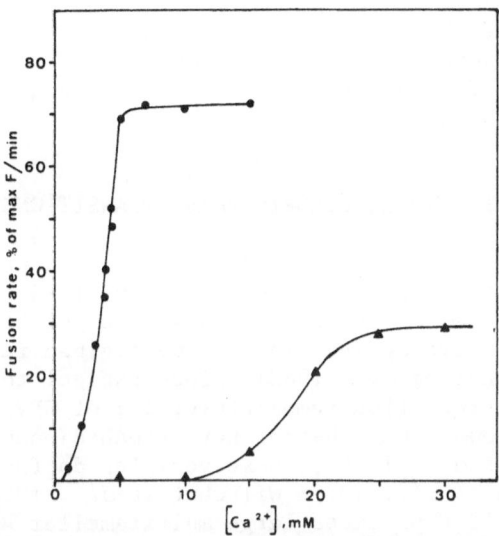

Fig. 6. Effect of substitution of phosphatidate (circles) for
phosphatidylinositol (triangles) on the Ca^{2+} concentration
dependence of the initial rate of fusion of vesicles
composed of phosphatidate (or phosphatidylinositol)/
phosphatidylserine/phosphatidylethanolamine/phosphatidyl-
choline (1:1:2:1). (From Sundler and Papahadjopoulos, 1981).

threshold concentration for the fusion of these membranes is shown
in Fig. 6. Clearly, membranes containing PA fuse at considerably
lower Ca^{2+} concentrations (Sundler and Papahadjopoulos, 1981). The
difference between the fusion characteristics of the two phospholipids
may be ascribed to the hydrated, bulky inositol group which would be
resistant to the formation of dehydrated complexes between the apposed
membranes.

Phospholipid vesicles are a very simple approximation of
biological membranes. Nevertheless, the physicochemical disposition
of certain mixtures of phospholipids for fusion in the presence of
Ca^{2+} and Mg^{2+} leads to the possibility that membrane fusion may be
induced between the phospholipid bilayers of exocytotic vesicles
and plasma membranes. Membranes containing only 10-25% PS can be
induced to fuse with Ca^{2+} or Mg^{2+} (Fraley et al., 1980; Düzgüneş
et al., 1981c). Furthermore, the presence of Mg^{2+} lowers the
$[Ca^{2+}]_t$ for certain types of membranes. Thus, millimolar
concentrations of Mg^{2+} in the cytoplasm may facilitate membrane
fusion by lowering the threshold for Ca^{2+} concentration at the sites
of fusion and increase the rate of the reaction. Creutz et al. (1982)

205

have suggested that the concentration of Ca^{2+} at the site of membrane interaction during exocytosis could be on the order of 400 μM–1 mM, a range which is sufficient to induce fusion in certain phospholipid membranes.

ROLE OF THERMOTROPIC AND IONOTROPIC PHASE TRANSITIONS

The physical state of the lipid bilayer has a profound effect on the rate and extent of Ca^{2+}-induced fusion of PS vesicles. At $0^{o}C$, when LUV composed of PS are mostly in the gel state, addition of Ca^{2+} induces massive aggregation but negligible fusion or release of contents (Wilschut et al., 1982). Since the gel to liquid-crystalline phase transition temperature, T_c, of SUV is expected to be significantly lower than that of LUV (Papahadjopoulos et al., 1976b; Suurkuusk et al., 1976), these vesicles do fuse to a limited extent at $0^{o}C$ (Sun et al., 1978; Wilschut et al., 1982). Addition of Mg^{2+} to PS SUV at $12^{o}C$ produces large multilamellar structures, in contrast to the aggregated vesicular structures slightly larger than the original vesicles obtained at $37^{o}C$. Since the T_c of the Mg^{2+}/PS complex is approx. $18^{o}C$ and that of the Na^{+}/PS complex approx. $8^{o}C$ (Jacobson and Papahadjopoulos, 1975), this suggests that at $12^{o}C$ an isothermal phase change from the liquid-crystalline to the gel state may be related to the extensive fusion of the bilayers (Papahadjopoulos et al., 1977). On the other hand, at $25^{o}C$ small PS vesicles do undergo limited fusion with Mg^{2+}, although the Mg^{2+}/PS complex is still in a fluid state. The correlation of ion-induced phase changes and induction of membrane fusion does hold for Ca^{2+}-induced fusion of PS vesicles since PS bilayers transform in the presence of Ca^{2+} from a fluid phase to a crystalline phase with a very high T_c ($>100^{o}C$). This isothermal phase change can be followed by batch or titration microcalorimetry (Portis et al., 1979; Rehfeld et al., 1981). With other divalent cations, such as Ba^{2+} and Sr^{2+}, fusion can be induced at temperatures both above and below the T_c of the metal ion/PS complex (N. Düzgüneş and K.B. Freeman, unpublished data). This also holds for La^{3+}-induced fusion of PS SUV, although in this case fusion is more extensive at the T_c of the La^{3+}/PS complex than at temperatures above or below it (Hammoudah et al., 1981). The above observations demonstrate that there is no strict correlation between the apparent occurrence of an ion-induced isothermal phase change and bilayer fusion. Importantly, it is not known whether an isothermal phase change occurs at the point of contact between fusing vesicles during their initial interaction, even under conditions where a transition does take place at a later stage. It is more likely that initially the dehydration of the bilayer surface at the point of contact is critically involved in the fusion of phospholipid vesicles (Portis et al., 1979; Wilschut et al., 1981; Düzgüneş et al., 1981c; Hoekstra, 1982a). The ability of Ca^{2+} to form an anhydrous trans complex between PS bilayers (Papahadjopoulos *et al.*, 1978; Portis *et al.*, 1979) may be crucial for the induction of fusion.

Another example of the difference between structures obtained at equilibrium in metal-ion/phospholipid complexes and those found during the early fusion events, is the formation of lapidic particles in certain phospholipid mixtures. These structures, observed in freeze-fracture electron microscopy of cardiolipin/PC (1:1) vesicles in the presence of divalent cations have been proposed to be membrane fusion intermediates (Verkleij *et al.*, 1979, 1980). When LUV made of this lipid mixture are rapid-frozen within seconds of Ca^{2+} addition no lipidic particles are evident at sites of membrane fusion in freeze-fracture preparations, although they are seen after long-term incubations (Düzgüneş *et al.*, 1982b; Bearer *et al.*, 1982). These observations do not exclude the possibility that non-bilayer intermediate structures occur during membrane fusion. However, these structural transitions may be too fast to be visualized with present techniques or they could be confined to a small area of the contact zone between vesicles.

We have pointed out above that the induction of fusion of PI/PE vesicles depends on the bilayer being in the fluid phase. A different dependence on the gel to liquid-crystalline phase transition is observed with LUV composed of PS/dipalmitoyl-PC (1:1). When they are above the T_c these vesicles aggregate in the presence of Ca^{2+}, but do not fuse, similar to the case of PS/egg-PC (1:1) LUV. However, within a narrow temperature range (15-20°C) which corresponds to the lower end of the endothermic phase transition considerable fusion is observed (Düzgüneş *et al.*, 1981d). In this temperature range PS may be partially phase separated from the dipalmitoyl-PC (Stewart *et al.*, 1979) such that the inhibitory effect of PC on fusion is diminished, allowing the PS domains in two adhering vesicles to fuse. That the fusion susceptibility of membranes may be controlled by the lateral distribution of phospholipids in the plane of the bilayer would have significant implications for the modulation of membrane fusion in different subcellular membranes.

MODULATION OF MEMBRANE FUSION BY CALCIUM-BINDING PROTEINS

Fusion of secretory vesicles *in vitro* can be induced by Ca^{2+} concentrations considerably lower than those required for the fusion of phospholipid vesicles, and indirect evidence suggests that proteins are involved in lowering the $[Ca^{2+}]_t$ for secretory vesicles (Dahl *et al.*, 1979; Gratzl *et al.*, 1980; Ekerdt *et al.*, 1981). It is likely that Ca^{2+} directly interacts with these proteins which then facilitate either the close apposition of the membranes or their fusion (or both).

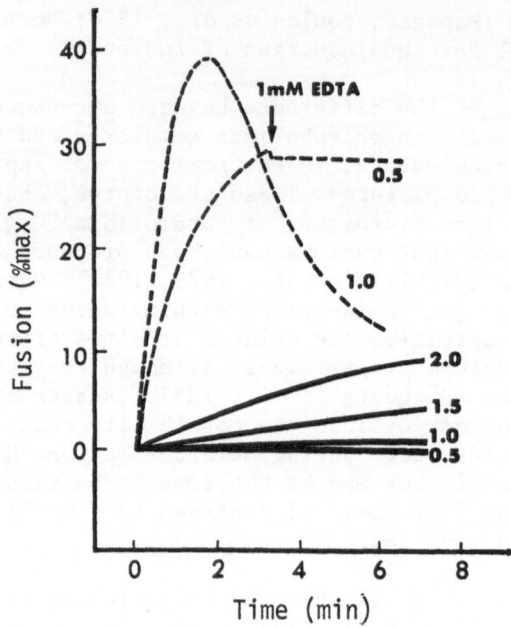

Fig. 7. Facilitation by synexin of Ca^{2+}-induced fusion of LUV made of phosphatidate/phosphatidylethanolamine (1:3). Ca^{2+} was added at t=0 at the indicated concentration (mM), in the presence (---) or absence (——) of synexin (6.2 μg/ml). The maximal fluorescence was determined as described in Fig. 1. (From Hong *et al.*, 1982a).

Synexin, a Ca^{2+}-binding protein originally isolated from bovine adrenal medulla and subsequently identified in several other tissues, induces the aggregation of isolated chromaffin granules at Ca^{2+} concentrations above 6 μM (Creutz *et al.*, 1978). Synexin also increases the initial rate of fusion of PS and PS/PE vesicles induced by Ca^{2+} and lowers the $[Ca^{2+}]_t$ (Düzgüneş *et al.*, 1980; Hong *et al.*, 1981). However, with PS/PC vesicles it only enhances the aggregation without promoting fusion (Hong *et al.*, 1981). The facilitation of membrane fusion is even more dramatic in the case of PA/PE vesicles, the initial rate of fusion being enhanced by three orders of magnitude and the $[Ca^{2+}]_t$ being reduced to < 0.1 mM (Hong *et al.*, 1982a, Fig. 7). In the presence of sub-threshold concentrations of Mg^{2+}, fusion can be induced by free Ca^{2+} concentrations as low as 10 μM (Fig. 8). Other Ca^{2+}-binding proteins, such as calmodulin (Fig. 8) or parvalbumin are slightly inhibitory to fusion, whereas prothrombin and its proteolytic fragment 1 have a strong inhibitory effect (Düzgüneş *et al.*, 1980; Hong *et al.*, 1981, 1982a). The

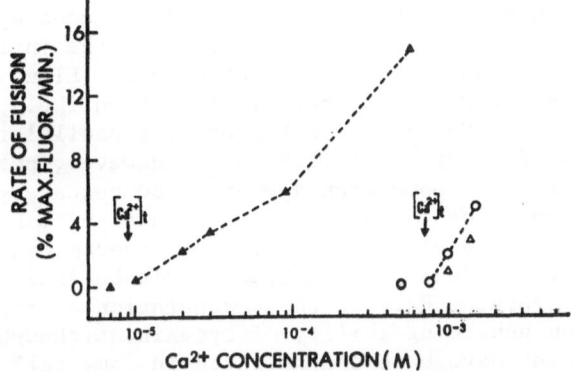

Fig. 8. Initial rate of fusion of phosphatidate/phosphatidyl-
ethanolamine (1:3) vesicles (LUV) as a function of the free
Ca^{2+} concentration, in the presence (solid triangles) or
absence (open circles) of 5 µg/ml synexin. 1 mM Mg^{2+} was
present in all experiments. The effect of 4 µg/ml calmodulin
is also shown (open triangles). (From Hong *et al.*, 1982a).

inhibitory effects of prothrombin and fragment 1 may be explained by
the possibility that the exposed surface of the protein after binding
to one vesicle is unable to interact with an oncoming vesicle. In
contrast, synexin, because of its bipolar nature (Creutz *et al.*,
1979), can presumably bind to two vesicles thus enhancing their
aggregation and fusion (Hong *et al.*, 1982a).

The presence of PC in the membrane in addition to PA and PE is
extremely inhibitory to the effect of synexin on fusion (Hong *et al.*,
1982b). Thus, lipid composition appears to be a crucial factor in
determining the action of synexin on the membrane (Table 3).

Table 3. The effects of synexin on different phospholipid
systems in the presence of calcium[a].

Phospholipid	Aggregation	Fusion
PS/PE, PA/PE, PS	Enhanced	Enhanced
PS/PC	Enhanced	No effect
PI/PE	Enhanced	Inhibited

[a] From Hong *et al.*, 1982a.

Since in the presence of synexin fusion can be induced by Ca^{2+} concentrations as low as 10 μM which is close to the intracellular free Ca^{2+} concentrations necessary for exocytosis (Llinas and Nicholson, 1975; Baker *et al.*, 1980), it is not unlikely that synexin is involved in Ca^{2+}-induced fusion of subcellular membranes (Creutz *et al.*, 1978; Hong *et al.*, 1982a,b). However, phospholipid domains of appropriate composition appear to be necessary for synexin to function as an intracellular modulator of membrane fusion (Hong *et al.*, 1982b). For example, the random distribution of even low amounts of PC in the membrane would be inhibitory for this function of synexin. Lateral phase separation of lipids may restore the fusion enhancing ability of synexin. Although phospholipid domains have been demonstrated in some cell membranes (Bearer and Friend, 1980; Klausner *et al.*, 1980), it is not known whether they exist in all membranes involved in exocytosis. The mechanism of action of synexin is also not understood well at present. It is possible that it perturbs the bilayer structure at points where it inserts into the membrane, or that it only enhances the close apposition of the bilayers by a bridging mechanism.

These findings on the role of synexin in membrane fusion suggest that other cytoplasmic or membrane proteins may also be involved in the enhancement and control of membrane fusion in the cell.

ACKNOWLEDGEMENTS

This work was supported by NIH Grant GM 28117 and NATO Research Grant 151.81. The expert secretarial work of Mrs. Karin van Wijk is gratefully acknowledged.

REFERENCES

Baker, P.F. and D.E. Knight, Calcium dependent exocytosis in bovine adrenal medullary cells with leaky plasma membranes, Nature Lond. 276:620 (1978).

Baker, P.F., D.E. Knight and M.J. Whitaker, Calcium and the control of exocytosis, in: "Calcium-Binding Proteins: Structur and Function", F.L. Siegel, E. Carafoli, R.H. Kretsinger, D.H. MacLennan and R.H. Wasserman, eds., Elsevier/North-Holland Biomedical Press, Amsterdam, The Netherlands, p. 47 (1980).

Bearer, E., N. Düzgüneş, D.S. Friend and D. Papahadjopoulos, Calcium induced fusion of phospholipid vesicles arrested by quick-freezing: The question of lipidic particles as membrane fusion intermediate, Biochim. Biophys. Acta, in press (1982).

Bearer, E.L. and D.S. Friend, Anionic lipid domains: Correlation with functional topography in a mammalian cell membrane, Proc. Natl. Acad. Sci. USA 77:6601 (1980).

Bentz, J., S. Nir and J. Wilschut, Mass action kinetics of vesicles
    aggregation and fusion, J. Coll. Interface Sci, in press (1982a).
Bentz, J., N. Düzgüneş and S. Nir, Kinetics of divalent cation
    induced vesicle fusion: Correlation between binding affinity
    and fusogenic capacity, submitted for publication (1982b).
Boggs, J.M., M.A. Moscarello and D. Papahadjopoulos, Phase separation
    of acidic and neutral phospholipids induced by human myelin
    basic protein, Biochemistry 16:5420 (1977a).
Boggs, J.M., D.D. Wood, M.A. Moscarello and D. Papahadjopoulos,
    Lipid phase separation induced by a hydrophobic protein in
    phosphatidylserine-phosphatidylcholine vesicles, Biochemistry
    16:2325 (1977b).
Ceccarelli, B., W.P. Hurlbut, and A. Mauro, Depletion of vesicles
    from frog neuromuscular junctions by prolonged tetanic
    stimulation, J. Cell Biol. 54:30 (1972).
Cochrane, D.E. and W.W. Douglas, Calcium-induced extrusion of
    secretory granules (exocytosis) in mast cells exposed to
    48/80 or the ionophores A-23187 and X-537A, Proc. Natl. Acad.
    Sci. USA 71:408 (1974).
Cowley, A.C., N.L. Fuller, R.P. Rand and V.A. Parsegian, Measurement
    of repulsive forces between charged phospholipid bilayers,
    Biochemistry 17:3163 (1978).
Creutz, C.E., C.J. Pazoles and H.B. Pollard, Identification and
    purification of an adrenal medullary protein (synexin) that
    causes calcium-dependent aggregation of isolated chromaffin
    granules, J. Biol. Chem. 253:2858 (1978).
Creutz, C.E., C.J. Pazoles and H.B. Pollard, Self-association of
    synexin in the presence of calcium. Correlation with synexin-
    induced membrane fusion and examination of the structure
    of synexin aggregates, J. Biol. Chem. 254:553 (1979).
Creutz, C.E., J.H. Scott, C.J. Pazoles and H.B. Pollard, Further
    characterization of the aggregation and fusion of chromaffin
    granules by synexin as a model for compound exocytosis,
    J. Cell. Biochem. 18:87 (1982).
Cullis, P.R. and B. de Kruijff, The polymorphic phase behaviour of
    phosphatidylethanolamines of natural and synthetic origin.
    A ^{31}P NMR study, Biochim. Biophys. Acta 513:31 (1978).
Dahl, G., R. Ekerdt and M. Gratzl, Models for exocytotic membrane
    fusion, Symp. Soc. Exp. Biol. 33:349 (1979).
Day, E.P., J.T. Ho, R.K. Kunze and S.T. Sun, Dynamic light scattering
    study of calcium-induced fusion in phospholipid vesicles,
    Biochim. Biophys. Acta 470:503 (1977).
Douglas, W.W., Stimulus-secretion coupling: The concept and clues
    from chromaffin and other cells, Brit. J. Pharmacol. 34:451
    (1968).
Düzgüneş, N. and S. Ohki, Calcium-induced interaction of phospholipid
    vesicles and bilayer lipid membranes, Biochim. Biophys. Acta
    467:301 (1977).

Düzgüneş, N. and S. Ohki, Fusion of small unilamellar liposomes with phospholipid bilayer membranes and large single-bilayer vesicles, Biochim. Biophys. Acta 640:734 (1981).

Düzgüneş, N., K. Hong and D. Papahadjopoulos, Membrane fusion: The involvement of phospholipids, proteins and calcium binding, in: "Calcium Binding Proteins: Structure and Function", F.L. Siegel, E. Carafoli, R.H. Kretsinger, D.H. MacLennan and R.H. Wasserman, eds., Elsevier/North-Holland Biomedical Press, Amsterdam, p. 17 (1980).

Düzgüneş, N., S. Nir, J. Wilschut, J. Bentz, C. Newton, A. Portis and D. Papahadjopoulos, Calcium and magnesium-induced fusion of mixed phosphatidylserine/phosphatidylcholine vesicles: Effect of ion binding, J. Membrane Biol. 59:115 (1981a).

Düzgüneş, N., S.J. Rehfeld, K.B. Freeman, C. Newton, D.J. Eatough and D. Papahadjopoulos, Microcalorimetric analysis of divalent cation interaction with phosphatidylserine vesicles: Correlation with binding and fusion, Abstr. VIIth Int. Biophys. Cong. Mexico City, p. 226 (1981b).

Düzgüneş, N., J. Wilschut, R. Fraley, and D. Papahadjopoulos, Studies on the mechanism of membrane fusion: Role of head-group composition in calcium- and magnesium-induced fusion of mixed phospholipid vesicles, Biochim. Biophys. Acta 642:734 (1981c).

Düzgüneş, N., K. Hong, J. Wilschut, N. Lopez and D. Papahadjopoulos, Modulation of Ca^{2+}-induced membrane fusion by phase transitions, glycolipids and lectins, Abstr. VIIth Int. Biophys. Cong. Mexico City, p. 108 (1981d).

Düzgüneş, N. and D. Papahadjopoulos, Ionotropic effects on phospholipid membranes: Calcium/magnesium specificity in binding, fluidity and fusion, in: "Membrane Fluidity in Biology", R.C. Aloia, ed., Academic Press, New York, Vol. I, in press (1982).

Düzgüneş, N., J. Wilschut, K. Hong, D. Hoekstra and D. Papahadjopoulos, Retention of aqueous contents during divalent cation-induced fusion of phospholipid vesicles, in preparation (1982a).

Düzgüneş, N., E. Bearer and D. Papahadjopoulos, Phospholipid vesicle fusion monitored by rapid-freezing and mixing of aqueous contents, Biophys. J. 37:25a (1982b).

Eisenberg, M., T. Gresalfi, T. Riccio and S. McLaughlin, Adsorption of monovalent cations to bilayer membranes containing negative phospholipids, Biochemistry 18:5213 (1979).

Ekerdt, R., G. Dahl and M. Gratzl, Membrane fusion of secretory vesicles and liposomes, Biochim. Biophys. Acta 646:10 (1981).

Ekerdt, R. and D. Papahadjopoulos, Intermembrane contact affects calcium binding to phospholipid vesicles, Proc. Natl. Acad. Sci. USA 79:2273 (1982).

Foreman, J.K., J.L. Mongar and B.D. Gomperts, Calcium ionophores and movement of calcium ions following physiological stimulus to a secretory process, Nature Lond. 245:249 (1973).

Fraley, R., J. Wilschut, N. Düzgüneş, C. Smith and D. Papahadjopoulos, Studies on the mechanism of membrane fusion: Role of phosphate in promoting calcium ion induced fusion of phospholipid vesicles, Biochemistry 19:6021 (1980).

Gratzl, M. and G. Dahl, Fusion of secretory vesicles isolated from rat liver, J. Membrane Biol. 40:343 (1978).

Gratzl, M., C. Schudt, R. Ekerdt and G. Dahl, Fusion of isolated biological membranes, in: "Membrane Structure and Function", E.E. Bittar, ed., Vol. 3, John Wiley New York, p. 59 (1980).

Gresh, N., Intermolecular chelation of two serine phosphates by Ca^{2+} and Mg^{2+}. A theoretical structural investigation, Biochim. Biophys. Acta 597:345 (1980).

Hammoudah, M.M., S. Nir, J. Bentz, E. Mayhew, T.P. Stewart, S.W. Hui and R.J. Kurland, Interactions of La^{3+} with phosphatidylserine vesicles. Binding, phase transition, leakage, ^{31}P-NMR and fusion, Biochim. Biophys. Acta 645:102 (1981).

Hauser, H., M.C. Phillips and M.D. Barratt, Differences in the interactions of inorganic and organic (hydrophobic) cations with phosphatidylserine membranes, Biochim. Biophys. Acta 413:341 (1975).

Hauser, H., I. Pascher, R.H. Pearson and S. Sundell, Preferred conformation and molecular packing of phosphatidylethanolamine and phosphatidylcholine, Biochim. Biophys. Acta 650:21 (1981).

Heuser, J.E., T.S. Reese, M.J. Dennis, Y.N. Jan, L.Y. Jan and L. Evans, Synaptic vesicle exocytosis captured by quick-freezing and correlated with quantal transmitter release, J. Cell Biol. 81:275 (1979).

Hoekstra, D., Role of lipid phase separations and membrane hydration in phospholipid vesicle fusion, Biochemistry 21:2833 (1982a).

Hoekstra, D., Kinetics of intermixing of lipids and mixing of aqueous contents during vesicle fusion, Biochim. Biophys. Acta, in press (1982b).

Hoekstra, D., A. Yaron, A. Carmel and G. Scherphof, Fusion of phospholipid vesicles containing a trypsin-sensitive fluorogenic substrate and trypsin, FEBS Lett. 106:176 (1979).

Holz, R.W. and C.A. Stratford, Effects of divalent ions on vesicle-vesicle fusion studied by a new luminescence assay for fusion, J. Membrane Biol. 46:331 (1979).

Hong, K., N. Düzgüneş and D. Papahadjopoulos, Role of synexin in membrane fusion. Enhancement of calcium-dependent fusion of phospholipid vesicles, J. Biol. Chem. 256:3641 (1981).

Hong, K., N. Düzgüneş and D. Papahadjopoulos, Modulation of membrane fusion by calcium-binding proteins, Biophys. J. 37:297 (1982a).

Hong, K., N. Düzgüneş, R. Ekerdt and D. Papahadjopoulos, Synexin facilitates fusion of specific phospholipid vesicles at divalent cation concentrations found intracellularly, Proc. Natl. Acad. Sci. USA 79:4642 (1982b).

Ingolia, T.D. and D.E. Koshland Jr., The role of calcium in fusion of artificial vesicles, J. Biol. Chem. 253:3821 (1978).

Jendrasiak, G.L. and J.H. Hasty, The hydration of phospholipids, Biochim. Biophys. Acta 337:79 (1974).

Jacobson, K. and D. Papahadjopoulos, Phase transitions and phase separations in phospholipid vesicles, induced by changes in temperature, pH and concentration of divalent metals, Biochemistry 14:152 (1975).

Kanno, T., D.E. Cochrane and W.W. Douglas, Exocytosis (secretory granule extrusion) induced by injection of calcium into mast cells, Can. J. Physiol. Pharmacol. 51:1001 (1973).

Kantor, H.L. and J. Prestegard, Fusion of fatty acid containing lecithin vesicles, Biochemistry 14:1790 (1975).

Klausner, R.D., A.M. Kleinfeld, R.L. Hoover and M.J. Karnovsky, Lipid domains in membranes. Evidence derived from structural perturbations induced by free fatty acids and lifetime heterogeneity analysis, J. Biol. Chem. 255:1286 (1980).

Koter, B., B. de Kruijff and L.L.M. van Deenen, Calcium-induced aggregation and fusion of mixed phosphatidylcholine-phosphatidic acid vesicles as studied by ^{31}P NMR, Biochim. Biophys. Acta 514:255 (1978).

Kurland, R.J., M. Hammoudah, S. Nir and D. Papahadjopoulos, Binding of Ca^{2+} and Mg^{2+} to phosphatidylserine vesicles: Different effects on P-31 NMR shifts and relaxation rates, Biochem. Biophys. Res. Commun. 88:927 (1979a).

Kurland, R., C. Newton, S. Nir and D. Papahadjopoulos, Specificity of Na^+ binding to phosphatidylserine vesicles from a ^{23}Na NMR relaxation rate study, Biochim. Biophys. Acta 551:137 (1979b).

Lagunoff, D., Membrane fusion during mast cell secretion, J. Cell Biol. 57:252 (1973).

Lawson, D., M.C. Raff, B. Gomperts, C. Fewtrell and N.B. Gilula, Molecular events during membrane fusion. A study of exocytosis in rat peritoneal mast cells, J. Cell Biol. 72:242 (1977).

Le Neveu, D.M., R.P. Rand and V.A. Parsegian, Measurement of forces between lecithin bilayers, Nature Lond. 259:601 (1976).

Liao, M-J. and J.H. Prestegard, Fusion of phosphatidic acid-phosphatidylcholine mixed lipid vesicles, Biochim. Biophys. Acta 550:157 (1979).

Llinás, R. and C. Nicholson, Calcium role in depolarization-secretion coupling: an aequorin study in squid giant synapse, Proc. Natl. Acad. Sci. USA 72:187 (1975).

Maeda, T. and S.-I. Ohnishi, Membrane fusion. Transfer of phospholipid molecules between phospholipid bilayer membranes, Biochem. Biophys. Res. Commun. 60:1509 (1974).

McIver, D.J.L., Control of mebrane fusion by interfacial water: A model for the actions of divalent cations, Physiol. Chem. Phys. 11:289 (1979).

McLaughlin, S., N. Mulrine, T. Grefalsi, G. Vaio and A. McLaughlin, The adsorption of divalent cations to bilayer membranes containing phosphatidylserine, J. Gen. Physiol. 77:445 (1981).

214

Michell, R.H., Inositol phospholipids and cell surface receptor function, Biochim. Biophys. Acta 415:81 (1975).

Miledi, R., Transmitter release induced by injection of calcium ions into nerve terminals, Proc. R. Soc. Lond. Ser. B. 183:421 (1973).

Miller, C. and E. Racker, Fusion of phospholipid vesicles reconstituted with cytochrome c oxidase and mitochondrial hydrophobic protein, J. Membrane Biol. 26:319 (1976).

Miller, D.C. and G.P. Dahl, Early events in calcium-induced liposome fusion, Biochim. Biophys. Acta 689:165 (1982).

Newton, C., W. Pangborn, S. Nir and D. Papahadjopoulos, Specificity of Ca^{2+} and Mg^{2+} binding to phosphatidylserine vesicles and resultant phase changes of bilayer membrane structure, Biochim. Biophys. Acta 506:281 (1978).

Nir, S., Van der Waals interactions between surfaces of biological interest, Prog. Surface Sci. 8:1 (1977).

Nir, S. and J. Bentz, On the forces between phospholipid bilayers, J. Colloid Interface Sci. 65:399 (1978).

Nir, S., C. Newton and D. Papahadjopoulos, Binding of cations to phosphatidylserine vesicles, Bioelectrochem. Bioenerg. 5:116 (1978).

Nir, S., J. Bentz and J. Wilschut, Mass action kinetics of phosphatidylserine fusion as monitored by coalescence of internal vesicle volumes, Biochemistry 19:6030 (1980).

Nir, S., J. Bentz and N. Düzgüneş, Two modes of reversible vesicle aggregation: Particle size and the DLVO theory, J. Coll. Interface Sci. 84:266 (1981).

Nir, S., J. Bentz, J. Wilschut and N. Düzgüneş, Aggregation and fusion of phospholipid vesicles, Prog. Surface Sci., in press (1982a).

Nir, S., J. Wilschut and J. Bentz, The rate of fusion of phospholipid vesicles and the role of bilayer curvature, Biochim. Biophys. Acta 688:275 (1982b).

Nir, S., N. Düzgüneş and J. Bentz, Binding of monovalent cations to phosphatidylserine and modulation of calcium- and magnesium-induced vesicle fusion, submitted for publication (1982c).

Ohki, S., A mechanism of divalent ion-induced phosphatidylserine membrane fusion, Biochim. Biophys. Acta 689:1 (1982).

Ohki, S. and N. Düzgüneş, Divalent cation-induced interaction of phospholipid vesicle and monolayer membranes, Biochim. Biophys. Acta 552:438 (1979).

Ohki, S. and R. Kurland, Surface potential of phosphatidylserine monolayers. II. Divalent and monovalent ion binding, Biochim. Biophys. Acta 645:170 (1981).

Ohki, S. and R. Sauvé, Surface potential of phosphatidylserine monolayers. I. Divalent ion binding effect, Biochim. Biophys. Acta 511:377 (1978).

Ohki, S., N. Düzgüneş and K. Leonards, Phospholipid vesicle aggregation: Effect of monovalent and divalent ions, Biochemistry 21:2127 (1982).

Orci, L., A. Perellet and D.S. Friend, Freeze-fracture of membrane fusions during exocytosis in pancreatic B-cells, J. Cell Biol 75:23 (1977).

Palade, G., Intercellular aspects of the process of protein synthesi: Science 189:347 (1975).

Papahadjopoulos, D. and A.D. Bangham, Biophysical properties of phospholipids. II. Permeability of phosphatidylserine liquid crystals to univalent ions, Biochim. Biophys. Acta 126:185 (1966).

Papahadjopoulos, D. and S. Ohki, Stability of asymmetric phospholipi bilayers, Science 164:1075 (1969).

Papahadjopoulos, D., G. Poste, B.E. Schaeffer and W.J. Vail, Membran fusion and molecular segregation in phospholipid vesicles, Biochim. Biophys. Acta 352:10 (1974).

Papahadjopoulos, D., W.J. Vail, K. Jacobson and G. Poste, Cochleate lipid cylinders: Formation by fusion of unilamellar lipid vesicles, Biochim. Biophys. Acta 394:483 (1975).

Papahadjopoulos, D., W.J. Vail, W.A. Pangborn and G. Poste, Studies on membrane fusion. II. Induction of fusion in pure phospholipid membranes by calcium and other divalent metals, Biochim. Biophys. Acta 448:265 (1976a).

Papahadjopoulos, D., S. Hui, W.J. Vail and G. Poste, Studies on membrane fusion. I. Interactions of pure phospholipid membrane and the effect of myristic acid, lysolecithin, proteins and DMSO, Biochim. Biophys. Acta 448:245 (1976b).

Papahadjopoulos, D., W.J. Vail, C. Newton, S. Nir, K. Jacobson, G. Poste and R. Lazo, Studies on membrane fusion. III. The role of calcium-induced phase changes, Biochim. Biophys. Acta 465:579 (1977).

Papahadjopoulos, D., A. Portis Jr., and W. Pangborn, Calcium-induced lipid phase transitions and membrane fusion, Ann. N.Y. Acad. Sci. 308:50 (1978).

Portis, A., C. Newton, W. Pangborn and D. Papahadjopoulos, Studies on the mechanism of membrane fusion: Evidence for an inter-membrane Ca^{2+} phospholipid complex, synergism with Mg^{2+}, and inhibition by spectrin, Biochemistry 18:780 (1979).

Rand, R.P., Interacting phospholipid bilayers: Measured forces and induced structural changes, Ann. Rev. Biophys. Bioeng. 10:277 (1981).

Rehfeld, S.J., N. Düzgünes, C. Newton, D. Papahadjopoulos and D.J. Eatough, The exothermic reaction of calcium with unilamellar phosphatidylserine vesicles: Titration microcalorimetry, FEBS Lett. 123:249 (1981).

Reiss-Husson, F., Structure des phases liquides-crystallines de différents phospholipides, monoglycerides, sphingolipides, anhydrides ou en présence d'eau, J. Mol. Biol. 25:363 (1967).

Satir, B., C. Schooley and P. Satir, Membrane fusion in a model system. Mucocyst secretion in *Tetrahymena*, J. Cell Biol. 56:153 (1973).

Schullery, S.E., C.F. Schmidt, P. Felgner, T.W. Tillack and
    T.E. Thompson, Fusion of dipalmitoylphosphatidylcholine
    vesicles, Biochemistry 19:3919 (1980).
Stewart, T.P., S.W. Hui, A.R. Portis and D. Papahadjopoulos,
    Complex phase mixing of phosphatidylcholine and
    phosphatidylserine in multilamellar membrane vesicles,
    Biochim. Biophys. Acta 556:1 (1979).
Struck, D.K., D. Hoekstra and R.E. Pagano, Use of resonance energy
    transfer to monitor membrane fusion, Biochemistry 20:4093
    (1981).
Sundler, R. and D. Papahadjopoulos, Control of membrane fusion by
    phospholipid head groups: I. Phoshatidate/phosphatidylinositol
    specificity, Biochim. Biophys. Acta 649:743 (1981).
Sundler, R., N. Düzgüneş and D. Papahadjopoulos, Control of membrane
    fusion by phospholipid head-groups. II. The role of
    phosphatidylethanolamine in mixtures with phosphatidate and
    phosphatidylinositol, Biochim. Biophys. Acta 649:751 (1981).
Sun, S.T., E.P. Day and J.T. Ho, Temperature dependence of calcium-
    induced fusion of sonicated phosphatidylserine vesicles,
    Proc. Natl. Acad. Sci. USA 75:4325 (1978).
Suurkuusk, J., B.R. Lentz, Y. Barenholz, R.L. Biltonen and
    T.E. Thompson, A calorimetric and fluorescent probe study
    of the gel-liquid cyrstalline phase transition in small,
    single-lamellar dipalmitoyl phosphatidylcholine vesicles,
    Biochemistry 15:1393 (1976).
Vanderwerf, P. and E.F. Ullman, Monitoring of phospholipid vesicle
    fusion by fluorescence energy transfer between membrane-bound
    dye labels, Biochim. Biophys. Acta 596:302 (1980).
Verkleij, A.J., C. Mombers, W.J. Gerritsen, L. Leunissen-Bijvelt and
    P.R. Cullis, Fusion of phospholipid vesicles in association
    with the appearance of lipidic particles as visualized by
    freeze-fracturing, Biochim. Biophys. Acta 555:358 (1979).
Verkleij, A.J., C.J.A. van Echteld, W.J. Gerritsen, P.R. Cullis and
    B. de Kruijff, The lipidic particle as an intermediate
    structure in membrane fusion processes and bilayer to
    hexagonal H_{II} transitions, Biochim. Biophys. Acta 600:620 (1980).
Verwey, E.J.A. and J.Th.G. Overbeek, "Theory of the Stability of
    Lyophobic Colloids", Elsevier, Amsterdam and New York (1949).
Weinstein, J.N., S. Yoshikami, P. Henkart, R. Blumenthal and
    W.A. Hagins, Liposome-cell interaction: Transfer and
    intracellular release of a trapped fluorescent marker,
    Science, 195:489 (1977).
Wilschut, J. and D. Papahadjopoulos, Ca^{2+}-induced fusion of
    phospholipid vesicles monitored by mixing of aqueous contents,
    Nature Lond. 281:690 (1979).
Wilschut, J., N. Düzgüneş, R. Fraley and D. Papahadjopoulos, Studies
    on the mechanism of membrane fusion: Kinetics of calcium ion
    induced fusion of phosphatidylserine vesicles followed by a
    new assay for mixing of aqueous vesicle contents, Biochemistry
    19:6011 (1980).

Wilschut, J., N. Düzgüneş and D. Papahadjopoulos, Calcium/ magnesium specificity in membrane fusion: Kinetics of aggregation and fusion of phosphatidylserine vesicles and the role of bilayer curvature, <u>Biochemistry</u> 20:3126 (1981).

Wilschut, J., N. Düzgüneş and D. Papahadjopoulos, Studies on the mechanism of membrane fusion: Temperature dependence of divalent cation-induced fusion of phosphatidylserine vesicles and the role of bilayer curvature, in preparation.

CORRELATION BETWEEN STEADY-STATE AND

TIME-RESOLVED FLUORESCENCE ANISOTROPY DATA

W. van der Meer*, H. Pottel** and W. Herreman**

*Physiological Laboratory, University of Leiden
P.O.Box 9604, NL-2300 RC Leiden, The Netherlands
**Interdisciplinary Research Center, Katholieke
Universiteit Leuven, Campus Kortrijk, B-8500
Kortrijk, Belgium

INTRODUCTION

The most common fluorescence depolarization measurement is a steady-state experiment. Continuous illumination with monochromatic polarized light is used to excite fluorescent probes embedded in the lipid regions of the membrane sample. One measures the fluorescence intensities parallel ($I_\parallel$) and perpendicular ($I_\perp$) to the polarization direction of the excitation light. The relevant parameter is the steady-state fluorescence anisotropy (FA), defined as

$$r_s = (I_\parallel - I_\perp)/(I_\parallel + 2I_\perp) \qquad (1)$$

Here we are concerned with the FA of the probe 1,6-diphenyl-1,3,5-hexatriene (DPH) in a membrane system which is macroscopically isotropic. In that case we have $0 \leq r_s \leq r_o$; theoretically r_o equals 0.4, experimental values lie between 0.362 [1] and 0.395 [2]. The steady-state FA, r_s, reflects the hindrance of the probe rotation; $r_s = r_o$ means that the probes do not rotate within the fluorescence lifetime and $r_s = 0$ means that the probe molecules rotate rapidly and without restriction within that time.

Recently, it has been shown[3,4] that r_s can be resolved into a static part r_∞, and a dynamic part r_f:

$$r_s = r_f + r_\infty \qquad (2)$$

The static part is proportional to the square of the order parameter for the probe[3,4]

$$S = (r_\infty/r_0)^{\frac{1}{2}} \tag{3}$$

The dynamic part reflects the rate of probe rotation[3,4]. It has been shown[3,4] that γ defined as

$$\gamma = \phi/\tau \tag{4}$$

where ϕ is the rotational relaxation time and τ is the fluorescence lifetime, is related to r_s and r_∞ according to

$$\gamma = (r_s - r_\infty)/(r_0 - r_s) \tag{5}$$

If the membrane were similar to an isotropic liquid such as a mineral oil, and $r_\infty = 0$ were a good approximation, then r_s could be translated into a "microviscosity" η[1,5] by applying classical hydrodynamic expressions of the Perrin type, that is:

isotropic case $\quad \eta = Vr_s/(r_0 - r_s) \tag{6}$

where V is $C\tau T$, C is a geometrical factor, and T is the absolute temperature. For DPH, V = 2.4 Poises is a good approximation[1].

However, a membrane is anisotropic, that is, the molecules tend to be oriented perpendicular to the membrane plane. Therefore, it is necessary to apply a correction. Recently an approximate expression has been proposed for the so-called "true microviscosity" or "corrected viscosity"[5,6]:

anisotropic case $\quad \eta_0 = Vr_0(r_s - r_\infty)/\{(r_0 - r_s)(r_0 - r_\infty)\} \tag{7}$

This viscosity parameter is smaller than that of equation (6) as becomes apparent by rearranging terms,

$$\eta_0 = \eta - Vr_\infty/(r_0 - r_\infty) \tag{8}$$

It is the purpose of this paper to investigate the correlation between the true microviscosity η_0 and the order parameter S with steady-state FA data for DPH. We present a comprehensive set of η_0 and S data of a large number of membranes, for which r_f and r_∞ values are known from the literature. It is shown that η_0 can be estimated accurately only from steady-state FA data for rather low r_s values, and that the inaccuracy in deriving η_0 from r_s increases strongly with increasing r_s. The opposite is true for the order parameter S; for low r_s the estimation of S from r_s is inaccurate and the evaluation of S from r_s can be done with confidence for larger r_s values. A more detailed discussion can be found elsewhere[7].

INTERPRETATION OF DATA

In the last few years r_s and r_∞ values for a large number of DPH labelled membranes have been determined using time-resolved FA decay measurements[2,10-17] or differential polarized phase fluorimetry[18,19] in combination with steady-state FA measurements. For these membranes the order parameter S and the true microviscosity η_o can be calculated according to equations (3) and (5) respectively as a function of r_s/r_o, which value can be obtained from the simple steady-state method alone.

The Order Parameter S

The S values for 125 membranes are plotted as a function of r_s/r_o in Figure 1a. These membranes belong to 3 different groups:

1) model membranes (66 points) from one lipid component [see references 10, 11, 18, and 19]
2) model membranes (39 points) containing various amounts of cholesterol [see references 12, 13, and 19]
3) various biological membranes [see references 2, 15, and 16]

The quoted r_o values were taken, except from reference 16 for which r_o values have been calculated from r_s, r_∞, ϕ, and τ using equations (2) and (5); r_∞ values of the membranes in reference 12 were calculated from cone angle data. If the decay of the total fluorescence was taken double-exponential, we used a weighted average for τ.

Although these membranes differ considerably in temperature and composition, they all follow closely one relation $S(r_s/r_o)$, which is well described by equation (13) below, the middle curve in Figure 1a. This relation can be derived from the following arguments:

1. The theory of rotational diffusion of a rodlike object in an anisotropic potential[20-22] provides an expression for the product $D\phi$ of the wobbling diffusion constant D times the effective rotational correlation time ϕ. In the cone model[9,20] $D\phi$ is approximately proportional to $1-S^2$,

$$D\phi \propto (1-S^2) \tag{9}$$

2. The wobbling diffusion constant will in general depend upon the order parameter S[22,23]. A plot of the D data versus $1/S$ suggests a linear relationship,

$$D \propto 1/S \tag{10}$$

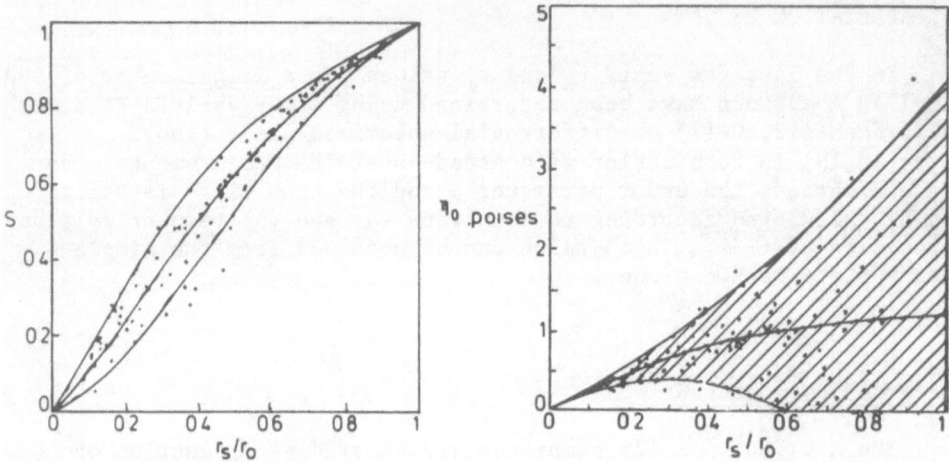

Fig. 1.   Time-resolved FA data as a function of r /r (see text).

3.   The fluorescence lifetime is also correlated with the order para-
     meter[8,24].   Following Van Blitterswijk et al.[8] we assume the
     proportionality

$$\tau \propto (1+S) \tag{11}$$

Combining equations (4), (9), (10), and (11) we obtain

$$\gamma = S(1-S) \tag{12}$$

where we have put the coefficient on the right-hand side equal to 1,
which is close to the average of $\gamma/\{S(1-S)\}$ for the 125 data points
used in Figure 1.   This average is 1.09±0.09 (standard deviation).
Substitution of equations (3) and (12) into (5) gives our $S(r_s/r_0)$
relation:

$$r_s/r_0 = S/(1+S-S^2) \tag{13}$$

The middle curve in Figure 1a corresponds to equation (13).   Combining equation (3) and (13), we obtain a $r_\infty(r_s)$ relation, a non-
linear function in r_s.   For $0.33<r_s/r_0<0.7$, r_∞ can be approximated by

$$r_\infty = \frac{4}{3}r_s - 0.28r_0 \tag{14}$$

in agreement with the result of Van Blitterswijk et al.[8]; for
$r_0=0.4$, the difference between our r_∞ and theirs is less than 0.01,
which is well within experimental error.

222

The True Microviscosity η_0

Substitution of equations (3) and (13) into (7) gives our $\eta_0(r_s/r_0)$ relation:

$$\eta_0 = VS/(1+S) \text{ with } r_s/r_0 = S/(1+S-S^2) \tag{15}$$

This is the middle curve in Figure 1b, where we have plotted η_0 values calculated from literature data for the same membranes as in Figure 1a. Three data points, however, have a value for η_0 greater than 5 Poises.

Estimation of Errors

Comparing Figures 1a and 1b, we see that the deviations from the $S(r_s)$ relation are much smaller than the deviations from the $\eta_0(r_s)$ curve. Since the location of a point in the η_0-r_s/r_0 plane follows from its corresponding coordinates in the S-r_s/r_0 plane, it is clear that these scattered errors are interrelated. To quantify this connection, we have studied the maximal deviation ΔS from the $S(r_s/r_0)$ curve as a function of r_s/r_0. The error ΔS is well described by

$$\Delta S/S = \tfrac{1}{2}\Delta r_\infty/r_\infty = A(1-r_s/r_0)^2 \tag{16}$$

We have chosen A=0.6, so that 90% of the data lie between the upper curve (corresponding to $S(r_s/r_0) + \Delta S$) and the lower curve (corresponding to $S(r_s/r_0) - \Delta S$). By differentiating equation (8) with respect to r_∞, we obtain $\Delta\eta_0$ at fixed r_s/r_0

$$\Delta\eta_0 = Vr_0\Delta r_\infty/(r_0-r_\infty)^2 = 2AV(r_s/r_0)^2 \tag{17}$$

$\eta_0(r_s/r_0) + \Delta\eta_0$ $(\eta_0(r_s/r_0) - \Delta\eta_0)$ is the upper (lower) curve in Figure 1b and corresponds to the lower (upper) curve of Figure 1a. In the region where r_s is high, a small error in r_∞ has a negligible effect upon the order parameter, but results in an enormous error in the true microviscosity.

DISCUSSION

We have confirmed the conclusion of Van Blitterswijk et al.[8] that there is a one-to-one correspondence between the order parameter S and the relative steady-state FA, r_s/r_0, allowing for an evaluation of S from r_s data. This would suggest that the dynamic contribution to r_s and therefore the true microviscosity η_0, could also be estimated from FA measurements[5,25]. However, the latter conclusion is wrong, because of inaccuracy: while the spreading of the data around the $S(r_s/r_0)$ relation is rather small, the inaccuracy due to this scattering is strongly amplified in the η_0-r_s/r_0 plane, as is appar-

ent from the figure. Adopting the criterium that an error of 20% is
the highest inaccuracy still acceptable, we find:

S from r_s/r_o accurate for $r_s/r_o > 0.4$, not reliable for $r_s/r_o < 0,4$
η_o from r_s/r_o accurate for $r_s/r_o < 0.2$, not reliable for $r_s/r_o > 0.2$

Since the great majority of biological membranes has $r_s/r_o > 0.4$, we
conclude that S can be reliably evaluated from stead-state FA data
for biomembranes, but η_o cannot. Even the more optimistic choice A =
0.4 does not alter the conclusion. In that case the evaluation of S
is accurate for $r_s/r_o > 0.3$ and the estimation of η_o is reliable for
$r_s/r_o < 0.3$.

The arguments leading to $\gamma = S(1-S)$ and from there to the $S(r_s/r_o)$
relation, have some shortcomings[23,26]. However, the relation works
surprisingly well for all types of membranes, except for those where
the florescence lifetime is considerably shortened due to quenching,
e.g. energy transfer, as is clearly the case for artificial membranes
containing large amounts of Cytochrome Oxidase[27] and the Purple
membrane which is rich in Bacteriorhodopsin[2]. For these membranes
the S values deviate from equation (13). Equations (9), (10), and
(11) apply if no quenching occurs. It should be stressed that these
equations are only rough approximations exhibiting certain trends
that follow from theory and experiment. It is possible to arrive at
improved relations by taking into account not only the variation in γ
due to the order parameter, but from other physical parameters as
well, as proposed by Hare[26]. The present approach has the great
advantage that it is simple and straightforward and only requires a
measurement of r_s. Moreover it is empirically justified as is shown
by Van Blitterswijk et al.[8] and in the present paper. Our $S(r_s)$
relation could be very convenient in practice, because it allows for
a direct evaluation of the order parameter from steady-state FA
measurements only. The accuracy in a particular S value can be
estimated from equation (16).

We have taken a constant V=2.4 poise. However, if one calcul-
ates for every membrane its own V value, essentially the same pattern
emerges. Kinosita et al., have calculated a "viscosity in the cone",
η_c [2,9,10-12]. The spreading in the η_c data is of the same order as
for the η_o data. There seems to be an interesting analogy between
the approach of Heyn[3] and Jähnig[4] (the concept of an order parameter) on one hand and the approach of Shinitzky and Yuli[5] (the
concept of a gradient in viscosity) on the other hand. Molecules
rotating in a gradient of viscosity will orient themselves in such a
way that the majority will "feel" the smallest component of the
viscosity tensor, which is assumed to be of second rank[5]. Therefore, the anisotropy in the viscosity R introduced by Shinitzky and
Yuli[5] should be related to the order parameter S. A possible
relation could be S=-2R, which is correct for S=0 and S=1, but
remains to be verified.

Acknowledgements

W. van der Meer is supported by the Queen Wilhelmina Foundation for Cancer Research.

Thanks are due to R. P. H. Kooyman, W. J. Van Blitterswijk, R. P. Van Hoeven, M. Shinitzky and A. H. Parola for stimulating discussions.

REFERENCES

1. M. Shinitzky and Y. Barenholz, Biochim.Biophys.Acta, 515:367 (1978).
2. K. Kinosita, R. Kataoka, Y. Kimura, O. Gotoh, and A. Ikegami, Biochemistry, 20:4270 (1981).
3. M. P. Hyen, FEBS Lett., 108:359.
4. F. Jähnig, Proc.Natl.Acad.Sci.USA, 76:6361.
5. M. Shinitzky and I. Yuli, Chem.Phys.Lipids, 30:261 (1982).
6. M. P. Heyn, R. J. Cherry, and N. A. Dencher, Biochemistry, 20:840 (1981).
7. H. Pottel, W. van der Meer, and W. Herreman, submitted to FEBS Lett.
8. W. J. Van Blitterswijk, R. P. Van Hoeven, and W. Van der Meer, Biochim.Biophys.Acta, 644:323 (1981).
9. K. Kinosita, S. Kawato, and A. Ikegami, Biophys.J., 20:289 1977).
10. S. Kawato, K. Kinosita, and A. Ikegami, Biochemistry, 16:2319 (1977).
11. C. D. Stubss, T. Kouyama, K. Kinosita, and A. Ikegami, Biochemistry, 20:4257 (1981).
12. S. Kawato, K. Kinosita, and A. Ikegami, Biochemistry, 17:5026 (1978).
13. K. Hildenbrand and C. Nicolau, Biochim.Biophys.Acta, 553:365 (1979).
14. C. Sené, D. Genest, A. Obrénvitch, P. Wahl, and M. Monsigny, FEBS Lett., 88:181 (1978).
15. C. E. Martin and D. C. Foyt, Biochemistry, 17:3587 (1978).
16. A. H. Parola, P. W. Robbins, and E. R. Blout, Exp.Cell Res., 118:205 (1979).
17. A. H. Parola, P. W. Robbins, and E. R. Blout, Israel J.Med.Sci., 12:1362 (1976).
18. J. R. Lakowicz and F. G. Prendergast, Science, 200:1399 (1978).
19. J. R. Lakowicz, F. G. Prendergast, and D. Hogen, Biochemistry, 18:508 (1979).
20. G. Lipari and A. Szabo, Biophys.J., 30:489 (1980).
21. C. Zannoni, Molec.Phys., 38:1813 (1979).
22. C. Zannoni, Molec.Phys., 42:1303 (1981).
23. G. Moro and P. L. Nordio, Chem.Phys., 43:303 (1979).
24. P. K. Wolber and B. S. Hudson, Biophys.J., 37:262 (1982).

25. D. Grünberger, R. Haimovitz, and M. Shinitzky, <u>Biochim.Biophys. Acta</u>, 688:764 (1982).
26. F. Hare, Workshop: Fluorescent techniques and membrane markers in cancerology - immunology, Montpellier, France, 14-15 December (1981).
27. K. Kinosita, S. Kawato, A. Ikegami, S. Yoshica, and Y. Orii, <u>Biochim.Biophys.Acta</u>, 647:7 (1981).

MEMBRANE LIPID FLUIDITY AT A PHYSIOLOGICALLY RELEVANT SCALE;

DETERMINATION BY FLUORESCENCE POLARIZATION

Meir Shinitzky

Department of Membrane Research
The Weizmann Institute of Science
76100 Rehovot, Israel

PREFACE

The intention of this chapter is to provide an outline of the current views and methodology relating to lipid fluidity of cell membranes, as determined by fluorescence polarization techniques, and its involvement in membranal activities.

Relevant material can be found in the following references:

Shinitzky and Barenholz, Biochim. Biophys. Acta 515:367(1978).
Shinitzky and Henkart, Intl. Rev. Cytol. 60:121 (1979).
Yuli et al., Biochemistry 20:4250 (1981).
Shinitzky and Yuli, Chem. Phys. Lipids 30:261 (1982).
Grunberger et al., Biochim. Biophys. Acta 688:764 (1982).

INTRODUCTION

The lipid fluidity is implicated in most membrane functions where the dynamic protein-lipid interplay serves as the control framework. Yet the analysis of the correlation between a membranal activity and its lipid fluidity is hampered by the complexity and heterogeneity of the lipid matrix. Another complication is that data on lipid fluidity can in principle, belong to either one of three distinct levels of resolution, microscopic, macroscopic or submacroscopic. The microscopic level provides information on individual atoms or molecular segments at various depths of the lipid layer. The most applicable technique for this level of resolution is nuclear magnetic resonance. The macroscopic level deals with lipid domains in bulk thermodynamic terms,

227

which are determined by mechanical or calorimetric means. The sub-macroscopic level provides low molecular resolution of lipid fluidity, generally expressed in macroscopic terms, with the aid of spectral probes. Data on lipid fluidity obtained by fluorescence depolarization, excimer formation and electron spin resonance, belong to this category. Intermediate levels of resolution are obtained in specialized methods where the reporting unit is introduced into selective lipid domains or into the inner or outer lipid monolayers.

Inasmuch as the physiologic relevance of a certain dynamic parameter lays or its effect on the overt activity, the submacroscopic level of resolution of membrane fluidity complies well with the majority of membranal processes. This physiologically relevant lipid fluidity parameters can be most conveniently derived by fluorescence polarization of a well defined probe. These parameters are of low microscopic resolution and can be expressed in macroscopic units (e.g. poise).

VISCOSITY OF WEAKLY-ASSOCIATED FLUIDS

The viscosity coefficient η, or "viscosity", is defined as the shear stress (the longitudinal force per unit area), f, which can induce a gradient of flow velocity, Δv, between two parallel layers separated by a distance of $\Delta \ell$:

$$\eta = \frac{f \cdot \Delta \ell}{\Delta v} \tag{1}$$

The cgs unit of η is poise (dyne.sec.cm^{-2}) and that of "fluidity" (the reciprocal of viscosity), is poise^{-1}. Viscosity is basically a dynamic term, which integrates the kinetic rate of thermal shuttling between flow-units, molecules or molecular segments, with intermolecular energy and volume.

For simple weakly-associated fluids (e.g., aliphatic hydrocarbons) an important empirical dependence of η on the specific volume, $\bar{V}$ (the reciprocal of density), was discovered by Batschinsky:

$$\eta = \frac{B}{\bar{V} - \bar{V}_\infty} = \frac{B}{\Delta \bar{V}} \tag{2}$$

where $\bar{V}_\infty$ is the limiting specific volume at infinite viscosity which, at low temperature, approaches the volume of the molecular backbone, and $\Delta \bar{V}$ is the intermolecular free volume. B is a constant given in units of erg.sec ("action") namely, a combination of energy and frequency. These could be attributed to the flow parameters which correspond to the intermolecular energy and to frequency of molecular shuttling which contribute to η. In a family of weakly-associated fluids like hydrocarbons, B and $\bar{V}_\infty$ are approximately constant and the factor which determines the viscosity is almost exclusively the density (or free volume). The implied inverse correlation between the fluid free-volume and viscosity can be used to translate the kinetic meaning of viscosity to the stationary thermodynamic parameter of free volume. This has a great operational value when the effect of lipid microviscosity on membranal activity is analyzed (see below).

The dependence of viscosity on temperature for simple weakly-associated fluids obeys the exponential relation

$$\eta = A \, e^{\,\Delta E_T/RT} \tag{3}$$

where T is the absolute temperature, ΔE_T is the flow activation energy and A is a constant which is formally the residual viscosity at the critical point of transition to gas. Relation 3 indicates that viscous flow obeys the common Boltzmann statistics, and that the flow can be considered as an exchange reaction between "flow-units". In a simple Newtonian fluid, the flow unit is the molecule itself but, in general, the flow unit can also be molecular segments as in the case of elongated molecules which flow through segmental sliding, or molecular aggregates which flow as a cooperative exchange. The effect of pressure (P) on η approximately obeys a similar dependence:

$$\eta_P = \eta_{P=o} e^{\,P\Delta V^{\neq}/RT} \tag{4}$$

where $\Delta V^{\neq}$ is the activation volume, namely the change in volume at the active state of the process. Other physical effectors (e.g. electric field) presumably obey similar exponential relations.

VISCOSITY AT THE CORE OF THE LIPID BILAYER - "MICROVISCOSITY"

The lipid bilayer can be considered as a two dimensional fluid which is fundamentally different than an isotropic fluid. In principle, the anisotropic nature of the lipid bilayer can be presented as a combination of two principal viscosity vectors - across ($\eta_{\shortparallel}$) and along ($\eta_{\perp}$) the plane of the bilayer. In addition, asymmetric distribution of lipids between the two monolayers, and possible segregation of specific lipid domains, add to the complexity of the term "lipid fluidity". Another important characteristic of the lipid bilayer is the gradient of thermal motion and of dielectric constant, extending from the head-group region to the hydrocarbon core. Yet, both the packing and the thermotropic features of the lipid bilayer are determined largely by the hydrocarbon region and, from the submacroscopic point of view, it can be treated analogously to an isotropic fluid, despite the complexities mentioned above.

Lipid microviscosity, $\bar{\eta}$, is the macroscopic simulation of the viscosity in the hydrocarbon core of lipid layers and is generally derived from steady-state fluorescence depolarization (see below). It complies well with both the Batschinsky relation (Eq. 2), namely constant B value and inverse correlation between $\bar{\eta}$ and $\Delta\bar{V}$, and with the exponential dependence of $\bar{\eta}$ on 1/T and on other physical effectors. The lipid microviscosity is about 2 orders of magnitude greater than a hydrocarbon fluid of a similar chain length. This difference is predominantly due to an increase in B by about 2 orders of magnitude and to a much lesser extent to a decrease in lipid free volume ($\Delta\bar{V}$, see Eq. 2). The marked increase in B in lipid layers as compared to an analogous hydrocarbon fluid is presumably a combination of an increase in intermolecular energy of association at the headgroup region and a decrease in frequency of shuttling between two adjacent segments. In fluid lipid layers $\bar{V}$ is only slightly smaller than an equivalent hydrocarbon and is significantly greater than V_∞ which imparts high compressibily on the system. Increase in lipid microviscosity (e.g. by incorporation of cholesterol) is almost exclusively due to decrease in $\bar{V}$ with only small change in B.

Compressibility of lipid bilayer is predominantly related to changes in surface area, A. The thickness of the bilayer, can be assumed to remain constant during changes in lipid microviscosity. Therefore between two states of lipid microviscosity the following approximate relation presumably holds:

$$\frac{\eta_1}{\eta_2} = \frac{\bar{V}_2 - \bar{V}_\infty}{\bar{V}_1 - \bar{V}_\infty} = \frac{A_2 - A_\infty}{A_1 - A_\infty} \qquad (5)$$

The change in specific area with lipid microviscosity implies that increase in lipid microviscosity is associated with increase in order. Microviscosity is therefore a special parameter which combines structure and dynamics of lipid bilayers. Upon increase in microviscosity fluorescent probes like DPH display an increase in the residual fluorescence anisotropy (r_∞) which is due to alignment along the lipid chains - a reflection of increase in order (see below).

The tendency of fluorescent probes to align along the lipid chains upon increase in microviscosity results in different weighing of order vs. dynamics with an apparently different microviscosity values, for different hydrocarbon probes. With DPH, the currently most used fluidity probe, the apparent microviscosity weighs considerably more order than dynamics. Since these parameters are actually interrelated one can, in principle, ignore this problem by relating to the apparent microviscosity of the same probe as a structure-dynamic scale for membranal activity. Alternatively, one can evaluate the "true" microviscosity of system according to

$$\bar{\eta}_o = \frac{1}{3} \eta_{\shortparallel} + \frac{2}{3} \eta_{\perp} \tag{6}$$

where $\eta_{\shortparallel}$ and $\eta_{\perp}$ are the vectorial viscosities along and across the lipid acyl chains. $\bar{\eta}_o$ should, in principle, be independent on the probe structure.

LIPID FLUIDITY AND MEMBRANE FUNCTION

Membranal processes can be grossly divided into those driven by metabolic energy (active processes) and those carried out through diffusion (passive processes). The latter are spontaneous and comply with the thermodynamics of protein diffusion and position where the membrane lipid fluidity is a critical determinant. The natural modulators of lipid fluidity can be divided into chemical modulators and physical effectors. The main modulators are the cholesterol level - presented as cholesterol/phospholipids, the degree of unsaturation of the phospholipid acyl chains, the level of sphingomyelin - generally presented as sphingomyelin/lecithin, and the level of membrane proteins - presented as protein/lipid. The stationary levels of these modulators in a biological membrane, can change in response to a regulatory signal or stress. The change can be either by a passive translocation of one of the chemical modulators with the external medium (e.g., serum) or by membrane biogenesis, both of which processes reach a new steady-state within hours or days. The physical effectors of lipid fluidity are temperature, pressure, pH, membrane potential and Ca^{+2}, and their effect is practically instantaneous.

The physiological relevance of the submacroscopic approach to membrane lipid fluidity can be demonstrated by the fact that a proper physiological function is maintained at a well-defined range of lipid fluidity, while this can - in principle - be attained by any combination of the above chemical or physical factors. For example, in response to changes in temperature (e.g. in poikilotherms or during hybernation) the level of cholesterol or the degree of phospholipid unsaturation changes to the levels required for restoration of the apparent lipid fluidity. Thus, the change in temperature is compensated for by changes in chemical composition, which are obviously different from the microscopic point of view. Similar compensatory mechanisms on a submacroscopic level operate when lipid fluidization induced by narcotics (e.g.,alcohol) is reversed by hydrostatic pressure to restore normal function. Also, when the biosynthesis of one of the chemical modulators (e.g. cholesterol) is impaired, it can be compensated for by another chemical modulator (e.g., degree of unsaturation) which preserves normal function.

In some specific membranal functions, however, microscopic details play nevertheless an important role. This is especially pertinent to cases where the functional unit is activated by a specific interaction with a lipid, which can also act as a fluidity modulator. In such cases, the effect on lipid fluidity - induced by this lipid - is only partially relevant.

The overt effect of the lipid fluidity on a passive membranal process is mainly mediated through the degree of accessibility of the functional site and its rates of rotational and lateral diffusions. Increase in the lipid microviscosity decreases the lipid free volume and in turn decreases the solubilization of the protein in the hydrocarbon core. In parallel, the energy of interactions between the protein residues and the lipid chains decreases and the net effect is a shift in the equilibrium position towards the aqueous domain in either sides of the membrane ("vertical displacement"). Changes in lipid microviscosity can alternatively be compensated for by protein-protein association through "lateral displacement" which in extreme cases, can create segregated domains of lipids and proteins.

As in other equilibria, thermal fluctuations around the median vertical position of the function site (e.g., a receptor) are expected. This can be simulated by a quasi-equilibrium between an operating and a cryptic form of the site which are at constant shuttling. For most cases the dependence of the fraction of operating sites, α, on the membrane lipid microviscosity, $\bar{\eta}$, is given by:

$$\alpha = \frac{1}{1+(\frac{\bar{\eta}}{\bar{\eta}_{\frac{1}{2}}})^{-m}} \tag{7}$$

where $\bar{\eta}_{\frac{1}{2}}$ is the specific microviscosity at which half of the sites are in the operating form. The power m is an expansion factor which characterizes the sensitivity of the site accessibility to changes in the lipid microviscosity. This feature relates to the site *plasticity* and is expressed in the slope of the function $\alpha = f(\bar{\eta}/\bar{\eta}_{\frac{1}{2}})$. Modulation of accessible fraction, α, of antigens receptors[2] and transport channels by changes in membrane lipid fluidity approximately obey the dependence given in Eq. 7.

The overt activity of a passive membranal process can be

presented in terms of Michaelis-Menten kinetics. Accordingly, V_{max} is a product of the number of operating sites C_+ and the average rate of activity, k, of an isolated unit:

$$V_{max} = k \cdot C_+ = k \cdot \alpha \cdot C_o \qquad (8)$$

C_o is the total operateable sites stored in the membrane (the site *capacity*). Since for almost all instances k is inversely proportional to $\bar{\eta}$, combination of Eqs. 7 and 8 yields

$$\frac{1}{V_{max}} = A \cdot \bar{\eta} [1 + (\frac{\bar{\eta}}{\bar{\eta}_{\frac{1}{2}}})^{-m}] \qquad (9)$$

This dependence of V_{max} on $\bar{\eta}$ has the important feature that when m > 1, V_{max} reaches a peak value at $\bar{\eta}_{max} > 4/3 \ \bar{\eta}_{\frac{1}{2}}$. Namely, in some cases a membranal function reaches maximal activity at a specific $\bar{\eta}$. These characteristics could be verified for glucose and amino acid transports in intact cells. A summary of such an experiment is shown in Figure 1.

Manipulation of passive membranal processes by in vitro or in vivo alteration of lipid composition and fluidity ("passive modulation") bears a great potential for restoration of impaired functions and for clinical treatments.

DETERMINATION OF LIPID FLUIDITY BY FLUORESCENCE POLARIZATION

The most direct and sensitive tool for evaluation of lipid fluidity parameters on a submacroscopic scale is fluorescence polarization. At this level of resolution the dynamics of the hydrocarbon region in lipid bilayers can be treated as a non-associated fluid though basic differences between the two-dimensional fluid structure of the lipid bilayer and its equivalent isotropic fluid still prevail even at this low resolution. Nonetheless, to a first approximation, the differences in microscopic and anisotropic thermal motions, which comprise the rotational motions monitored by fluorescence polarization, may be assumed to be averaged out.

Steady-State Measurements

Since lipids are non-fluorescent, labelling with a fluorescence probe is required. An efficient probe should have a series of spectral characteristics and could be either free (e.g., DPH or perylene) or bound to a lipid backbone (e.g., dansyl-PE or anthroyl-stearic acid). The most important characteristics of the probe are

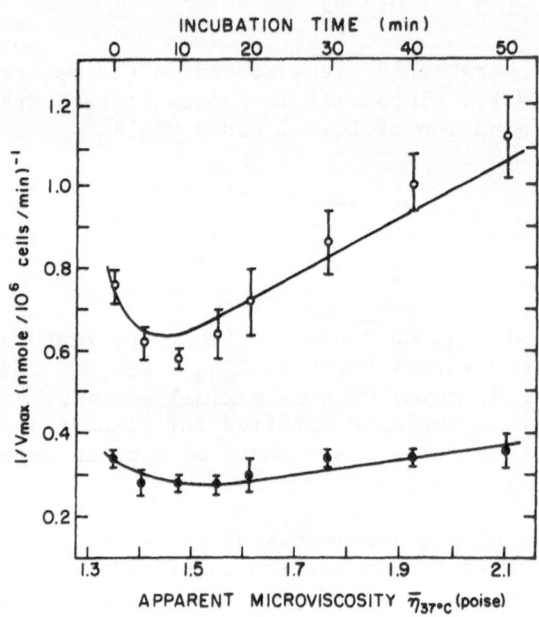

Figure 1: The effect of membrane microviscosity on the basal (open
symbols) and the insulin stimulated (100 ng/ml, solid symbols) maximal rate (V_{max}) of ^{3}H-aminoisobutyric acid (AIB) uptake by 3T3
mouse fibroblasts at 37°C. The plasma membrane microviscosity was
selectively increased by incorporation of cholesteryl hemisuccinate
(50 µg/ml) in PVP (see last section) at increasing time of incubation. The plasma membrane microviscosity $\bar{\eta}$ was determined by
steady-state fluorescence depolarization of DPH and resolved by the
quenching method with TNBS (see last section). The experimental
points represent the mean and standard deviation of 6 independent
experiments. The lines were computed for data fitting according
to Eq. 9. For both curves $\bar{\eta}_{\frac{1}{2}}$ = 1,32 ± 0,03 poise at 37°C and m>8,
indicating that less than 50% of the transport channels are
available at the native state ($\bar{\eta} < \bar{\eta}_{\frac{1}{2}}$), its accessibility can
increase markedly upon slight increase in $\bar{\eta}$ (high m value) and
that insulin has no effect on these parameters. The factor A,
however, decreases about 3-fold upon insulin binding which can
be due to either recruitment of new transport carriers (increase
in C_0) or to reduction of the energy of activation of AIB transport or to a combination of these mechanisms.

a well-defined region of intercalation and even distribution in various lipid domains or a known pattern of distribution (e.g. exclusion from gel phases). The structural perturbation, induced by the probe molecules, seems to be of secondary importance and can be assessed by the effect of probe concentration on the result or by comparing the same parameters obtained with different probes.

The evaluation of the lipid viscosity by steady-state fluorescence polarization is based on the Perrin equation, which combines the measured fluorescence parameters of the probe with the hydrodynamic parameters of the rotational correlation time (ϕ). The convenient form of it is:

$$\frac{r_o}{r} = 1 + \frac{\tau}{\phi} = 1 + C(r) \frac{T \cdot \tau}{\eta} \tag{10}$$

where r is the fluorescence anisotropy, a parameter which is measured directly in the experiment, r_o is the limiting value of r (at infinite viscosity), T is the absolute temperature, τ is the excited state lifetime of the probe, and C(r) is a complex structural parameter of the probe which is slightly dependent on r. η is defined as the "microviscosity" of the system and can be regarded as the submacroscopic term for the lipid viscosity. It is, therefore, an average term given in macroscopic unites (e.g., poise), which describes the bulk fluidity property of the lipid. As stressed before, this term is of direct physiological relevance even though it does not disclose the microscopic details of the system.

Time-Resolved Measurements

The averaged results obtained by the steady-state measurements can be partially resolved by the shape of decay of the fluorescence intensity (F_t) and the anisotropy (r_t) after pulse excitation. A homogeneous system will yield single exponential decays for both F_t and r_t, characterized by a single excited state lifetime (τ) and a single rotational correlation time (ϕ):

$$F_t = F_o\, e^{-t/\tau} \tag{11}$$

$$r_t = r_o\, e^{-t/\phi} \tag{12}$$

For homogeneous systems averaging of r_t over F_t should yield the steady-state value of r:

$$r = \frac{\int_o^\infty r_t \cdot F_t \cdot dt}{\int_o^\infty F_t \cdot dt} \tag{13}$$

In a heterogeneous system each homogeneous compartment, which is fluorescently-labelled, can possess specific excited state lifetime (τ_i) and rotational correlation time (ϕ_i), and the shapes of F_t and r_t will deviate from single exponential decays.

$$F_t = F_o \, \Sigma_i \, f_i \, e^{-t/\tau_i} \tag{14}$$

$$r_t = r_o \, \Sigma_i \, r_i \, e^{-t/\phi_i} \tag{15}$$

where f_i is the relative contribution of compartment i. In reality, resolution of such decays to more than two exponential decays is impractical. Therefore, the analysis of F_t and r_t is commonly approached by simulation to two compartments which provides a general information on the degree of heterogeneity of the system.

In some membranes, the decay of r_t does not reach zero as expected, but approaches a residual fluorescence anisotropy of r_∞. Such a decay can be presented by:

$$r_t = r_\infty + (r_o - r_\infty) \, e^{-t/\phi} \tag{16}$$

and indicates a residual order in the distribution of the probe molecular after an extensive rotational motion, steming from preferential orientation with respect to the lipid chains. The rod-like probe DPH tends to be oriented in parallel to the lipid chains. The degree of alignment of DPH increases with increase in lipid density (i.e. viscosity), namely increase in r (see Eq.17), and is reflected in an increase in r_∞ which obeys the empirical relation:

$$r_\infty = 1.33 \, r - 0.1; \quad \text{for} \quad 0.13 < r < 0.28 \tag{17}$$

It was therefore suggested to use the term r_∞/r_o of DPH as the square of an order parameter, analogously to the order parameters used in spin resonance techniques.

Microviscosity in Restricted Rotation

Although the actual geometrical boundaries of the restricted rotation are hard to evaluate, it could be simulated by a wobbling motion restricted to a cone of an angle $\Theta_c < 90°$. The symmetry axis is presumably perpendicular to the plane of the membrane and the viscosity within the boundaries is homogeneous. The relation between r_∞ and Θ_c in this model is given by:

$$\frac{r_\infty}{r_o} = [\frac{\cos\Theta_c (1 + \cos\Theta_c)}{2}]^2 \tag{18}$$

A more realistic simulation is of a motion in a gradient of viscosity between the two principal viscosities $\eta_\perp$ and $\eta_{||}$ (see Eq.6). Operationally, this leads to definition of 3 types of microviscosities:

(i) The "true" microviscosity - $\bar{\eta}_o = \frac{1}{3} \eta_{||} + \frac{2}{3} \eta_\perp$.

(ii) The microviscosity estimated through Eq. 11 (namely, assuming that r_∞ = o) - $\bar{\eta}$.

(iii) The apparent microviscosity sensed by the probe in its rotational boundaries - $\bar{\eta}_a$

Introduction of r_∞ into Eq. 10 leads to

$$r = r_\infty + \frac{r_o - r}{\tau} \phi_a = r_\infty + \frac{(r_o - r)\bar{\eta}_a}{C_{(r)} \cdot T \cdot \tau} \tag{19}$$

Where ϕ_a is the apparent rotational correlation time of the restricted rotation. Comparing Eq. 10 with Eq. 19 shows that

$$\frac{\bar{\eta}_a}{\bar{\eta}} = 1 - \frac{r_\infty}{r} \tag{20}$$

Since r_∞ is of the same sign as r, Eq. 20 indicates that the probe orientation favours alignment with a network which is less viscous than the bulk. Therefore, ignoring r_∞ actually compensates for the pre-selection of the less viscous environment. Yet, this compensation is not full and $\bar{\eta}$ is higher than η_o as shown in the following. The two other correlations between the 3 microviscosities are

$$\bar{\eta} = \bar{\eta}_a + \frac{2}{3} \frac{r_\infty}{r_o} (\eta_\perp - \eta_{||}) \tag{21}$$

and the approximate expression

237

$$\frac{\bar{\eta}}{\bar{\eta}_0} = \frac{1 - \dfrac{r_\infty}{r_0}}{1 - \dfrac{r_\infty}{r}} \tag{22}$$

Introducing the empirical relation for DPH given in Eq. 17 into the linear part of Eq. 22, together with the value of $r_0 = 0.362$, leads to the approximate relation

$$\bar{\eta}_0 = (0.089/r - 0.116)\ \bar{\eta} \tag{23}$$

which enables to evaluate $\bar{\eta}_0$ without measuring r_∞.

Equations 20-23 indicate that the magnitude of the 3 viscosities follows the order $\bar{\eta}_a < \bar{\eta}_0 < \bar{\eta}$ suggesting that the probe orients itself in favour of more fluid coordinates ($\bar{\eta}_a < \bar{\eta}_0$). Yet, the artificial increase in r measured by steady-state fluorescence depolarization caused by the probe orientation increases the apparent $\bar{\eta}$ beyond the compensation for the preferential orientation which senses more fluid environment ($\bar{\eta} > \bar{\eta}_0$). This over-compensation increases with r_∞ and is therefore more pronounced in probe shapes which are remote from a spherical structure (e.g. the rode-like structure of DPH). Operationally, one can relate to the value of $\bar{\eta}$ derived by steady-state fluorescence depolarization as a microviscosity parameter in which $\bar{V}$ in Eq. 2 is artificially reduced and the contribution of order to the apparent $\bar{\eta}$ value is more than expressed in Eq. 2. Nonetheless, the inverse correlation between $\bar{\eta}$ and free volume (ΔV in Eq. 2) is still preserved and for all practical purposes $\bar{\eta}$ derived from DPH fluorescende depolarization can serve as a useful parameter of lipid fluidity.

SOME PRACTICAL OUTLINES

The apparent $\bar{\eta}$ determined by fluorescence depolarization can be presented at different levels of accuracy which are discussed below and summed up in Table 1.

Qualitative Estimation of $\bar{\eta}$

Fluorescence polarization probes generally possess a high-emission quantum yield and environmental-insensitive lifetime of excited state. Therefore, unless uncommon fluorescence quenching occurs in the system, the fluorescence anisotropy, r, itself can serve as a qualitative scale for fluidity: the higher the r, the lower is the fluidity. It is important to note that the scale of

Table 1. Evaluation of Lipid Microviscosity by Steady-State Fluorescence Polarization

Level of $\bar{\eta}$ presentation	Fluorescence polarization relation	Comments
Qualitative	r	Highly non-linear
Quantitative on a relative scale	$(\frac{r_o}{r} - 1)^{-1}$	Approximate; useful in measurements of the same system at different conditions
Approximate quantitative for DPH (in poise)	$2.4(\frac{r_o}{r} - 1)^{-1}$	Apparent value $-\bar{\eta}$
Apparent quantitative (in poise)	$C_{(r)} \cdot T \cdot \tau (\frac{r_o}{r} - 1)^{-1}$	For freely rotating probes. A combination of the true viscosity and the degree of order $-\bar{\eta}$
"True" quantitative for DPH($\bar{\eta}_o$)	$\bar{\eta} \cdot (\frac{0.089}{r} - 0.116)$ for $0.13 < r < 0.28$	Defined as $\frac{1}{3}\eta_{\shortparallel} + \frac{2}{3}\eta_{\perp}$, considerably smaller than the apparent $\bar{\eta}$

The "true" microviscosity ($\bar{\eta}_o$) weighs appropriately the action and the free volume determinants (B and ΔV in Eq. 2). The apparent microviscosity $\bar{\eta}$, which is derived from steady-state fluorescence depolarization (ignoring r_∞) overweighs ΔV. Since B is much less sensitive to changes in lipid composition both $\bar{\eta}$ and $\bar{\eta}_o$ can be regarded as scales of free volume. For measurements with DPH at high fluidity (r < 0.13) r_∞ is negligible and the apparent $\bar{\eta}$ value weighs appropriately B and ΔV. At higher r values the contribution of r_∞ increases and hence the order parameter becomes a significant part of $\bar{\eta}$. At high r values (r > 0.28) the order parameter is the main constituent of the apparent $\bar{\eta}$. Evaluation of $\bar{\eta}_o$ at this range is inaccurate since r_∞ approaches the value of r_o.

r versus $\bar{\eta}$ is non-linear (see Eq. 10) and around $r \to 0$ or $r \to r_o$, where small changes in r reflect large changes in $\bar{\eta}$, this presentation becomes insensitive. Only around $r \sim r_o/2$, small changes in $\bar{\eta}$ can be detected and monitored accurately.

Relative Scales for $\bar{\eta}$

According to Eq. 10, the term $(\frac{r_o}{r} - 1)^{-1}$ is proportional to $\bar{\eta}$ and inversely proportional to $C(r)T \cdot \tau$. Since, for most determinations, changes in $C(r)$, T and τ are relatively small, and for most cases, in directions opposite to each other, the product $C(r) \cdot T \cdot \tau$ can be assumed to be approximately constant. The term $(\frac{r_o}{r} - 1)^{-1}$ can, therefore, serve as a useful relative scale for $\bar{\eta}$. For DPH $C(r) \cdot T \cdot \tau \sim 2,4$ poise and therefore the product $2,4(\frac{r_o}{r} - 1)^{-}$ can be used as an approximate absolute scale for $\bar{\eta}$ presented in macroscopic units (poise). A more accurate scale for $\bar{\eta}$ is obtained when changes in τ are taken into account: $\bar{\eta} \propto (\frac{r_o}{r} - 1)^{-1} \cdot \tau$. Since direct measurement of τ is complicated, one can express τ on a relative scale of fluorescence intensity, F/F_o, which is the corresponding fluorescence intensity, F, divided by the fluorescence intensity, F_o, at quantum yield of 1. A practical way of evaluation of F/F_o is by measuring F as a function of reduced temperature. For probes with high quantum yield, a plateau of F (which corresponds to F_o) can be reached at temperatures around or slightly below room temperature. The product $(\frac{r_o}{r} - 1)^{-1} \cdot \frac{F}{F_o}$ thus obtained represents $\bar{\eta}$ more accurately. When temperature changes are also involved in the measurements, the scale for $\bar{\eta}$ can be further extended to $(\frac{r_o}{r} - 1)^{-1} \cdot \frac{F}{F_o} \cdot T$.

The Absolute Apparent $\bar{\eta}$ Value

The submacroscopic approach allows, in principle, to evaluate the absolute $\bar{\eta}$ when all other parameters appearing in Eq. 10 are known. The most problematic parameter is $C_{(r)}$ and only in cases of simple symmetry (e.g., a sphere) it can be calculated directly from hydrodynamic equations. In practice, it is possible to evaluate $C_{(r)}$ empirically, by using an equivalent macroscopic fluid for calibration profiles. Yet, this approach may introduce some basic uncertainties, which are due to differences between the calibration fluid and the simulated lipid layer. Nonetheless, this

240

approach is of practical advantage since it provides a way to bridge between the microscopic system and an equivalent macroscopic fluid and thus allowing to employ basic hydrodynamic equations like Eqs. 2 and 3. For DPH, the calibration curve was obtained with heavy paraffin oil (American Oil USP35)which is presented by

$$\frac{21}{r^2} + \frac{242}{r} - 827 = \frac{T \cdot \tau}{\bar{\eta}} \ (^\circ K \cdot nsec \cdot poise^{-1}); \ for \ r > 0,16 \qquad (24)$$

$$\frac{510}{r} - 1680 = \frac{T \cdot \tau}{\bar{\eta}} \ (^\circ K \cdot nsec \cdot poise^{-1}); \qquad for \ r < 0,16 \qquad (25)$$

The empirical translation of the fluorescence polarization parameters into lipid microviscosity has led to a vast amount of absolute parameters of lipid viscosity which, despite the possible deviation from the true figures are of great practical value.

Resolution of Plasma Membrane Lipid Fluidity in Intact Cells

Molecular probes tend to partion between the plasma membrane and intracellular organelles. In cells labelled with DPH indirect evaluation of the plasma membrane fluidity can be approached by selective quenching by nonradiative energy transfer of the fluorescence emitted from the plasma membrane after tagging the cell-surface with a suitable impermeable electron acceptor. This can be achieved by chemical binding of 2, 4, 6 trinitrobenzene sulfonate (TNBS) or by incorporation of N-bixinoyl glucosamine (BGA). This approach has an advantage over membrane isolation since the latter requires cell disruption which may lead to elimination of scpecific lipid regions from the final preparation [Grunberger et al. Biochim. Biophys. Acta 688: 764 (1982)].

a. The TNBS method. DPH in tetrahydrofuran(2×10^{-3}M) is dispersed in aqueous buffer (e.g. phosphate buffered saline, PBS) by 1:1000 dilution with vigorous stirring. 10^7 cells are incubated in this dispersion (10^6/ml) sufficiently long to allow complete partitioning of DPH between the various lipid compartments (up to 1 hr at 37°C). The labelled cells are divided accurately to 5 identical samples, each containing 1 ml of 2×10^6 cells, and placed at 4°C. To each of the samples 1 ml of PBS containing 4, 3, 2, 1 and 0 mg TNBS is added, respectively. After 1 hr of incubation the cells are washed and resuspended in 2ml PBS. Cells should be then counted. The fluorescence anisotropy (r) and fluorescence intensity (F) are recorded, and the latter is corrected for cell number. The results are plotted as r vs. 1/F where the intercept corresponds to the fluorescence anisotropy of the plasma membrane (r_m).

b. The BGA method. Cells are labelled with DPH as described above. A stock solution of 3×10^{-5} BGA in tetrahydroform is first diluted 1:30 into the aqueous buffer (e.g. PBS), generally 10 μl into 0.3 ml and then diluted 1:50 or 1:25 into a sample of DPH labelled cells ($\sim 2 \times 10^6$ cell in 2 ml buffer) while in the fluorescence instrument. The fluorescence intensity and fluorescence anisotropy are recorded within a few seconds and then at time intervals which correspond to significant decrease in F (i.e. incorporation of BGA). In general, the fluorescence quenching reaches a plateau after about 20 min. The results are processed as above.

In vitro Manipulation of Membrane ¯ipid Fluidity

Under physiological physical conditions membrane lipid composition, and hence the fluidity, can be altered through either acute or systemic approaches. In the acute approach cells (or membranes) are incubated in a medium which contains a large excess of a fludizing or a rigidifying lipid for minutes or hours. The cells are washed and transferred into the tissue culture medium where the experiment continues. In general, restoration of the membrane fluidity of the modified cells takes more than 12 hours, namely past the triggering of most physiological signals. In the systemic approach the incubation medium of the cells is supplemented with a relatively small amount of lipid modifying mixture and the experiment is conducted in the modified medium. Alteration of membrane fluidity in this approach is a combination of passive translocation of lipids between the membrane and the external pool, as in the acute method, in addition to changes due to intracellular lipid synthesis and membrane biogenesis. In this approach the apparent change in membrane lipid fluidity reaches a peak level at around 24 hours and then reverses towards restoration through homeostasis mechanisms. In the following, 3 procedures for acute treatment are given. In principle, each of the treating media can be mixed with the culture media for systemic treatment.

a. The PVP method. Polyvinyl pyrrolidone, PVP, is an uncharged polyamide which is highly soluble in water and can thus provide a hydrophobic support for lipophilic materials when dispersed in water. A solution of 3.5% PVP (40.000 MW), 1% serum albumin and 0.5% glucose is a medium applicable for almost all types of lipid manipulations. For lipid rigidification cholesterol, or preferably cholesteryl hemisuccinate (CHS), at a final concentration of 50 μg/ml in the PVP medium is recommended. This is obtained by 1:100 dilution of 5 mg/ml cholesteror or CHS in ethanol into the PVP medium. For fluidization egg lecitanin or a phospholipid mixture at a final concentration of 100 μg/ml in PVP medium is prepared analogously. Cells (10^6-10^7 per ml) are incubated with gentle shaking in the PVP medium at room temperature up to 90 minutes. In some cells incubation of only a few minutes results in significant lipid alteration. Overtime incubation can result in cell death.

b. Lyophilized lipid enriched serum. Serum or serum-medium for cell culture is supplemented with 1 mg cholesterol (or CHS) per ml serum or 2 mg lecithin (or phospholipid mixture) per ml serum. A solution of 10 mg/ml cholesterol or 20 mg/ml lecithin in tetrahydrofuran is diluted into the serum or serum-medium with good mixing and then lyophilized to complete dryness. The dry mixture is reconstituted with sterile distilled water before use. Modulation of membrane fluidity follows a similar procedure as above but in general it is milder and slower than with the PVP.

c. Incorporation of fatty acids. Fatty acids incorporate very rapidly into cell membranes and induce changes in lipid fluidity according to their degree of unsaturation. However, shortly after incorporation the fatty acid are internalized and metabolized. Therefore, manipulation of lipid fluidity by fatty acids can be of practical use only for isolated membranes or when the cell metabolism is suppressed. Another important factor which should be considered is the critical micellar concentration (CMC) of the fatty acid. Above the CMC the fatty acids act as detergents which are harmful to the membranes. In general, a stock solution of $2x10^{-2}$ M fatty acid in ethanol is diluted into an aqueous buffer or medium (e.g. PVP) and immediately applied to the cells for a few minutes. A convenient couple for lipid fluidity modulation is stearic acid (18:0) for rigidification and linoleic acid (18:2) for fluidization (CMC~10^{-4} M). A potent adequate incubation medium is made by 1:500 dilution of the $2x10^{-2}$ M ethanolic stock solution.

DYNAMIC INTERACTIONS OF FLUORESCENTLY LABELED
ADENOSINE DEAMINASE IN INTACT HUMAN ERYTHROCYTES:
RELATION TO ABNORMAL ACTIVITY IN MALIGNANT CELLS

A. H. Parola, G. Skorka, P. Shuker, and D. Gill

Department of Chemistry and Physics
Ben-Gurion University of the Negev
Beer-Sheva 84105, Israel

INTRODUCTION

Membrane modifications associated with cell transformation and malignancy are the subject of our study[1,2]. Specifically, we explore changes due to lipid-protein and protein-protein dynamic interactions, as manifested through fluorescence depolarization of the tagged membrane component[3,4].

In the present study a membrane enzyme, the specific activity of which is known to be modified in malignancy, was chosen to be specifically labeled. Given the relatively minute number of such labeling sites, fluorescence techniques, which are highly sensitive and informative of the molecular dynamics, were the method of choice. Specific tagging of the enzyme may be accomplished by photoaffinity labeling with a fluorescent substrate analogue. This approach complements and sharpens our previous work, in which labeled membrane lipids and glycoproteins showed strong interdependence[3,4].

Our search for a suitable enzyme resulted in the choice of adenosine deaminase (ADase), which has the following attractive features:

1. It is a key enzyme in the metabolism of adenoise[5]
2. It regulates cell differentiation and proliferation[6]
3. Its specific activity in malignant cells is abnormal[6,7]
4. It is an attractive model enzyme: a) in having a specific, tightly binding inhibitor (2'-deoxycoformycine)[8], b) in the evidence for its presence in the cell membrane[9-11], c) in its flexible quaternary structure[11,12].

In spite of the clinical importance of ADase (e.g. in the fatal immunodeficiency[13] and in its potential to inactivate potent anti-tumor nucleosides)[14], relatively little is known about this enzyme (e.g. primary structure). Human ADase (E.C. 3.5.4.4) exists in multiple forms[11]. Two soluble forms have estimated M.W. of 36,000-38,000 (small form) and 222,000-298,000 (large form) and are inter-convertible. The purified small form ADase from human erythrocytes was shown to be a single polypeptide with less than 10% carbohydrate and to be identical with the small form from other tissues[15]. The small form ADase combines with the binding protein in a molar ratio of 2:1 to produce the large form of ADase, present in some tissues[11,12]. The enzyme thus exhibits variable activity in either the small form (monomer or dimer) or in the large form, and both activity and quaternary structure may be modified with malignant transformation[6,16]. For example this is the case in colon adeno-carcinoma[6], where the observed increase in ADase specific activity coincided with a dissociation of the large form.

ADase in transformed cells may be chemically altered in its primary and secondary structure. Alternatively, milder perturbation, e.g., the straining of ADase quaternary structure due to lipid-protein interactions, would suffice to account for the observed changes[3,4,17]. We undertook the study of the latter possibility, which may be simulated by artificially modifying membrane lipid properties and monitoring the concomitant changes in ADase behavior. Accordingly, we have synthesized 3'-0-[5-dimenthylamino-naphthalene-1-sulfonyl]adenosine-(dansyl-3'-0-adenosine) and established its unique binding to ADase and its role as a reversible competitive inhibitor of the deamination of adenosine[18]. Subsequently, a photoaffinity label variant of this substrate analogue - 8-azido-dansyl-3'-0-adenosine - was prepared. The spectroscopic character-istics of the covalently labeled enzyme were determined. Erythro-cytes labeled by 8-azido-dansyl-adenosine allowed the study of lipid-protein interaction as modified by added cholesterol. This methodology proved its applicability to studies of transformed cells.

RESULTS

8-Azido-dansyl-3'-0-adenosine was prepared from 8-azido adeno-sine (according to Holmes and Robins)[19] and dansyl chloride[18], and was characterized by TLC, NMR, IR, U.V. and elemental analy-sis[20]. Table 1 summarizes the fluorescence properties of this label when free in solution, prior and after photolysis and when bound to ADase, noncovalently and covalently.

The probe exhibits the typical dansyl solvent dependence[18], i.e. large blue shift and increased quantum yield in nonpolar sol-vents. Upon non-covalent binding to ADase, a 65 nm blue shift in emission maxima and an increase in P up to 0.21 are observed. Inhi-

Table 1. Fluorescence Characteristics of Free and Bound Probe[a]

	Photo-lysis	max λ_{ext}, nm	max λ_{em}, nm	ϕ	τ, nsec	P[b]
Free	–	326	503–507(514–518)[c]	0.46	$12.6\pm.2$[d]	0.00
	+	326	503–507(514–518)[c]	–	$13.2\pm.2$[d]	0.00
Bound	–	340	438–440	–	–	0.21
Dialysed	+	340	405, shoulder at 440	0.65	$18.2\pm.2$[e]	0.37

[a] In phosphate buffer, pH 7.6 at 10°C.
[b] At maximum excitation and emission wavelengths.
[c] Corrected fluorescence spectra.
[d] Single exponential decay analysis.
[e] Double exponential decay analysis.

bition kinetics[18] indicate that the probe is a reversible competitive inhibitor of ADase, $K_I = (2.3\pm0.4)$ x 10^{-6}M, i.e. a more effective inhibitor than dansyl-3'-0-adenosine, for which $K_I = (1.6\pm0.1)$ x 10^{-5}M [18]. Binding specificity of dansyl-3'-0-adenosine was evident from a) the almost quantitative decrease in the fluorescence of the ADase-bound probe upon addition of adenosine in excess, b) p-hydroxy-mercuriphenyl-sulfonic acid (as SH-blocker) and c) 2'-deoxycoformycine ($K_I = 2.5$ x 10^{-12}M).

The covalent complex between ADase and our photoaffinity label was obtained after equilibration in the dark followed by photolysis at 365 nm at 25° under N_2. While no significant changes in fluorescence were noted after photolysis of the free label (Table 1), the dialysed covalent ADase-probe complex showed pronounced variations in all the fluorescence parameters examined. Amino acid analysis identified the photolabeled glutamic acid and alanine, presumably present at the active site.

Active-site directed binding was further probed by its initial blocking with 2'-deoxycoformycine in the presence of our photoaffinity label. Subsequent photolysis and dialysis resulted in reduced fluorescence intensity, and a P = 0.24. Furthermore, the contribution of ADase tertiary structure to the observed fluorescence polarization was evaluated by denaturing ADase in 8M urea. A decrease from P = 0.37±0.03 to P = 0.15±0.03 was observed. Upon dialysis, P = 0.19±0.03 was measured, indicating partial renaturation. Renaturation in the presence of free photolyzed, 8-azido-dansyl-3'-0-adenosine showed further increase to P = 0.24±0.03, presumably due to non-covalent adsorption.

Studies with human erythrocytes were in the following sequence:
1) Determination of noncovalent binding specificity, 2) Determinatic
of covalent binding specificity of the photoaffinity label by gel
electrophoresis, and 3) Spectroscopic characterization of the labele
erythrocytes prior to and after the addition of cholesterol.

Labeled intact human erythrocytes showed upon excitation at 340
nm an emission maximum at 500 nm when measured by frontal surface
fluorometry[20]. Upon addition of excess adenosine, a 40% decrease
in this fluorescence emission was noted. Covalent attachment was
achieved by photolysis, after which maximum fluorescence was shifted
to 490±5 nm, with P = 0.06. Fluorescent protein fractions obtained
from DEAE-cellulose column separation of erythrocyte hemolysate
supernatant, were shown on SDS gels by electrophoresis[12].
Coummasie blue staining showed a single band with R_f = 0.22 coincident with the single fluorescence maximum of fractions extracted fro
a sliced gel. This R_f value is in good agreement with that reported
from purified erythrocyte ADase.

Covalent labeling of cholesterol-enriched erythrocytes[21]
resulted in 60% increase in fluorescence emission intensity, a red
shift in emission maximum from 490 nm to 505 nm and an increase in
fluorescence polarization value to P = 0.30. Similar increase in
fluorescence intensity was observed for non-covalently labeled eryth
rocytes upon addition of cholesterol, which diminished upon addition
of adenosine. Remarkably, erythrocyte ghosts were found to be unlabeled, presumably due to shedding of the relatively loosely-bound
ADase.

DISCUSSION

The newly prepared fluorescent photoaffinity label, 8-azido-
dansyl-3'-0-adenosine was shown to preserve the attractive features
of its constituents. In exhibits dansylic sensitivity to the polarity of its environment and binding affinity even superior to adenosine[18]. Furthermore, it selectively binds to the active site of
ADase. After photolysis the covalent complex with ADase indicates a
tight rigid binding to the hydrophobic active site. The high quantu
yield, the relatively long lifetime (τ = 18.2 nsec) and high P value
(P = 0.37) render it suitable for application to studies of protein
rotational dynamics of loosely bound (peripheral) enzymes in living
cells. Our studies identified cysteine, alanine and glutamic acid
residues at the active site. Specific covalent binding to the activ
site was evident both from the reduced fluorescence emission obtaine
from attempts to label ADase blocked by 2'-deoxycoformycin and from
the reduction to P = 0.15 upon partial denaturation by 8M urea; the
competition with adenosine further indicates binding site specificity.

248

The 40% decrease in fluorescence intensity upon addition of adenosine to non-covalently labeled erythrocytes indicate that in equilibrium that fraction is specifically bound. After photolysis and washing, background due to free label is eliminated. In labeling whole erythrocytes, it should be considered that specific binding sites other than those of ADase are present, e.g., adenosie kinase. We observed that most of the specific binding is due to ADase as indicated by the gel electrophoresis results. Quenching by hemoglobin excludes any signal from cytosolic ADase (if present) and positively localizes the fluorescent binding sites to the plasma membrane.

Upon incorporation of cholesterol in the erythrocyte membrane, a 60% fluorescence intensity enhancement was observed. This could possibly arise from increased binding to ADase due to its enhanced exposure. The accompanying red shifted fluorescence emission to 505 nm would indicate a more hydrophilic environment. The remarkable increase in fluorescence polarization values (from 0.06 to 0.30) indicates restricted rotational motion, as would be expected in a more rigid membrane. These results are indicative of remarkable non-covalent lipid-protein interactions, to which our method is highly responsive. Similar effects in the membranes of malignant cells may be responsible for the abnormal ADase activity.

Acknowledgement

This work was supported, in part, by the Office of the Head Scientist, Ministry of Health, State of Israel.

REFERENCES

1. A. H. Parola, in: "Membranes in Tumor Growth", T. Galeotti, ed., Elsevier, North-Holland Biomedical Press, Amsterdam, in print (1982).
2. A. H. Parola, P. W. Robbin, and E. R. Blout, Exp.Cell Res., 118:205 (1979).
3. G. Fleischer, I. Nathan, A. Livne, A. Divilansky, and A. H. Parola, in: "Platelets: Cellular Response Mechanisms and their Biological Significance", A. Rothman, ed., John Wiley & Sons (1980).
4. I. Nathan, G. Fleischer, A. Divilansky, A. Livne, and A. H. Parola, Biochim.Biophys.Acta, 598:417 (1980).
5. R. E. Parks and R. P. Agarwal, "The Enzymes", P. O. Boyer, ed., 3rd Edition, 7:483 (1972).
6. P. P. Trotta and M. E. Balis, Biochemistry, 17:270 (1978).
7. P. K. Chiang, G. L. Cantoni, D. A. Ray, and J. P. Bader, Biochem.Biophys.Res.Commun., 78:336 (1977).
8. R. P. Agarwal and R. E. Parks, Jr., Method Enzymol., 51:502 (1978).

9. R. P. Agarwal and R. E. Parks, Jr., <u>Biochem.Pharmacol.</u>, 24:547 (1975).
10. C. H. M. M. de Bruyn and T. L. Oei, <u>Adv.Exp.Med.Biol.</u>, 41A:233 (1974).
11. M. B. Van de Weyden and W. N. Kelley, <u>J.Biol.Chem.</u>, 251:5418 (1976).
12. P. E. Daddona and W. N. Kelley, <u>Biochim.Biophys.Acta</u>, 580:302 (1979).
13. M. S. Coleman, J. Donofrio, J. J. Hutton, L. Hahn, A. Daud, B. Lampkin, and J. Dyminski, <u>J.Biol.Chem.</u>, 253:1619 (1978).
14. W. Plunkett and S. S. Cohen, <u>Cancer Res.</u>, 35:1547 (1975).
15. W. P. Schroder, A. R. Stacy, and B. Pollara, <u>J.Biol.Chem.</u>, 251:4026 (1976).
16. J. Meier, M. S. Coleman, and J. J. Hutton, <u>Br.J.Cancer</u>, 33:312 (1976).
17. M. Shinitzky, <u>Physiol.Rev.</u>, in press (1982).
18. G. Skorka, P. Shuker, D. Gill, J. Zabicky, and A. H. Parola, <u>Biochemistry</u>, 20:3103 (1981).
19. R. E. Holmes and R. K. Robins, <u>J.Amer.Chem.Soc.</u>, 87:1772 (1965).
20. G. Skorka, PhD. thesis, Ben-Gurion University (1983).
21. M. Shinitzky, in this book.

PHASE SEPARATION IN PHOSPHOLIPID MIXTURES

BY FLUORESCENCE ANISOTROPY

T. Parasassi*, N. Rosato**, E. De Felip*
and F. Conti*

* Istituto di Chimica Fisica, Università di Roma
**Istituto di Biofisica, CNR, Pisa

The study of phospholipid mixtures can give information on biologically important processes occurring in membranes such as phase separation and fusion.

A powerful tool for this kind of study is the fluorescence anisotropy behavior of probes located in different positions within the bilayer, such as DPH and ANS.*

In this paper we report some results obtained with mixtures of DLPC-DPPC, DMPC-DPPC, DPPE-DPPC, also in the presence of perturbing agents, namely AA, PA and calcium ions, able to influence the structure and the functions of biological membranes.

MATERIALS AND METHODS

All lipids were purchased from Sigma and used without further purification. Standard solutions were prepared in chloroform, stored at -20°C and their purity was checked periodically by two-dimensional thin-layer chromatography. DPH and ANS were from Sigma. The 2 mM standard solution of DPH in tetrahydrofuran was stored in the dark at -20°C; the 1 mM standard solution of ANS in water was stored in the dark at +4°C. Solvents were of spectroscopic purity grade.

---

*Abbreviations used are: DPH, 1,6-diphenyl-1,3,6-hexatriene; ANS, 1-anilinonaphthalene-8-sulfonic acid; DLPC, dilauroyl-L-α-phosphatidylcholine; DMPC, dimyristoyl-L-α-phosphatidylcholine; DPPC, dipalmitoyl-L-α-phosphatidylcholine; DPPE, dipalmitoyl-L-α-phosphatidyl ethanolamine; AA, arachidonic acid; PA, palmitic acid.

Sample Preparation

Appropriate amounts of chloroform solutions of phospholipids and fatty acids and of DPH tetrahydrofuran solution were mixed, evaporated under N_2 stream and finally under high vacuum for one hour. Dulbecco's PBS was then added.

For ANS labeling, after the evaporation of the chloroform solutions of phospholipids and fatty acids, the appropriate amount of ANS water solution was added together with Dulbecco's PBS. Water-suspended labeled lipids were vortex stirred at intervals, above their transition temperature, during a period of 20 min for DPH-labeled samples and of 45 min for ANS-labeled samples. Final concentrations were: 1 mM total lipids, 1.5 μM DPH, 10 μM ANS.

All measurements were carried out by a double-edged SLM 4800 spectrofluorimeter; the temperature was controlled by a circulating water bath and a magnetic stirrer was inserted under the cell holder. Fluorescence anisotropy measurements of DPH were carried out in the ratio mode, with the excitation at 360±2 nm and the emission was analyzed through Skott kV cut-off filters, cutting below 418 nm. In ANS experiments the excitation was 390±4 nm and the emission was observed with the same cut-off filters described above, or with a monochromator set at 470±4 nm. The polarization values were corrected for the intrinsic polarization of the monochromator.

RESULTS

DLPC-DPPC Mixtures

In Figure 1 is reported the DPH fluorescence anisotropy plot of an equimolar mixture of dilauroyl- and dipalmitoyl-phosphatidyl-choline as a function of temperature. Two separate phase transitions occur: the first at higher temperature (0°-6°C) with respect to the pure DLPC (T_c = 0°C), and the second at lower temperature (18°-30°C) with respect to the pure DPPC (T_c = 41.5°C).

Arachidonic or palmitic acid was then added to the mixture, both at concentrations below their critical micellar concentrations, i.e. 5 mol% of the total phospholipids. Fluorescence anisotropy plots versus temperature (Figure 1) show that with AA the low-temperature phase transition is broadened and flattened while PA affects mainly the high-temperature phase transition (Figure 2).

These results are in agreement with a preferential partition of AA and PA in two hypothetic distinct domains[2] as expected from phospholipids differing of four carbon atoms in their acyl chains[1].

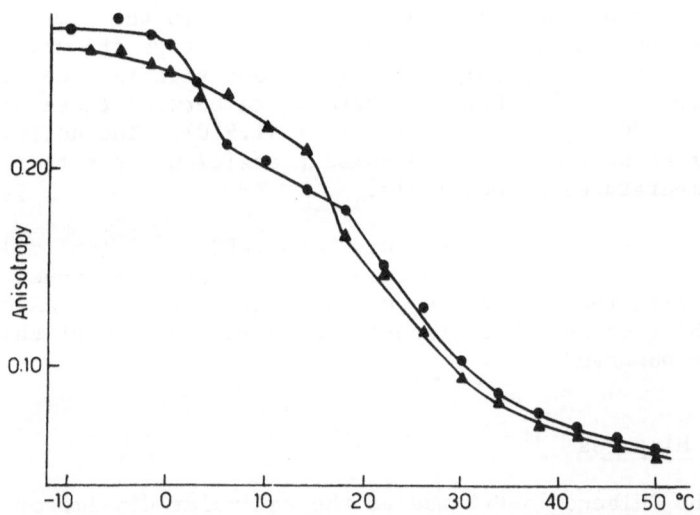

Fig. 1. DPH fluorescence anisotropy plots vs. temperature of (●)
DLPC–DPPC equimolar mixture and (▲) DLPC–DPPC equimolar
mixture with 5 mol% arachidonic acid.

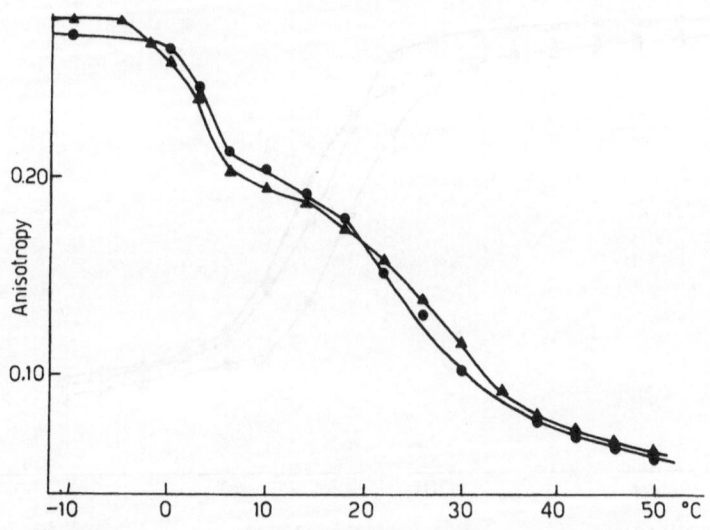

Fig. 2. DPH fluorescence anisotropy, plots vs. temperature of (●)
DLPC–DPPC equimolar and (▲) DLPC–DPPC equimolar mixture with
5 mol% palmitic acid.

DMPC-DPPC Mixtures

The DPH fluorescence anisotropy profile in the dimyristoyl- and dipalmitoyl-phosphatidylcholine equimolar mixture (Figure 3) versus temperature shows that only one phase transition is present at a temperature of 35.6°C that is intermediate between those of the pure compounds (DMPC: T_C = 24°C, DPPC: T_C = 41.5°C). The addition of 5 moles % of AA or PA shifts the phase transition range to lower or higher temperatures respectively.

The same plot for the mixture DMPC:DPPC = 7:3 (mol:mol) shows (Figure 4) a phase transition temperature shifted to lower values (T_C = 26.6°C) with respect to the equimolar mixture. When AA or PA is added a shift of the phase transition as in the case of the equimolar mixture is observed.

DPPC-DPPE Mixtures

In the Dulbecco's PBS medium the equimolar dipalmitoylphosphatidyl-choline and dipalmitoylphosphatidylethanolamine mixture showed the same behavior as the DMPC-DPPC mixture, with an intermediate phase transition temperature (T_C = 58°C using DPH and T_C = 48°C using ANS) between those of the pure compounds that are 41.5°C for DPPC and 63°C for DPPE[7]. The addition of calcium ions to this mixture

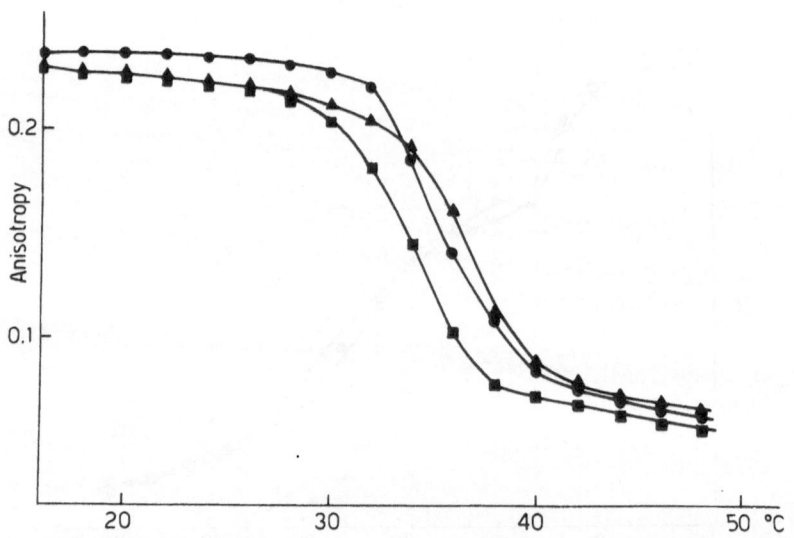

Fig. 3.   DPH fluorescence anisotropy plots vs. temperature of (●) DMPC-DPPC equimolar mixture; (■) DMPC-DPPC equimolar mixture with 5 mol% arachidonic acid; (▲) DMPC-DPPC equimolar mixture with 5 mol% palmitic acid.

produces opposite results on the fluorescence anisotropy of DPH (Figure 5) and of ANS (Figure 6). DPH fluorescence anisotropy values increase in the range of the phase transition as a function of calcium concentration, while ANS fluorescence anisotropy values decrease in the same range. At the molar ratio Ca^{2+}: phospholipids = 1 the ANS fluorescence anisotropy plot shows a T_c = 41.5°C that corresponds to the pure DPPC phase transition temperature. No relevant further decrease was noticed in the anisotropy plot at the molar ratio Ca^{2+}: phospholipids = 0.5. The same plots based on DPH fluorescence anisotropy show the increase of T_c with the increase of calcium concentrations (Figure 6). In the ANS labeled DDPC-DPPE mixture, however, the addition of different amounts of Ca^{2+} produces a small decrease of the ANS fluorescence intensity (Figure 7) without any shift in the emission spectrum.

From the plots of the fluorescence anisotropy versus temperature it is possible to derive the "fluid" to "solid" molar ratio within the studied temperature range. Considering the DPH equally partitioned in the solid and in the fluid lipid domains and considering the total fluorescence anisotropy as a sum of only two components:

$$\bar{A}_t = A_s\, x_s + \quad A_f\, x_f \qquad \text{with } x_s + x_f = 1$$

then

$$x_s = \frac{\bar{A}_t - A_f}{A_s - A_f} \qquad\qquad x_f = \frac{A_f - \bar{A}_t}{A_f - A_s} \qquad\qquad (1)$$

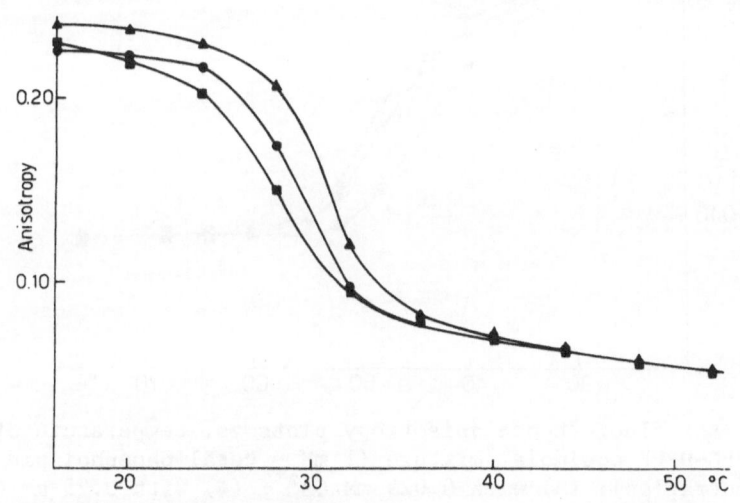

Fig. 4. DPH fluorescence anisotropy plots vs. temperature of (●) DMPC:DPPC = 7:3 (mol:mol); (■) with 5 mol% arachidonic acid; (▲) with 5 mol% palmitic acid.

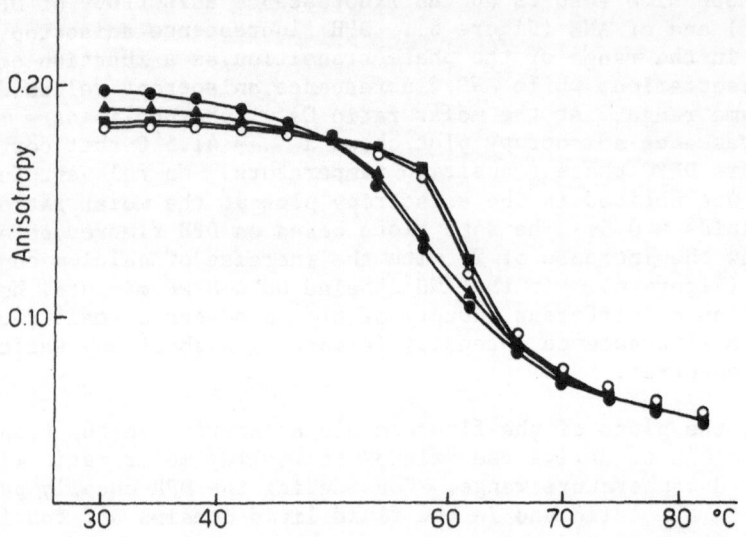

Fig. 5. DPH fluorescence anisotropy plots vs. temperature of (●)
DPPC-DPPE equimolar mixture (total phospholipid concentration = 1 mM); (▲) with 0.025 mM Ca^{2+}; (○) with 0.25 mM
Ca^{2+}; (■) with 0.5 mM Ca^{2+}.

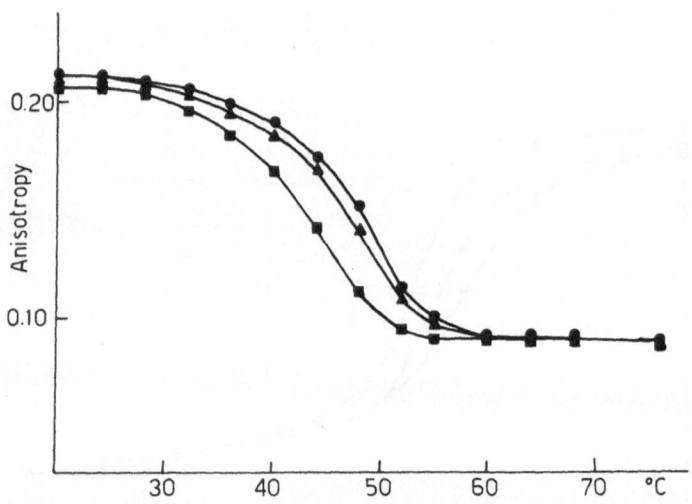

Fig. 6. ANS fluorescence anisotropy plots vs. temperature of (●)
DPPC-DPPE equimolar mixture (1 mM = total phospholipid concentration); (▲) with 0.025 mM Ca^{2+}; (■) with 0.25 mM Ca^{2+}.

with: $\bar{A}_t$ = experimental fluorescence anisotropy determined at given
temperature; A_s = fluorescence anisotropy of the "solid" phase; A_f =
fluorescence anisotropy of the "fluid" phase; x_s = "solid" molar

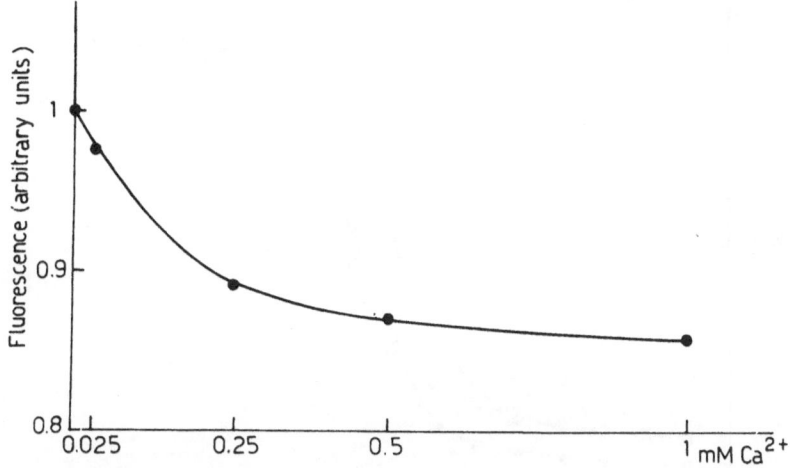

Fig. 7. ANS fluorescence intensity decrease as a function of Ca^{2+} concentration in the equimolar DPPC-DPPE mixture. 1 mM total phospholipid concentration.

fraction at given temperature; x_f = "fluid" molar fraction at given temperature. The small DPH quantum yield variation as a function of temperature it is not reported in the calculations. These calculations are carried out by two different procedures. Method 1: assuming for A_s and A_f the maximum and the minimum fluorescence anisotropy values, obtained at temperatures below and above the phase transition range. Method 2: evaluating A_s and A_f at each temperature within the phase transition range by linear extrapolations of the fluorescence anisotropy values, below and above the phase transition, respectively. Both the obtained results are reported in Figure 8 with the mixture DMPC:DPPC = 1:1.

DISCUSSION

It has been previously inferred[2], using the solution theory, that fatty acids preferentially partition in "fluid" or "solid" lipid domains of a pure phospholipid, namely DMPC, depending on their chemical characteristics. In the present work we show that the study of the whole temperature range of the phase transition of phospholipid mixtures can give a better knowledge of these systems. This is especially true when different phospholipids are present and they mix with the whole phase transition range, and therefore the increase or the decrease of the probe fluorescence anisotropy cannot be definitely attributed to a domain separation between the different phospholipids. However the previously reported interpretation[2] based on the solution theory can be applied to the fatty acids partition between the coexisting phases.

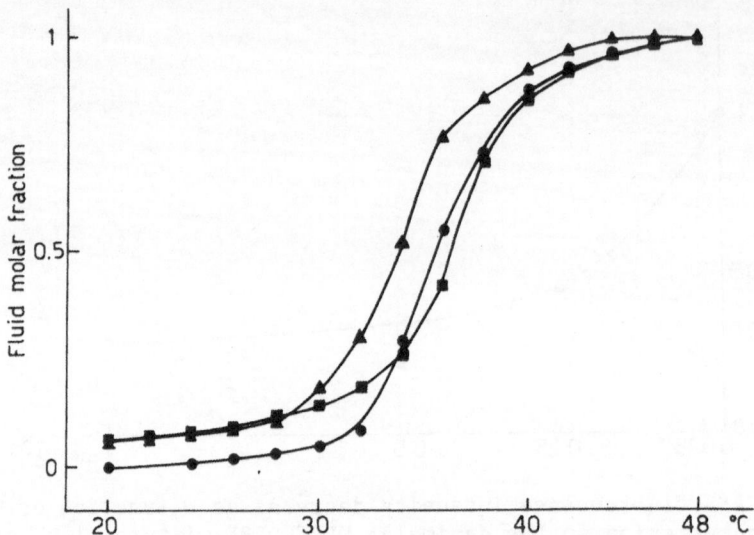

Fig. 8A. Fluid molar fraction vs. temperature calculated by
equation (1) of DPH fluorescence anisotropy plots of DMPC-
DPPC equimolar mixture (●); with 5 mol% arachidonic acid
(▲); with 5 mol% palmitic acid (■). Calculations were done
as sub method 1: (see Results).

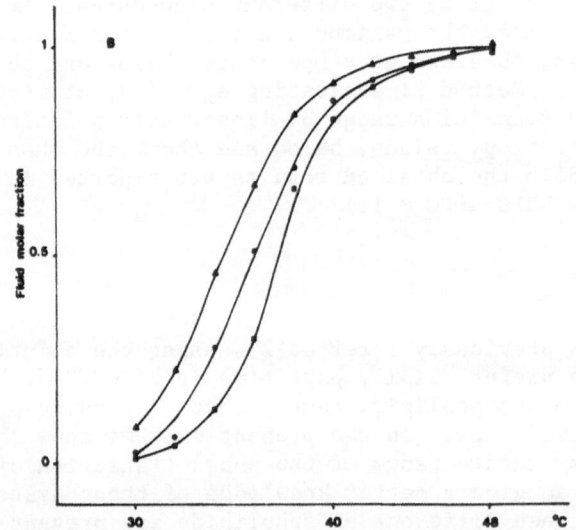

Fig. 8B. Fluid molar fractions vs. temperature calculated by
equation (1) of DPH fluorescence anisotropy plots of DMPC-
DPPC equimolar mixtures (●); with 5 mol% arachidonic acid
(▲); with 5 mol% palmitic acid (■). Calculations were done
as sub method 2: (see Results).

258

Generally, two kinds of behavior are possible: binary mixtures where two separate phase transitions occur and binary mixtures where only one intermediate phase transition is present. In the first case are included mixtures of phospholipids differing from at least four carbon atoms: in this case the addition of fatty acids can give an immediate answer due to the modification of only one of the pre-existing phase transitions. This is the case of DLPC-DPPC equimolar mixture that shows the first phase transition between 0° and 6°C and the second between 18° and 30°C. The results agree with those obtained by DSC (present volume, H. De Gier et al.) for the same phospholipid mixture. The possible explanation of this behavior is then the coexistence of two separate phases, one richer in DLPC that is the lower melting, the other with a higher DPPC concentration that is the higher melting. This is confirmed by the effects of the addition of the two fatty acids. Thus, according to this hypothesis, AA partitions preferentially in the "fluid" domain and its addition flattens and broadens the low temperature phase transition. On the contrary PA addition affects the high temperature phase transition, according to its supposed preferential partition in the "solid" domain.

As for the second kind of behavior, this includes phospholipids which mix in the "solid" phase as well as in the "fluid" one. Only one phase transition occurs at an intermediate temperature between those of the pure components. In this case the addition of AA shifts the transition plot to lower temperatures, whereas the addition of PA shifts the transition to higher temperatures.

Consequently the fluorescence anisotropy variation cannot be related to a domain segregation between the different phospholipids but only to the phase separation in the transition interval.

By comparison of the anisotropy behavior of DPH and ANS in the DPPE-DPPC mixture it is possible to show a domain formation also by means of Ca^{2+} ions. The increase of DPH fluorescence anisotropy due to Ca^{2+} addition in this DPPE-DPPC mixture can be explained by the general decreased mobility of the acyl chains due to Ca^{2+} binding to the polar head-groups of phospholipids. The decrease of the ANS fluorescence anisotropy in the transition range following the Ca^{2+} addition can be related to a migration of the ANS to the DPPC region, which is more fluid than DPPE migration due to the binding of Ca^{2+} to DPPE. It is known, in fact, that ANS locates itself in a pre-existing pocket at the hydrophobichydrophylic interface of the bilayer[3,4]. In the absence of Ca^{2+} ions the DPPC-DPPE transition plot obtained by ANS fluorescence anisotropy shows $T_c = 48.2°C$ (Figure 6) that is an intermediate value between the phase transitions of the two phospholipids. However, this temperature is closer to the transition temperature of DPPC indicating, as previously reported[4], a preferential partition of ANS in DPPC domains, which were also present before the Ca^{2+} addition. Increasing the Ca^{2+}

concentration in the mixture originates a progressive shift to lower values of the phase transition temperature. At the molar ratio Ca^{2+}: DPPE = 2 the transition temperature of the mixture coincides with that of the pure DPPC (Figure 6).

These data are consistent with the hypothesis of a preferential partition of ANS in the domain which is richer in DPPC; at the highest studied Ca^{2+} concentrations the ANS is only present in the DPPC domain. The Ca^{2+} addition causes the progressive exclusion of ANS from the DPPE richer domains, which is consistent with the presence of a binding between Ca^{2+} and DPPE. Another support to this explanation is given by the small decrease of ANS fluorescence intensity as a function of the Ca^{2+} concentration (Figure 7) indicating an ANS migration to DPPC richer domain more than to water.

As a conclusion the obtained results have shown:

1) the existence of a phase separation and of a domain formation, in the DLPC-DPPC mixture, in the studied temperature range;
2) the effect and the partition of fatty acids on different lipid phases can be determined using the solution theory[2] also in phospholipid mixtures; the segregation of different phospholipid domains can be univocally determined when two different transitions are present;
3) a phase separation and a domain formation is present in DPPC-DPPE mixtures;
4) a binding of Ca^{2+} ions with the DPPE component has been evidenced in DPPC-DPPE mixtures.

Such an interaction[5,6], is expected to occur only if negatively charged phospholipids are present.

REFERENCES

1. E. J. Shimshick and H. M. McConnel, Biochemistry, 12:2351-2360 (1973).
2. R. D. Klausner, A. M. Kleinfeld, R. L. Hoover, and M. J. Karnovsky, J.Biol.Chem., 255(4):1286-1295 (1980).
3. D. H. Haynes and H. Staerk, J.Membrane Biol., 17:313-340 (1974).
4. J. Slavik, Biochim.Biophys.Acta, 694:1-25 (1982).
5. N. Düzgünes, J. Wilschut, R. Fraley, and D. Papahadjopoulos, Biochim.Biophys.Acta, 642:182-195 (1981).
6. K. Houg, N. Düzgünes, and D. Papahadjopoulos, Biophys.J., 37: 297-305 (1982).
7. A. Blume and T. Ackermann, FEBS Lett., 43(1):71-74 (1974).

MEMBRANE FLUIDITY AND LIPID COMPOSITION

H.K. Kimelberg

Div. of Neurosurgery, Depts. of Anatomy and Biochemistry
Albany Medical College and Dept. of Biology, State University of New York at Albany, Albany, New York U.S.A.

The fluidity of a fluid is the reciprocal of its viscosity
which in turn is a measure of the resistance of the fluid to movement within it [1,2]. The unit of viscosity is the Poise (P) and,
measured by the motion of various physical probes, membranes at ambient temperatures seem to be far more viscous or far less fluid
than ordinary liquids which are of the order of 10^{-2} poise or a
centipoise (cP). The viscosity of red blood cell membranes was
found to vary from 2.0 to 6.0 P with increasing cholesterol/phos-
pholipid ratios [3]. Fluid lecithin phospholipid membranes have viscosities around 1 P at temperatures of 30° [4], which is approximately the viscosity of light oil [5].

While the concept of viscosity or fluidity has a precise meaning when applied to homogeneous isotropic fluids, biological membranes are anisotropic and far from homogeneous. The real value
of the concept of fluidity as applied to membranes is not that of
a precise physicochemical concept but that changes in membrane
fluidity or viscosity can be conveniently measured and compared by
a number of physical methods, as described and discussed in other
chapters of this volume. In simplified model phospholipid membranes in which proteins can be reconstituted, changes in fluidity
often relate directly to functional changes in these proteins. In
more complex natural membranes such direct correlations may or may
not occur, but some effect on membrane function is usually seen.

The molecular basis of membrane fluidity is thought to be predominantly a property of membrane lipids because changes in these
constituents change membrane fluidity as measured by physical techniques in a predictable way. However, changes in membrane lipids
are only the most obvious way in which membrane fluidity can be

altered and the complexities of protein-lipid interactions may well produce more subtle but biologically significant effects. In this chapter I will discuss some basic concepts that appear to pertain to membrane fluidity and how such fluidity can be altered by manipulation of phospholipid composition, as well as the effects of addition of divalent cations and proteins. In the following chapter I will describe some selected examples of the functional consequences of such changes drawn from studies on membrane enzymes, transmembrane ion transport and membrane receptors for hormones.

Membrane Fluidity and Viscosity

Lands [7] following the treatment of Hildebrand [1,2] has recently postulated a unifying physicochemical basis for membrane fluidity. He considers this to be the density or mass per unit volume (m/V). A more dense environment results in more energy-absorbing collisions per minute. Viscosity (η) is a quantitative index of the absorption of kinetic energy, whereas fluidity ($\emptyset$) represents a lack of that absorption and is the reciprocal of viscosity. The concept of fluidity being related to the density of a material by the absorption of kinetic energy was expressed by Hildebrand and Lamoreaux as;

$$\emptyset = B \frac{(V-Vo)}{(Vo)}$$

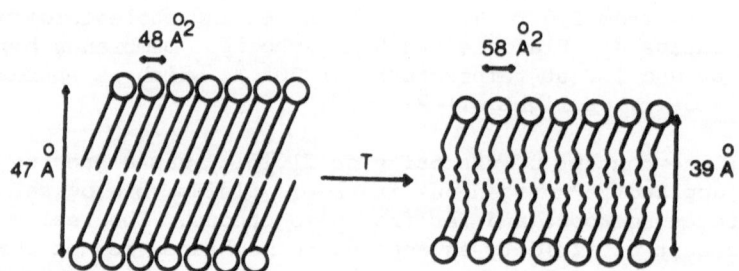

Fig. 1. Schematic representation of molecular orientation in phospholipid bilayers below (left hand side) and above the phase transition. The head-groups of the phospholipid molecules are represented by open circles and the two fatty acyl chains by the 2 solid lines. Tightly packed, solid acyl chains with inhibited motion are represented by straight lines. Mobile, liquid acyl chains are represented by wiggly lines. The dimensions shown relate to the data obtained with dipalmitoyl phosphatidylcholine bilayers. From ref. 8.

where Vo is the molal volume when there is no viscous flow and $\emptyset =$ O and can be obtained by plotting $\emptyset$ at different temperatures against V. When $\emptyset = $ O, V = Vo. B/Vo = slope. However, while the area per phospholipid molecule in a bilayer increases markedly at the transition temperature the effective width can also decrease due to increased motion of the fatty acyl chains, thus reducing the potential increase in molal volume. This is shown diagramatically in figure 1.

It would appear then that the phase transition in phospholipid bilayers may be mainly associated with an increase in molecular motion or decrease in order rather than large changes in molal volume. In some cases, however, the latter effect may predominate.

Phase Transitions of Phospholipids

As pointed out above a large increase in the motion of the fatty acyl chains occurs for membrane phospholipids marking the transition from a relatively solid to a far more fluid state, which may also be associated with an increased molal volume or decreased density. Such phase transitions are usually studied in aqueous suspensions of phospholipids in the form of liposomes consisting of spherical bilayers which form spontaneously when dried phospholipids are exposed to aqueous media [8].

Phospholipids are compounds in which two fatty acyl groups of usually 16-18, but also up to 26, carbons are esterified to the 1 and 2 positions of a glycerol molecule, with a phosphate group esterified to the 3 position. The head group specificity of the phospholipid is determined by the nature of the group (e.g. choline, ethanolamine or serine) which is esterified in turn to this phosphate group. The major phase transition of a phospholipid in the form of a liposome in excess water, represents the transition from the solid all-trans state of its fatty acyl group, to one with considerable trans-gauche rotational isomerism (see next section). This transition occurs within a characteristic temperature range. The enthalpy of this transition is, in part, the energy required for the total gauche isomerizations, the disruption of Van der Waals forces between neighboring alkyl chains and for the structure-breaking of ordered solvent around phospholipid head groups. This transition can be quite abrupt for some phospholipids suggesting a highly cooperative change between the two states. Thus, for dipalmitoylphosphatidylcholine, in which the 16 carbon palmitic acid is esterified to both the 1 and 2 position of the glycerol backbone and a choline group is esterified to the phosphate group, the mid-point (T_m) of the transition has been found to occur at values from 42° to 45°C, within a narrow temperature spread of < to several degrees (see Table 1).

It should be noted that most studies have been done on non-sonicated dispersions. For dipalmitoylphosphatidylcholine it has been reported, however, that sonication broadens the transition and lowers the T_m by 3-5°C [9].

Table 1. Phase Transition Data of Various Phospholipid Dispersions
          Measured by Differential Scanning Calorimetry

Phospholipid	T_1(°C)	T_m(°C)	T_2(°C)	ΔH (kcal/mole)	Ref.
Dimyristoyl (C14) PC	23	25.5	28	6.7	10
		23.7		6.2	11
Dipalmitoyl (C16) PC	41			8.7	12
	41.4	42.4	45	8.9	13
	41.8			9.7	11
Distearyl (C18) PC	58			10.7	12
		58.8		10.8	11
Dioleoyl (C18:1) PC	-22			7.6	12
Egg PC	-5				12
Dimyristoyl (C14) PE	48	51	55		10
Dimyristoyl (C14) PG	21.4	23.1	27		14
Dipalmitoyl (C16) PG	39.9	41.0	43.9	7.9	9
Distearyl (C18) PG	52	53.7	56		14
Dipalmitoyl (C16) PA	63	67	70		9

T_1, temperature at which slope of arm of main endothermic peak intercepts the base line (beginning of melt). T_2, corresponding point on high temperature side (end of melt). T_m is temperature of maximum peak height. The peak may or may not be symmetric (see diagram in Fig. 2). These values are either those quoted by the authors or obtained by inspection of their data. ΔH is the total enthalpy of the transition (major + pre-transition) and is equivalent to the area under the two peaks and is given if the value is quoted by the authors. All at pH 7.4 unless indicated otherwise. PC, phosphatidylcholine; PE, phosphatidylethanolamine; PG, phosphatidylglycerol; PA, phosphatidic acid.

264

Values for transition temperatures obtained from differential scanning calorimetry studies for a variety of phospholipids are shown in Table 1. It can be seen that alterations in the structure of phospholipid molecules alters the T_m and temperature range of these transitions. Thus, changing the choline of the head group in dimyristoly-phosphatidylcholine (DMPC) to dimyristoylphosphatidylethanolamine (DMPE) increases the T_m from 24-25°C to 51°C. Dipalmitoyl PC melts with a mid-point at 42.4°C, while dipalmitoyl-phosphatidic acid melts at 67°C at pH 6.5. These data indicate that specific head group interactions can markedly alter the melting behavior of the hydrocarbon chains. However, it should also be noted that the replacement of the PC with the PG head group has little effect on the phase transition (see Table 1). A more predictable determinant of the melting behavior of phospholipids are alterations in the fatty acyl chains themselves. With saturated chains the T_m shifts to higher temperatures with increasing chain length. Phospholipids in biological membranes, however, usually contain fatty acyl chains with one or more unsaturated <u>cis</u> C-C bonds. This has the extremely important effect of markedly lowering the T_m by putting a permanent kink in the chain and interfering with the regular all-<u>trans</u> packing of the chains since the bond angle between double-bonded carbon is 16° greater than that for single bonded carbons (109°28'). Figure 2A is a diagram of a typical differential scanning calorimeter heating curve for an endothermic type phospholipid transition, with the beginning, mid-point and end of the melt curve designated as T_1, T_m and T_2 respectively. Chapman and his co-workers (e.g. ref. 15) usually quote T_1 (beginning of melt) as the temperature of the transition, since they feel that the mid-point simply represents the maximal rate of change while the slope representing the completion of the change on the high temperature side is partly a reflection of instrument performance. Other authors either quote the mid-point, T_m or quote the entire temperature range occupied by the transition.

Naturally occurring phospholipids generally have one unsaturated and one saturated chain, with the unsaturated chain on the 2 position of the glycerol backbone. The melting behavior of phospholipid bilayers is affected when simple mixtures of phospholipids or phospholipids with different fatty acyl chains on the same glycerol backbone (intramolecular mixing) are measured. This is analogous to the situation actually found in biological membranes which consist of different phospholipids with heterogeneous fatty acyl chains. With intramolecular mixtures, sharp transitions intermediate between those that would be found for phospholipids consisting of either fatty acid alone are seen. When phospholipids with different head groups are mixed, an intermediate behavior is found, mixtures of DMPC and DMPE showing very broad intermediate transitions [15,16]. Thus, biological membranes may consist of both fluid and solid patches of lipid bilayers. However, the influence of other major components of biological membranes, such as cholesterol and proteins further complicates the type of states that will occur. Such states can be measured by a variety of physical techniques. In addition to differential scanning calorimetry which only measures the transition itself, there are a number of other techniques which

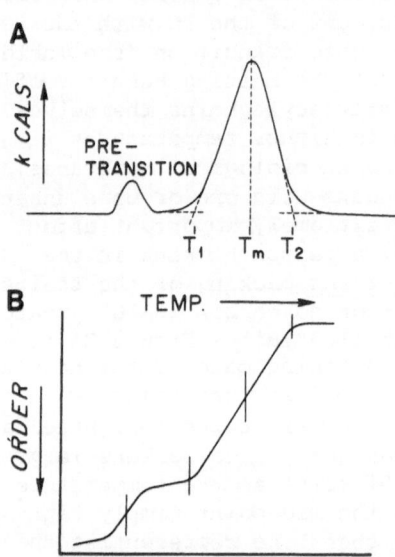

Fig. 2(A). A differential scanning calorimeter heating curve of a phospholipid liposome suspension. The absorption of heat is indicated in the upward direction and represents calories/unit time. The pretransition sometimes occurring on the low temperature side of the main endothermic transition is shown. The positions of the usually quoted temperatures of the main transition are also indicated. T_m represents the mid-point of the transition and T_1 and T_2 the lower and upper limits respectively. (B) Shows the comparable behavior of a fluorescence or spin-labelled probe in a phospholipid liposome. The degree or order is arbitrarily represented on the ordinate, increasing with decreasing temperature. It can be seen that the features of the main endothermic melt and the pretransition are easily detectable as sharp changes in the slope of the curve or as mid-points of the sharp rate of change in the order parameters between such changes in slope. From ref. 17.

can quantitatively measure the degree of fluidity of the membrane. Such techniques include: electron spin resonance and nuclear magnetic resonance, X-ray diffraction, fluorescence polarization, light scattering, dilatometry, and viscometry, many of which are dealt with in this volume. These techniques measure the overall bulk fluidity or the fluidity in the environment of a probe and may thus be limited in accurately analyzing the heterogeneity likely to be found in biological membranes. These techniques, however, when coupled with work on simple model systems do at least allow one to predict the basic type of behavior to be expected and to set certain limits for experiments with natural membranes.

As mentioned above, the presence of unsaturated bonds will lower the phospholipid transition temperature markedly. A single cis double bond disturbs the all-trans packing, and the solid to fluid transition will occur at a much lower temperature. Thus, the T_m for dioleoyl PC (two 18 carbon fatty acyl chains with a single cis double bond) is -22°C, whereas for the corresponding phospholipid with no unsaturated bonds, distearyl-PC, the T_m is 59°C (see Table 1), both values having been measured by differential scanning calorimetry. It is of interest that for phosphatidylserine from beef brain a broad transition from 5-20°C has been found [14]. This phospholipid has been reported to consist of 49% stearic (18:0), 37% oleic (18:1) and about 4% each of (20:1), (20:4) and (22:6) fatty acids [18]. The first figure in the brackets represents the number of carbons in the fatty acid and the second after the colon refers to the number of C=C bonds. Egg phosphatidylcholine with 34% (16:0), 20% (19:1) and 11.2% (18:2) [18] shows a transition at -5°C (Table 1). Thus, naturally occurring phospholipids show low melting temperatures as well as broad endothermic melts, as predicted from studies using synthetic phospholipids.

In model bilayers electron spin resonance techniques have been useful in detecting variations in fluidity along the fatty acyl chain progressing from the polar head group towards the interior of the bilayer. Using N-oxyloxazolidine labelled fatty acids as ESR probes [19], a profile of increasing motion or fluidity of the hydrocarbon chain has been detected from the polar head group region towards the terminal methylene group. It has been suggested that the gradual increase in fluidity detected by the spin label is in fact a result of the large ESR probe perturbing the packing of the bilayer and reporting an average fluidity. Deuterium magnetic resonance studies have found that in decanoic acid there is a constant order parameter until the last 3 carbons, where there was an abrupt increase in fluidity [20]. A spin label probe in the same molecule showed a gradual non-discontinuous increase in fluidity. A similar profile was also seen for naturally occurring egg lecithin [21] and, in addition, some motion of the quaternary nitrogen of the choline group was also detected. Other chapters in this book should be consulted for more recent references on this subject.

It has been mentioned above that mixtures of phospholipids with widely different T_m values can undergo two separate solid to fluid melts thus producing a bilayer consisting of separate solid and fluid areas. With mixtures of different head groups or binary mixtures of the same phospholipid whose saturated chains only differ by 2 carbon atoms, broad intermediate transitions indicating varying proportions of solid and fluid states producing an overall varying intermediate fluidity are seen. Both these phenomena have been referred to as phase separations [22]. Such broad transitions are also seen by differential scanning calorimetry for phospholipids with homogeneous saturated fatty acyl chains but possessing head groups with a net negative charge (see Table 1), and for natural phospholipids with heterogeneous fatty acyl chains and charged or neutral head groups [9,12,14,16].

Physicochemical Basis of Lipid Fluidity

Fluidity in phospholipid bilayers is generally considered to be due to rotational isomerization around the single C-C bonds of the fatty acyl chains [19,20,23]. Thus, a <u>trans</u> conformation can be converted to a <u>gauche</u> conformation by rotating a C-C bond by 120°. The continuous and rapid occurrence of these motions, the life time of any one conformation may be around 10^{-6} sec [6,20], will result in overall motion of the chain and a fluid condition or disordered state of the lipids. Below the phase transition the fatty acyl chains are in the all-<u>trans</u> conformation and are thus tightly and regularly packed in a crystalline array.

Effects of Cholesterol, Divalent Cations and pH on Lipid Fluidity

Apart from cold-blooded animals or homeotherms which hibernate, changes in temperature over and within the ranges indicated in Table 1 simply do not occur in homeotherms. Also, such temperature changes usually result in adaptive changes in lipid composition which are long-term. More short-term physiological effects will require more rapid isothermal changes in membrane fluidity. Changes in membrane cholesterol, divalent cations or local pH have the potential of effecting relatively rapid changes.

The effects of cholesterol on the properties of model membranes have attracted considerable interest in recent years and have been reviewed extensively (e.g. 10,24,25). In brief, cholesterol consists of a rigid sterol ring region approximately 11 Å long and about 25 Å^2 cross-sectional area, with a 3β-OH group at one end. Attached to the other end is an approximately 8 Å long 8 carbon hydrocarbon chain with about half the cross-sectional area of the sterol ring [26]. This molecule inserts into phospholipid bilayers with the -OH group in the region of the polar head groups and the remainder of the molecule parallel to the phospholipids [27,28]. The effect of cholesterol is to condense the average area per molecule and reduce the motion of the hydrocarbon chains of fluid phospholipid bilayers. On the other hand, by disrupting the all-trans crystalline packing of solid phospholipid bilayers it increases the fluidity of solid bilayers. This effect results in what has been termed an "intermediate fluid state" [10,29]. As seen by differential scanning calorimetry increasing amounts of cholesterol produce a progressive decrease in the height and some broadening of the transition peak, resulting in no effect on the T_m but a gradual decrease in the ΔH of the transition and a decreasing T_1 and increasing T_2 (see Fig. 3). Finally, a state of intermediate fluidity is obtained at 30 to 40 mole per cent cholesterol, at which point no transition can be detected by differential scanning calorimetry. Associated with the condensation effect on fluid phospholipid bilayers is a small increase in thickness, which in egg lecithin multilayers is from 39 Å to 42 Å.

Divalent cations, such as Ca^{2+} or Mg^{2+} are, like cholesterol, ubiquitous components of biological systems, and are likely to undergo rapid and reversible changes in their binding to membranes and membrane-associated macromolecules. Possible localized changes in pH at the membrane interface might also fit into this category. It is therefore of considerable interest that both divalent cations and pH can have significant effects on the fluidity of phospholipid bilayers. Ca^{2+} has been reported to increase the cation permeability of liposomes made from the negatively charged phospholipid, phosphatidylserine [30]. It has been found that Ca^{2+} added to one side of a membrane destabilizes it, thus explaining the increase in membrane permeability in liposomes, whereas Ca^{2+} added to both sides of the bilayer produces a highly stable membrane [30,31]. Consistent with this effect it has been found that Ca^{2+} concentrations > 1mM abolish the phase transition of phosphatidylserine and dipalmitoyl-phosphatidylglycerol [9] due to the formation, as identified by X-ray difrraction studies, of solid fatty acyl chains. Ca^{2+} at lower concentrations of 0.5mM, and Mg^{2+} at concentrations of 5-10 mM cause an increase in the T_m of 10-20°C with negatively charged phospholipids such as dipalmitoyl-phosphatidylglycerol and brain phosphatidylserine (fig. 4). Mg^{2+} is only effective at 10 x higher

269

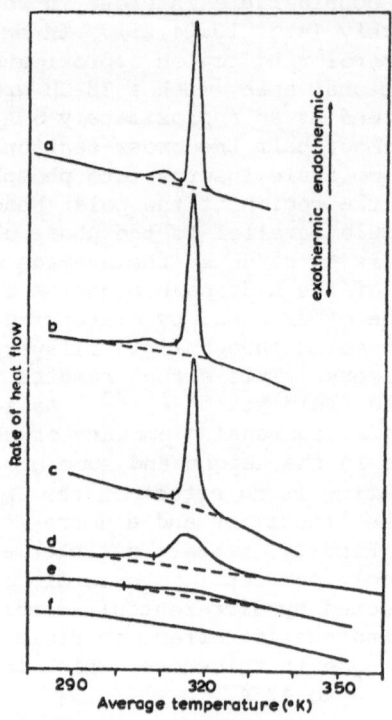

Fig. 3. Differential scanning calorimetry curves of
50% weight % dispersion in water of 1,2 dipalmitoyl-
L-lecithin-cholesterol mixtures of a) 0.0 moles %, b)
5.0 mole %, c) 12.5%, d) 20.0 mole %, e) 32.0 mole %
and f) 50.0 mole % cholesterol (from ref. 29).

concentration than Ca^{2+} and appears to form less solid membranes.
These effects of Ca^{2+} and Mg^{2+} have the fundamentally important
implication of enabling phase transitions or fluidity changes to
occur isothermally. Another aspect of potentially equal biological importance is the phenomenon of divalent cation-induced lateral
phase separation of acidic lipids when present in a mixture with
neutral lipids. Thus, 10 mM Ca^{2+} can induce the formation of
solid patches of Ca^{2+}-phosphatidylserine complexes in a mixture of
fluid phosphatidylserine-phosphatidylcholine mixtures as determined by ESR measurements [33] or differential scanning calorimetry [34].
These effects were not produced by Mg^{2+}. In addition, Ca^{2+} and to

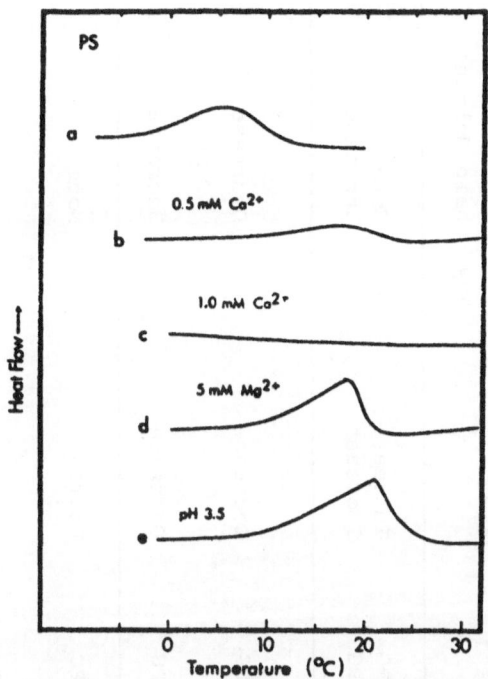

Fig. 4. Differential scanning calorimetry of non-
sonicated dispersions of beef brain phosphatidylserine in 100 mM NaCl at pH 7.4 under the different
conditions indicated (from ref. 9).

a lesser extent Mg^{2+} induces a similar lateral separation in egg
phosphatidylcholine-phosphaticid acid mixtures, but not with other
charged phospholipids such as phosphatidylethanolamine and dipalmitoylphosphatidylglycerol [35].

Consistent with divalent metal effects being partly mediated
through charge neutralization of the polar head groups, is the
finding that alteration in pH has a marked effect on the phase
transitions of charged phospholipids when it occurs in the ionization range of their charged groups [9,32]. Thus, the T_m of dipalmitoyl phosphatidic acid is altered from 67°C at pH 6.5 when the
free phosphate has only one negative charge to 58°C at pH 9.1 when
the secondary phosphate is completely ionized [9]. Fluorescence

Table 2. Effects of Several Proteins on Phase Transition of Liposomes as Measured by Differential Scanning Calorimetry

Type of Interaction	Protein	Lipid	(ΔH)	(T_m) Temperature of Acyl Chain Melting
1. Simple adsorption	Ribonuclease	DPPG	Increase	None
	Polylysine	DPPG	Increase	Increase
2. Adsorption + penetration	Cytochrome c	DPPG	Decrease	Decrease
	Myelin basic protein	DPPG	Decrease	Decrease
3. Penetration	Myelin	DPPC	Decrease	None
	Proteolipid gramidicin A	DPPC	Decrease	None

From ref. 13

DPPG, dipalmitoylphosphatidylglycerol; DPPC, dipalmitoylphosphatidylcholine.

272

polarization data also show that phase transitions can be induced
to occur isothermally by altering the pH [9],[32] (see fig. 4).

Effects of Proteins on Membrane Fluidity

The effect of proteins on the fluidity of membrane lipids is
of great importance, but is a very complex question which will be
dealt with in some detail in other chapters of this volume. Papahadjopoulos et al [13] using differential scanning calorimetry,
studied the effects of different classes of membrane proteins on
the phase transition of acyl chain melting in liposomes made from
different phospholipids. The results are summarized qualitatively
in Table 2. It is to be noted that the effects seen in groups 1
and 2 are for positively charged basic proteins and only occur with
phospholipids that have negative head groups. These are examples
of soluble or peripheral membrane proteins. The integral membrane
proteins in group 3 have a cholesterol-like effect. They lead to
a decrease in the area under the curve, indicating a decrease in the
enthalpy, but no change in the temperature of the transition. Similar data have been published for a Ca^{2+} ATPase-DPPC reconstituted
system. In addition, fluorescent polarization studies indicated
that the presence of proteins produced an "intermediate" fluid
state [36]. This could indicate that fewer phospholipids are parti-
cipating in the cooperative melting phenomenon due to immobilization of phospholipid acyl chains adjacent to the protein. This
effect is analogous to the boundary lipid concept for membrane proteins in which phospholipids adjacent to integral membrane proteins
are immobilized relative to phospholipids in the remaining part of
the membrane [37]. This concept has become somewhat controversial
since the layer of boundary lipids seems to disappear when measured
with certain techniques such as nuclear magnetic resonance (NMR),
but is detectable with electron spin resonance (ESR). At frequencies of around 10^6/sec NMR measurements will show a sharp spectrum
indicative of relatively free tumbling of the probe while ESR will
show a broad immobilized spectrum [6]. Thus, it appears that the
boundary lipids are immobile only on the ESR time scale of around
10^{-8} sec [6]. Thus, data on mobilities of membrane components should
not be interpreted on an all-or-none or qualitative basis, but
from a quantitative viewpoint. The activity of membrane proteins
can be a more sensitive indicator of the state of these boundary
lipids since the protein molecule will always be seeing its own
micro-environment of lipids with altered properties. Thus, phase
transitions, which will not be detected by physical averaging
techniques, might be detected by measuring the activity of such
proteins. This approach will be discussed in the following chapter.

REFERENCES

1. J.H. Hildebrand, Motions of molecules in liquids: viscosity and diffusivity, Science 174:490 (1971).
2. J.H. Hildebrand and R.H. Lamoreaux, Fluidity: a general theory, Proc. Nat. Acad. Sci. USA 69:3428 (1972).
3. R.A. Cooper, Influence of increased membrane cholesterol on membrane fluidity and cell function in human red blood cells, J. Supramolecular Structure 8:413 (1978).
4. U. Cogan, M. Shinitzky, G. Weber and T. Nishida, Microviscosity and order in the hydrocarbon region of phospholipid and phospholipid-cholesterol dispersions determined with fluorescent probes, Biochemistry 12:521 (1973).
5. D. Marsh, Spectroscopic studies of membrane stucture, in: Essays in Biochemistry, P.N. Campbell and W.N. Aldridge, eds. Academic Press, London, New York and San Francisco (1975).
6. M. Edidin, Molecular motions and membrane organization and function, in: Membrane Structure, J.B. Finean and R.H. Michell, eds., Elsevier/North Holland, Amsterdam, New York, Oxford (1981).
7. W.E.M. Lands, Fluidity of membrane lipids, in: Membrane Fluidity Biophysical and Cellular Regulation, M. Kates and A. Kuksis, eds., The Humana Press, Clifton, New Jersey (1980).
8. D. Papahadjopoulos and H.K. Kimelberg, Phospholipid vesicles (liposomes) as models for biological membranes: Their properties and interactions with cholesterol and proteins, in: Progress in Surface Science, S.G. Davison, ed., Pergamon, Oxford (1973).
9. K. Jacobson and D. Papahadjopoulos, Phase transitions and phase separations in phospholipid membranes induced by changes in temperature, pH, and concentration of bivalent cations. Biochemistry 14:142 (1975).
10. E. Oldfield and D. Chapman, Dynamics of lipids in membranes: Heterogeneity and the role of cholesterol, FEBS Lett. 23: 285 (1972).
11. H-J. Hinz and J.M. Sturtevant, Calorimetric studies of dilute aqueous suspensions of bilayers formed from synthetic L-α-lecithins, J. Biol. Chem. 247:6071 (1972).
12. B.D. Ladbrooke and D. Chapman, Thermal analysis of lipids, proteins and biological membranes. A review and summary of some recent studies. Chem. Phys. Lipids 3:304 (1969).
13. D. Papahadjopoulos, M. Moscarello, E.H. Eylar and T. Isac, Effects of proteins on thermotropic phase transitions of phospholipid membranes. Biochim. Biophys. Acta 401:317 (1975).
14. H.K. Kimelberg and D. Papahadjopoulos, Effects of phospholipid acyl chain fluidity, phase transitions and cholesterol on $(Na^+ + K^+)$-stimulated adenosine triphosphatase. J. Biol. Chem. 249:1071 (1974).

15. M.C. Phillips, D.B. Ladbrooke and D. Chapman, Molecular inter-
    actions in mixed lecithin systems, Biochim. Biophys. Acta
    196:35 (1970).
16. D. Chapman, J. Urbina and K.M. Keough, Biomembrane phase trans-
    itions. Studies of lipid-water systems using differential
    scanning calorimetry. J. Biol. Chem. 249:2512 (1974).
17. H.K. Kimelberg, The influence of membrane fluidity on the
    activity of membrane-bound enzymes, in: Cell Surface Re-
    views, G. Poste and G.L. Nicolson eds., ASP Biological and
    Medical Press (1977).
18. D. Papahadjopoulos and N. Miller, Phospholipid model membranes.
    I. Structural characteristics of hydrated liquid crystals.
    Biochim. Biophys. Acta 135:624 (1967).
19. W.L. Hubbell and H.M. McConnell, Molecular motion in spin-
    labeled phospholipids and membranes, J. Am. Chem. Soc. 93:
    314 (1971).
20. J. Seelig and W. Niederberger, Two pictures of a lipid bilayer.
    A comparison between deuterium label and spin-label exper-
    iments, Biochemistry 13:1585 (1974).
21. P.E. Godici and F.R. Landsberger, The dynamic structure of
    lipid membranes. A ^{13}C nuclear magnetic resonance study
    using spin labels. Biochemistry 13:362 (1974).
22. E.J. Shimshick and H.M. McConnell, Lateral phase separation in
    phospholipid membranes, Biochemistry 12:2351 (1973).
23. H. Träuble, The movement of molecules across lipid membranes:
    A molecular theory. J. Membrane Biol. 4:193 (1971).
24. M.C. Phillips, The physical state of phospholipids and chol-
    esterol in monolayers, bilayers and membranes, in: Progress
    in Surface and Membrane Science, J.F. Danielli, M.D. Rosen-
    berg and D.A. Cadenhead, eds., Academic Press, New York
    (1972).
25. M.K. Jain, Role of cholesterol in biomembranes and related
    systems, in: Current Topics in:Membranes and Transport,
    F. Bronner and A. Kleinzeller, eds., Academic Press, New
    York (1975).
26. J.E. Rothman and D.M. Engelman, Molecular mechanism for the
    interaction of phospholipid with cholesterol, Nature
    (London) New Biol. 237:42 (1972).
27. D. Chapman, Recent physical studies of phospholipids and
    natural membranes, in: Biological Membranes, D. Chapman,
    ed., Academic Press, London (1968).
28. R.P. Rand and V. Luzzati, X-Ray diffraction study in water of
    lipids extracted from human erythrocytes. The position of
    cholesterol in the lipid lamellae, Biophys. J. 8:125 (1968).
29. B.D. Ladbrooke, R.M. Williams and D. Chapman, Studies on leci-
    thin-cholesterol-water interactions by differential scan-
    ning calorimetry and X-ray diffraction. Biochim. Biophys.
    Acta 150:333 (1968).

30. D. Papahadjopoulos, W.J. Vail, W.A. Pangborn and G. Poste, Studies on membrane fusion. II. Induction of fusion in pure phospholipid membranes by Ca^{2+} and other divalent metals, Biochim. Biophys. Acta 448:265 (1976b).

31. D. Papahadjopoulos and S. Ohki, Stability of asymmetric phospholipid membranes. Science 164:1075 (1969).

32. H. Träuble and H. Eibl, Electrostatic effects on lipid phase transitions: Membrane structure and ionic environment, Proc. Nat. Acad. Sci. USA 71:214 (1972).

33. T. Ohnishi and H. Kawamura, Clustering of lecithin molecules in phosphatidylserine membranes induced by calcium ion binding to phosphatidylserine, Biochem. Biophys. Res. Commun. 51:132 (1973).

34. D. Papahadjopoulos, K. Jacobson, G. Poste and G. Shepherd, Effects of local anesthetics on membrane properties. I. Changes in the fluidity of phospholipid bilayers. Biochim. Biophys. Acta 394:504 (1975).

35. T. Ito and S.-I. Ohnishi, Ca^{2+}-induced lateral phase separations in phosphatidic acid-phosphatidylcholine membranes. Biochim. Biophys. Acta 352:29 (1974).

36. J.C. Gomez-Fernandez, F.M. Gini, D. Bach, C.J. Restall and D. Chapman, Biophysical studies of $(Ca^{2+} + Mg^{2+})$-ATPase reconstituted systems, Biochim. Biophys. Acta 598:502 (1980).

37. P. Jost, O.H. Griffith, R.A. Capaldi and G. Vanderkooi, Evidence for boundary lipid in membranes. Proc. Nat. Acad. Sci. USA 70:480 (1973).

MEMBRANE FLUIDITY AND MEMBRANE ACTIVITIES

H.K. Kimelberg

Div. of Neurosurgery, Depts. of Anatomy and Biochemistry
Albany Medical College and Dept. of Biology, State University of New York at Albany, Albany, New York U.S.A.

Observations that changes in membrane fluidity affect a wide variety of membrane and cellular activities [1-5] greatly stimulated interest in the study of membrane fluidity. It also suggested that such effects could be used to monitor such changes. In this chapter I will discuss selected examples of cases in which membrane fluidity seems to affect membrane enzymes, ion transport and hormone-receptor interactions in both reconstituted and natural membranes.

Arrhenius Plots and Enzyme Activity

The fluidity of both model and natural membranes can be altered by changing the temperature. The resulting changes can then be conveniently represented as Arrhenius plots [6]. The Arrhenius equation relates temperature and reaction velocity as follows:

$$\frac{d \ln k}{dT} = \frac{Ea}{RT^2} \tag{1}$$

where k is the rate constant, R the gas constant, T the absolute temperature in degrees Kelvin and Ea is known as the activation energy.

Equation (1) can be integrated to a more useful form for graphical representation:

$$\ln k = \frac{-Ea}{RT} + \text{constant} \tag{2}$$

Thus, if 1/T is plotted against the natural logarithm of the rate

277

constant, a straight line with a negative slope (-Ea/R) with respect to 1/T is usually obtained, since reaction rates generally increase with increasing temperature. The slope of this line thus gives the activation energy Ea, of the reaction. Ea is related to the enthalpy of the activated complex by:

$$Ea = \Delta H^* + RT \qquad\qquad\qquad (3)$$

where ΔH^* is the enthalpy of the activated state according to the Eyring rate theory. RT is only around 600 cal mole^{-1}.

For many membrane activities a plot of the logarithm of the reaction velocity against 1/T does not give a straight line but shows a sharp change in slope. The departures from linearity can take the form of actual breaks in the curve, or two or more linear lines with points of intersection as shown in figure 1 [3,6]. Since all these effects involve discontinuous changes in the curve they will all be referred to as "discontinuities". For enzymes the lower temperature lines in the majority of cases have larger slopes than the upper temperature lines. The two most likely reasons for a sharp, rather than a curvilinear change in slope are: (1) a sharp

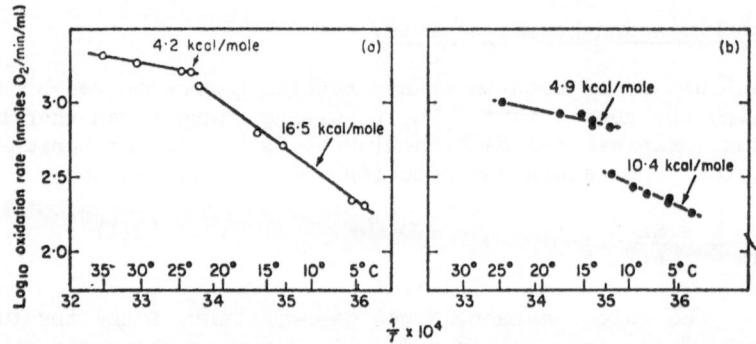

Fig. 1. Arrhenius plots of succinate oxidation by a) rat-liver mitochondria and b) chilling-sensitive cucumber fruit. Activation energies above and below the transition temperature (23 to 24°C for rat liver mitochondria, and 12°C for cucumber mitochondria) are as indicated on the graph. (From ref. 8).

phase change in the solvent (i.e. water or the membrane environment); or (2) a change from one conformation of the enzyme to another with differing activation energies (for a discussion on this point see ref. 7). It has been argued that at the point of intersection, where the reaction velocity is only marginally different, both phases are coexisting with about the same free energy of activation [6,8]. Since ΔH^* for the two phases, derived from the two slopes, is clearly different, a compensating change in the entropy of activation (ΔS^*) must occur to give the same ΔF^*. If this does not occur, two non-intersecting lines of different slopes with a gap between them indicating a significant change in ΔF^* should be obtained, which is in fact sometimes found [6,8] (see fig. 1). Thermodynamic calculations for both the Ca^{2+} ATPase from sarcoplasmic reticulum and reconstituted (Na+K) ATPase [9,10,11] show large changes in ΔS^* for differing values of ΔH^*. ΔH^* can be obtained simply from the measured Ea according to equation (3). F* can be obtained from the measured reaction rate (k) at any temperature from:

$$\Delta F^* = -RT \ln \frac{(k \cdot h)}{(k_B T)} \qquad (4)$$

where k_B is Bolzmann's constant, and h is Planck's constant.

ΔS^* is then simply calculated from the usual relationship

$$\Delta F^* = \Delta H^* - T \cdot \Delta S^*$$

The various values of these quantities for the Ca^{2+} ATPase and (Na+K) ATPase are shown in Table 1. It is clear that the decrease in Ea and therefore ΔH^* in the higher temperature range is compensated for by a decrease in ΔS^* thus keeping the free energy of activation (F*) relatively constant with changing temperature. A positive ΔS^* value indicates a gain of entropy by the activated complex, and implies that the activated complex is less ordered than the reactants. The values shown for ΔS^* in Table 1 are what one might expect if the transition temperature corresponds to a solid to fluid melt, since the ΔS^* values are greater (more positive) below the transition temperature suggesting that the reactants are more ordered than the activated complex, especially if the activated complex has a constant degree of order throughout the temperature range. From the data with (Na+K) ATPase reconstituted with phospholipids of differing fatty acyl groups it can be seen that for DMPG above its transition temperature values of both Ea and ΔS^* are about the same as DOPC which is fluid throughout the temperature range and shows a single Ea and ΔS^* value. Similar values are seen for PS above its upper Td value.

In most studies in which Arrhenius plots of enzyme activity have been used, the logarithm of the reaction velocity is usually

Table 1. Free Energy and Entropy of Activation for (Na⁺+K⁺) ATPase and Ca²⁺ ATPase Activity

	TRANSITION TEMP. (T_d)		Ea or ΔH^* (kcal mole⁻¹)	ΔS^* (cal deg⁻¹ mole⁻¹)	ΔG^* (kcal mole⁻¹)	Ref.
	Enzyme Activity (°C)	Lipid (°C)				
Ca²⁺ ATPase (in SR)	18	17 (ESR)	22 (t >T_d) / 27 (t <T_d)	16 / 35	17	9
Ca²⁺ ATPase - DOPC	--	--	26	29	17	
" - DPPC	29	29 (ESR)	23 (t >T_d) / 44 (t <T_d)	20 / 86	17 / 18	
Ca²⁺ ATPase (in SR)	20	20 (ESR)	17 (t >T_d) / 28 (t <T_d)	3 / 35	16 / 18	10
(Na⁺+K⁺) ATPase - DOPG	--	--	16	-19	22	11
" - DMPG	20	21–28 (DSC)	13 (t >T_d) / 29 (t <T_d)	-28 / 26	22 / 22	
" - PS	8,17	7–20 (DSC)	16 (t >17°) / 27 (t <17°)	-25 / 20	23 / 22	

(in SR) means a membrane preparation prepared from the original tissue. Other preparations are delipidated enzymes reconstituted with phospholipids. DPPC, dipalmitoyl phosphatidylcholine; DOPC, dioleoyl phosphatidylcholine; DOPG, dioleoylphosphatidylglycerol; DMPG, dimyristoylphosphatidylglycerol; PS, beef brain phosphatidylserine. ESR means using an electron spin resonance probe. DSC - differential scanning calorimetry. t > T_d or t < T_d values calculated at temperatures greater or less than the transition temperature respecitvely.

plotted against 1/T. It has been pointed out, however, that the measured velocity of an enzyme is usually the result of a number of different reaction rates, all of which may have a different temperature dependence [7]. The simplest procedure is to ensure that one is always measuring the maximal velocity of the enzyme reaction (Vmax). Under these conditions $V = k_{+2}e$ [7], so that one might assume that one was measuring the effect of temperature on the rate constant k_{+2} alone. k_{+2} is the rate at which the final ES complex breaks down to give product, and e is the total enzyme concentration. This would be true only if this remained the rate-limiting step throughout the temperature range studied which need not necessarily be the case. The Vmax should be accurately determined at each temperature value at infinite substrate concentration by Lineweaver-Burk plots, since one critical kinetic parameter, the substrate affinity of K_m, might vary markedly with temperature and the measured rate may no longer approximate Vmax [7]. Current practice has been, however, to determine initial rates at substrate levels which are saturating at the highest temperature measured. This is a reasonable operational technique as long as one is only asking the basic question of whether enzyme activity is affected by membrane fluidity.

As pointed out above the discontinuous changes in slopes in Arrhenius plots of enzyme activities or the physical properties of probes embedded within the membrane indicate a sharp change in both the activation energy and entropy of activation of the reaction. This could be due to a sharp change occurring in either the enzyme itself or in its environment. It is well-known that many soluble enzymes such as fumarase [12] and D-amino acid oxidase [13] also show discontinuities when their temperature dependence is plotted as Arrhenius plots. This has been explained in terms of two stable conformation states of the enzymes with different Ea values, which undergo sharp transition from one form to the other at the discontinuity temperature (T_d). Indeed, early studies which showed definite discontinuities in Arrhenius plots of the enzyme activity of membrane-bound enzymes such as the (Na+K) ATPase [14,15] were interpreted in terms of changes in protein conformation. Clearly, effects on enzyme activity basically involve changes in the active sites via local or general changes in the conformation of the protein molecule and the term "viscotropic" has been suggested as a descriptive term for these effects of membrane fluidity on enzyme activity [16].

Natural membranes often show broad phase transitions and when enzyme activity is measured over a sufficiently large range more than one discontinuity is seen [2,14]. We have also found this with delipidated (Na+K) ATPase reconstituted with beef brain phosphatidylserine (PS), as shown in figure 2. The Arrhenius plot of enzyme activity shows two transitions at 17 and 8°C giving three values for Ea of 15, 30 and > 100 kcal mole. Also shown in the

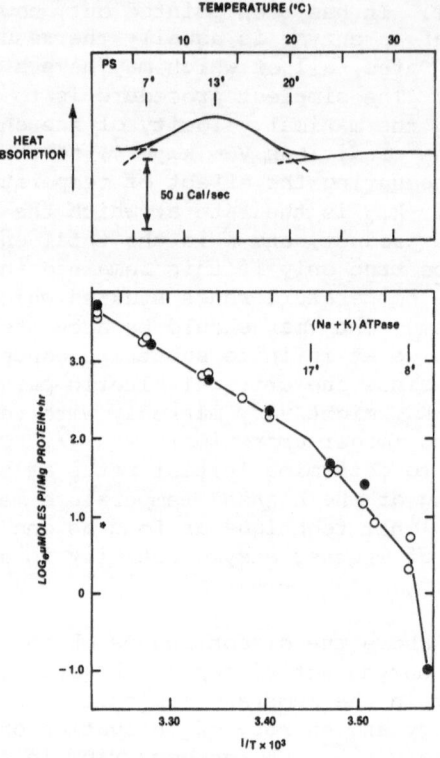

Fig. 2. <u>Upper figure</u>: DSC heating curve of non-sonicated beef brain phosphatidylserine (PS) liposomes. 1 µmole PS in 0.1M NaCl at pH 7.4. Sample volume 15 µl, scanning rate 2.5°/min. <u>Lower figure</u>: (Na+K) ATPase activities of rabbit kidney outer medulla microsomes delipidated with deoxycholate. Asterisk indicates activity with no added lipid at 37°C. Arrhenius plot of specific activity of 0.08 (●) or 0.13 (○) mg protein in presence of 0.5 µmol sonicated PS at pH 6.9 - 6.1. From ref. 17 and 33.

upper part of this figure is a DSC tracing of PS showing a broad transition between 7 and 20°C. Thus the two transitions seen for (Na+K) ATPase activity seem to correspond to the beginning and end of the phospholipid melt.

Hepatocyte plasma membranes from rat also show a broad transition as detected by DSC (summarized in Table 2) [18]. Two measured enzyme activities, however, showed discontinuities in Arrhenius plots corresponding only to the upper or lower limit of the transition. This was also found for other enzymes in these membranes by other workers as discussed in ref. 18. In another study from hamster hepatocyte membranes however, two transitions in Arrhenius plots were found for a variety of enzymes, although not for all [19]. These transitions were all in the range of 25-26 and 12-13°C. Transitions at these two temperatures were also found for the fluorescence of a membrane probe, 4 anilinonaphthalene-1-sulphonic acid (ANS) and were interpreted by Houslay and Palmer [19] as being the phase transitions of the phospholipids in the two halves of the membrane bilayer.

Table 2. Enzyme Activities and Membrane Transitions in Rat Hepatocyte Plasma Membranes

Measurement	Transition (°C)	
DSC	18 (lower)	31 (upper)
Fluorescence polarization	18	
5'nucleotidase	17	
Alkaline phosphate		26

Data from ref. 18

It should be noted at this point that although most membrane activities appear to depend on a fluid lipid membrane environment this is not always the case. A few enzyme activities are actually activated by a decrease in membrane fluidity. I found that the ouabain-insensitive component of delipidated (Na+K) ATPase was increased by addition of cholesterol while the ouabain-sensitive component of the (Na+K) ATPase was decreased [17]. This does not seem to be due to the decreased sensitivity of (Na+K) ATPase to ouabain in less fluid membranes [20], since the Mg ATPase activity of phospholipase treated rat liver plasma membranes measured in a 50 mM Tris.HCl, 10mM $MgCl_2$ buffer was increased by saturated phospholipids or when cholesterol was present, while fluid phospholipids inhibited this activity [21]. Basal adenylate cyclase activity also seems to be increased by decreased membrane fluidity [22].

Lateral Diffusion of Proteins

The above effects of membrane fluidity on the activity of membrane proteins are presumably mediated through effects of the motion of membrane lipids on conformational changes in membrane proteins required for their functioning [2,23]. As Edidin [23] points out these changes occur over time scales of 10^{-12} to 10^{-6} sec and distances of 10^{-9} to 10^{-6} cm and motions on this scale are probed directly in magnetic resonance experiments and indirectly when enzyme or receptor activity is changed by changes in membrane lipid composition and apparent viscosity [23]. Lateral diffusion of proteins is apparently affected to a greater extent than predicted by the effects of membrane viscosity on rotational movement of proteins and may occur in the range of 10^{-6} to 10^{-4} cm and 10^{-6} to 10^{-3} sec [23]. The work of Strittmatter and Rogers [24] using a mixture of NADH reductase and cytochrome b_5 reconstituted in dimyristoylphosphatidylcholine (DMPC) liposomes showed that the reduction of cytochrome b_5 by NADH reductase was markedly dependent on the fluidity of the membrane and a major sharp change in activity was found at 19-23°C and a smaller one at 10-12°, as shown in figure 3. Also, this effect was greatest at the lowest cyt b_5:NADH reductase ratios. Based on the necessity of collision between the two proteins for reduction of b_5 and that DMPC shows a main fatty acyl chain transition at 23-25° and a pretransition at 12-13°C these data suggest that this collision rate is affected by the fluidity of the lipid bilayers and this effect is most marked under conditions where the two enzymes are likely to be separated by the greatest distances.

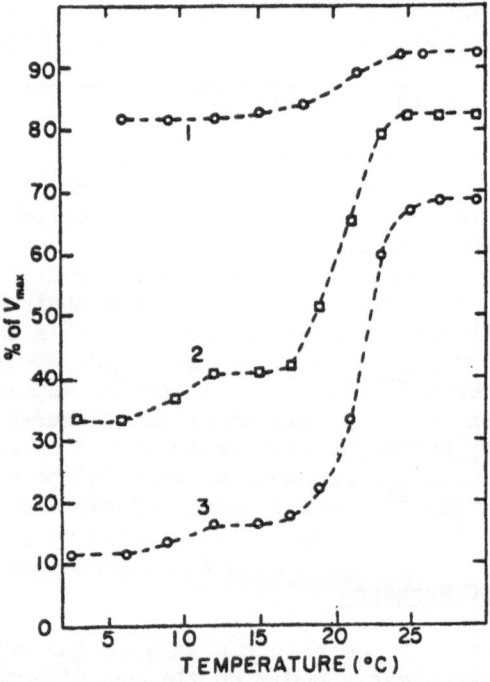

Fig. 3. Dependence of NADH: cytochrome b_5 reductase (measured by final transfer of electrons from cytochrome b_5 to soluble cytochrome c) in reconstituted DMPC liposomes, on temperature and cytochrome b_5 concentration. The cyt. b_5: NADH mole ratios for the different curves are approximately: 1) 50:1, 2) 5:1, 3) 2:1. Activity expressed as % of Vmax measured with NADH reductase: ferricyanide (from ref. 24).

The dependence of activity on translational diffusion due to collision between proteins may play an important role in the functioning of multienzyme complexes in natural membranes and the coupling of receptor molecules to effector proteins such as enzymes or transport proteins. The latter may have considerable significance in the functioning of receptor activated ion channels during synaptic transmission. Such effects have been observed for the activation of adenyl cyclase by adenosine or epinephrine in turkey erythrocytes [25]. In this case the rate constant for the activation of adenyl cyclase by 1-norepinephrine was markedly dependent on the fluidity of the membrane, altered by treating the cells with <u>cis</u>-vaccenic acid and measured directly by fluorescence polarization. In contrast activation of the adenyl cyclase by the adenosine-linked receptor was unaffected by fluidity (see fig. 4). It was important to measure the activation constant since adenyl cyclase activity itself will also be directly affected by membrane fluidity. The activation by the 1-norepinephrine receptor is thus an example of "collision coupling" while adenosine activation is an example of "permanent coupling".

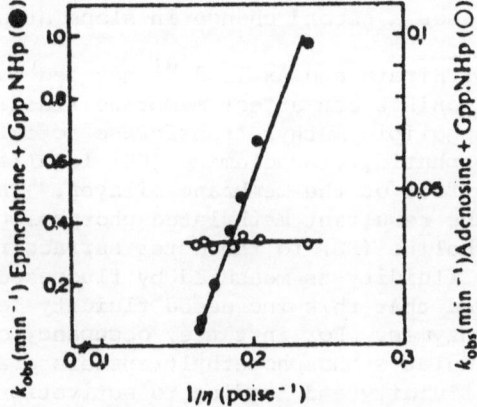

Fig. 4. Dependence on membrane fluidity of the rate constant of adenyl cyclase activation in turkey erythrocyte membranes by 1-norepinephrine or adenosine. First order processes of adenylate cyclase activation were measured as a function of membrane fluidity altered by <u>cis</u>-vaccenic acid treatment. $1/\eta$ measured by DPH fluorescence polarization. (●) 1-norepinephrine activation; (O) adenosine activation. From ref. 25.

285

Membrane Transport

Transport phenomena are usually markedly affected by membrane fluidity. In a classic study on transport of K^+ in lipid planar bilayers sharp jumps in conductance within the narrow temperature range causing a transition from a solid to a fluid membrane, similar to those seen in fig. 3 for NADH:cytochrome b_5 reductase, were found for the shuttle carriers valinomycin and nonactin [26]. In Arrhenius plots this behavior also gave a discontinuous jump in activity at the transition temperature for the solid to fluid melt. In contrast, the channel carrier gramicidin gave a temperature dependence for the conductance increase which showed an intersecting Arrhenius plot similar to the behavior of most membrane enzymes,as already discussed in this chapter. A similar temperature dependence with intersecting Arrhenius plots and greater Ea values below the T_d compared to Ea values at temperatures above the T_d, is found for most transport phenomena. For example, the anion exchange system [27] and glucose transport [28] in human red blood cells show discontinuities at 17° and 20°C respectively. A stearic acid spin label study of human red blood cells gave a discontinuity for the order parameter at 17°C, and alkaline phosphatase activity also showed a discontinuity at 17°-19°C[29] All these data suggest a generalized fluid to solid melt at 17°-20°C in the membranes of human red blood cells affecting both transport and enzyme proteins. Further examples of transport phenomena showing discontinuities in Arrhenius plots are from the giant neuron of the sea slug Aplysia and the lobster giant axon [30]. These data are shown in figure 5 and in all cases a discontinuous change in slope was found at 17-20°C and in some cases a second change in slope at around 7°C.

Recent work of Hirata and Axelrod [31] has led to the suggestion that a number of agonists can affect membrane fluidity by stimulating membrane phospholipid methyl transferase activity leading to methylation of phosphatidylethanolamine (PE) localized on the inner cytoplasmic surface of the membrane bilayer. This leads to translocation of the resultant methylated phospholipids and finally phosphatidylcholine (PC) to the outer surface resulting in increased membrane fluidity as measured by fluorescent polarization [32]. It is envisaged that this increased fluidity leads to activation of membrane enzymes. For instance, occupancy of the β-adrenergic receptor stimulates phosphomethyltransferase activity, increasing membrane fluidity and leading to activation of adenyl cyclase by increased lateral movement of the β-receptor to activate adenyl cyclase.

From all the selected examples mentioned it is clear that changes in membrane fluidity induced by changes in membrane phospholipids cause large changes in the functioning of membrane proteins. Correlations between the average membrane fluidity and activities are seen, especially in simple systems. There are now a very large

number of reports of such behavior for both model and biological
membranes. In some cases, however, correlations are not seen.
This implies either that the activity is responsive to a localized
but not the average membrane fluidity of the membrane [33,34] and
that the physical measurement is not probing the microenvironment
sensed by the protein, or the protein is not sensitive to its membrane environment. It does appear reasonable to conclude, however,
that generally changes in membrane fluidity play an important role
in the normal and abnormal functioning of cells.

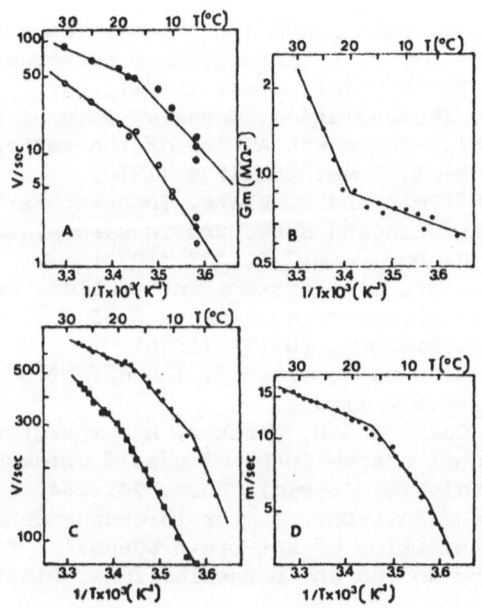

Fig. 5. A. Maximum rate of depolarization (●) and re
polarization (O) of action potential in Aplysia giant
neuron. B. Resting membrane conductance (Gm) of soma-
tic membrane of Aplysia giant neuron. C. Maximum rate
of depolarization (●) and repolarization (■) of action
potential of lobster giant axon. D. Speed of propagation of action potential in lobster giant axon. From
ref. 30.

REFERENCES

1. G. Lenaz, The role of lipids in the structure and function of membranes, in: Subcellular Biochemistry, D.B. Roodyn, ed., Plenum Press, New York and London (1979).
2. H.K. Kimelberg, The influence of membrane fluidity on the activity of membrane-bound enzymes, in: Cell Surface Reviews, G. Poste and G.L. Nicolson, eds., ASP Biological and Medical Press (1977).
3. R.B. Freedman, Membrane-bound enzymes, in: Membrane Structure, J.B. Finean and R.H. Michell, eds., Elsevier/North Holland, Amsterdam, New York and Oxford (1981).
4. Membrane Fluidity. Biophysical and Cellular Regulation, M. Kates and A. Kuksis, eds., The Humana Press, Clifton, N.J. (1980).
5. H. Sandermann, Jr., Regulation of membrane enzymes by lipids, Biochim. Biophys. Acta 515:209 (1978).
6. J.K. Raison, The influence of temperature-induced phase changes on the kinetics of respiratory and other membrane-associated enzyme systems, Bioenergetics 4:285 (1973).
7. M. Dixon and E.C. Webb, Enzymes, Academic Press, New York (1964).
8. J. Kumamoto, J.K. Raison and J.M. Lyons, Temperature "breaks" in Arrhenius plots: A thermodynamic consequence of a phase change, J. Theor. Biol. 31:47 (1971).
9. C. Hidalgo, N. Ikemoto and J. Gergely, Role of phospholipids in the calcium-dependent ATPase of the sarcoplasmic reticulum, J. Biol. Chem. 251:4224 (1976).
10. G. Inesi, M. Millman and S. Eletr, Temperature-induced transitions of function and structure in sarcoplasmic reticulum membranes, J. Mol. Biol. 81:483 (1973).
11. H.K. Kimelberg, Protein-liposome interactions and their relevance to the structure and function of cell membranes, Molec. Cell. Biochem. 10:171 (1976).
12. V. Massey, Studies on fumerase 3. The effect of temperature, Biochem. J. 53:72 (1953).
13. V. Massey, B. Curti and H. Ganther, A temperature-dependent conformational change in D-amino acid oxidase and its effect on catalysis, J. Biol. Chem. 241:2347 (1966).
14. N. Gruener and Y. Avi-Dor, Temperature-dependence of activation and inhibition of rat-brain adenosine triphosphatase activated by sodium and potassium ions, Biochem. J. 100:762 (1966).
15. K. Bowler and C.J. Duncan, The effect of temperature on the Mg^{2+}-dependent and Na^+-K^+ ATPase of a rat brain microsomal preparation, Comp. Biochem. Physiol. 24:1043 (1968).
16. H.K. Kimelberg and D. Papahadjopoulos, Phospholipid requirements for (Na^++K^+)-ATPase activity: Head-group specificity and fatty acid fluidity, Biochim. Biophys. Acta 282:277 (1972).

17. H.K. Kimelberg, Alterations in phospholipid-dependent (Na+K) ATPase activity due to lipid fluidity. Effects of cholesterol and Mg^{2+}, Biochim. Biophys. Acta 413:142 (1975).

18. C.J. Livingstone and D. Schachter, Lipid dynamics and lipid-protein interactions in rat hepatocyte plasma membranes, J. Biol. Chem. 255:10902 (1980).

19. M.D. Houslay and R.W. Palmer, Changes in the form of Arrhenius plots of the activity of glucagon-stimulated adenylate cyclase and other hamster liver plasma-membrane enzymes occurring on hibernation, Biochem. J. 184:909 (1978).

20. J.S. Charnock, Effect of lipid composition on activity of membrane bound enzymes in: Advances in Experimental Medicine and Biology, S. Wolf and A.K. Murray, eds. Plenum Press, New York and London (1981).

21. J.R. Riordan, Plasma membrane Mg^{2+} ATPase activity is inversely related to lipid fluidity in: Membrane Fluidity. Biophysical and Cellular Regulation, M. Kates and A. Kuksis, eds., The Humana Press, Clifton, New Jersey (1980).

22. M. Sinensky, K.P. Minneman and P.B. Molinoff, Increased membrane acyl chain ordering activates adenylate cyclase, J. Biol. Chem. 254:9135 (1979).

23. M. Edidin, Molecular motions and membrane organization and function in: Membrane Structure, J.B. Finean and R.H. Michell, eds., Elsevier/North Holland, Amsterdam, New York and Oxford (1981).

24. P. Strittmatter and M.J. Rogers, Apparent dependence of interactions between cytochrome b5 and cytochrome b5 reductase upon translational diffusion in dimyristoyl lecithin liposomes, Proc. Nat. Acad. Sci. USA 72:2658 (1975).

25. G. Rimon, E. Hanski, S. Braun and A. Levitzki, Mode of coupling between hormone receptors and adenylate cyclase elucidated by modulation of membrane fluidity, Nature 276: 394 (1978).

26. S. Krasne, G. Eisenman and G. Szabo, Freezing and melting of lipid bilayers and the mode of action of nonactin, valinomycin and gramicidin, Science 174:412 (1971).

27. A.L. Obaid and E.D. Crandall, HCO_3^-/Cl^- exchange across the human erythrocyte membrane: Effects of pH and temperature, J. Membrane Biol. 50:23 (1979).

28. L. Lacko, B. Wittke and P. Geck, The temperature dependence of the exchange transport of glucose in human erythrocytes, J. Cell. Physiol. 82:213 (1973).

29. C. Ziemann and G. Zimmer, Alkaline phosphatase in red cell membrane: Interconnection of activities and membrane lipid fluidity in: Membrane Fluidity. Biophysical and Cellular Regulation, M. Kates and A. Kuksis, eds., The Humana Press, Clifton, New Jersey (1980).

30. G. Romey, R. Chicheportiche and M. Lazdunski, Transition temperatures of the electrical activity of ion channels in the nerve membrane, Biochim. Biophys. Acta 602:610 (1980).

31. F. Hirata and J. Axelrod, Phospholipid methylation and biological signal transmission, Science 209:1082 (1980).

32. F. Hirata and J. Axelrod, Enzymatic methylation of phosphatidylethanolamine increases erythrocyte membrane fluidity, Nature 275:219 (1978).

33. H.K. Kimelberg and D. Papahadjopoulos, Effects of phospholipid acyl chain fluidity, phase transitions and cholesterol on $(Na^+ + K^+)$-stimulated adenosine triphosphatase, J. Biol. Chem. 249:1071 (1974).

34. J.C. Gomez-Fernandez, F.M. Gini, D. Bach, C.J. Restall and D. Chapman, Biophysical studies of $(Ca^{2+} + Mg^{2+})$-ATPase reconstituted systems, Biochim. Biophys. Acta 597:502 (1980).

MEMBRANE FLUORESCENCE ANISOTROPY

BEHAVIOR DURING CELL CYCLE

O. Sapora*, T. Parasassi**, L. M. Padovani*
and F. Conti**

* Laboratorio di Tossicologia Comparata, Istituto
  Superiore di Sanità, Viale Regina Elena 299, Roma, Italy
**Istituto di Chimica Fisica, Università degli Studi di
  Roma, Piazza Aldo Moro 5, Roma, Italy

Several studies have indicated that the cell surface could play an important role in the control of cell cycle, growth, differentiation and transformation[1,2]. Furthermore such biological events appear to be related to variations in the structure and chemical-physical behavior of the lipid-protein bilayer matrix. The characteristics of the bilayer are dependent on the composition as well as on the interactions between the different components[3,4,5].

One of the methods in use in such studies is the fluorescence depolarization of proper probes.

In the present paper we report preliminary results of our study on plasma membrane structural features during the cell cycle and cell differentiation of a proerythroblast cell line, measuring the steady-state fluorescence anisotropy of DPH.

Cellular System

K 562 cells grow in suspension in RPMI medium supplemented with 10% Fetal Calf Serum and antibiotics. This cell line was established by Lozzio and Lozzio[6] from a patient with chronic myeloid leukemia in the acute phase. These cells were thought to represent stem cell precursors of the myeloid lineage. Anderson[7], using specific antibodies and the benzidine reaction, has shown that these cells could be induced to synthesize hemoglobin by butyric acid or hemin treatment. Previous data suggest that different steps of differentiation can be identified by specific cell surface glycoproteins[8].

Synchronization

Cultured cells have an active cell replication cycle consisting
of pre- and post-DNA synthetic periods (G_1 and G_2), a DNA synthesis
period (S) and mitosis (M). The time period required to complete
this cycle varies with the cell type. At present there are several
methods available for the production and collection of synchronized
cultures[9]. A widely-used method to synchronize a large population
of cells is the use of a metabolic agent such as 2 mM thymidine (TdR)
to block the synthesis of DNA. However the inhibition of the cell
cycle progression, by the use of a relatively high concentration of
inhibitor, may change the ratio between different macromolecules in
the cell and induce unbalanced cell growth[10]. Such changes may
affect the mechanisms which control the content of different cell
components during the cycle and can be important in the evaluation of
experimental observations.

In order to mitigate this effect we chose a procedure suitable
for the particular cell line, taking the advantage of the fact that a
partially synchronized population requires a shorter period of inhibition of DNA synthesis to achieve maximal synchrony than do asyn-
chronous cultures[11]. Figure 1a shows the distribution of cell ages
over the cell cycle. A period of DNA synthesis inhibition equal to
the duration of the M and G_1 phases (14 hours) leads to a distribution like that shown in Figure 1b. If the block is then removed,
the cells traverse the S phase at the normal rate, and if the S phase
is less than one-half of the total cell cycle time, a distribution
such as that shown in Figure 1c, in which very few cells are found in
the S phase, will occur. At that moment, i.e. 24 hours after the
first block, the application of a second block for 10-12 hours will
collect the cells in a distribution similar to the one reported in
Figure 1d. As a reversible inhibitor of DNA synthesis we chose
thymidine at the concentration of 0.5 mM, far below the concentration
normally used. The results obtained using the described technique
are shown in Figure 2 where the cell number at the time zero, immediately after the inhibitor removal, was normalized. The time length
of S phase is measured incubating the synchronized cells, at different times after the removal of inhibitor, for 10 minutes in the
presence of tritium labelled thymidine (10 microCi/ml) and then the
radioactivity incorporated into the DNA is counted.

DPH Labeling

Cells collected at various times during the cycle were washed
and resuspended in Dublecco's PBS supplemented with 1% glucose, 1 mM
Ca^{++} and 1 μM DPH. The proper aliquot of 2 mM solution of DPH in
tetrahydrofuran was previously evaporated in PBS by nitrogen bub-
bling. The cell labeling was carried out for 45 minutes at room
temperature, with periodic stirring; during that time DPH fluor-

292

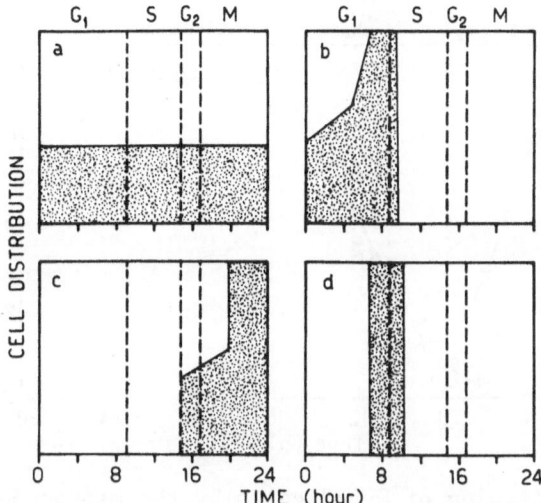

Fig. 1. Distribution of K 562 cells over the cell cycle during the
different phases of synchronization procedure. (a) Asyn-
chronous cell population, (b) cell distribution after the
first DNA synthesis inhibition, (c) cell distribution after
the reversal of inhibition, (d) cell distribution after the
application of the second period of inhibition.

escence intensity increases, as can be seen from Figure 3. Polarization measurements were done by the ratio mode with the emission
observed through Schott cut-off filters type KV cutting below 418 nm.
Excitation was 360±2 nm. Unlabeled cell scattering was always below
5% of the fluorescence. The temperature was controlled by a circulating water bath and the cells were maintained in suspension by
continuous low speed stirring.

The fluorescence anisotropy, r, was calculated from the polarization values according to: $r = 2P/3 - P$. We did not derive the
microviscosity values because of the large approximations needed for
its calculation. Furthermore the physical meaning of the fluorescence anisotropy is univocally defined, which this is not the case
of "η". From the constancy of anisotropy values during the incorporation of the probe into the cells we considered it segregated only in
the first encountered compartment, i.e. the plasma membrane. There
was little or no diffusion in other cellular compartments.

RESULTS AND DISCUSSION

Fluorescence anisotropy measurements on synchronized K562 cells
were carried out at 30°, 37° and 42°C. The averaged values, repro-

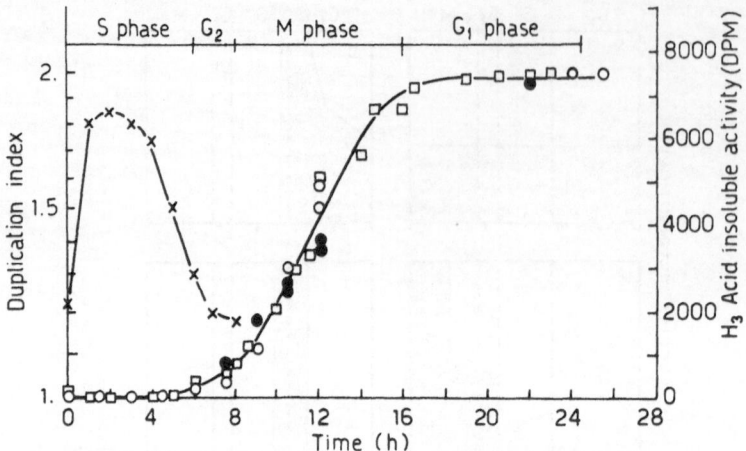

Fig. 2. Synchronization of K 562 cells by the method described in
the text. The initial cell number is normalized to 1. (x)
Acid insoluble radioactivity expressed in DPM. The symbols
(○, ●, □) represent data from different experiments.

ducible within 5%, from three independent synchronization experiments, are reported in Figure 4. A relative minimum value is present
during the G_2 and the early mitotic phase in all the temperatures
tested. During the G_1 phase a relative maximum appears followed by a
decrease to the initial value shown in S phase. With the increase of
the temperature a general decrease of the fluorescence anisotropy
values in all the cell cycle phases and a decrease in the fluctuations of the values, within the cycle, can be observed.

Similar measurements, but only at 37°C have been previously
carried out by other authors on a different cell line, the neuroblastoma cells[12]. A general disagreement with our results is
evident. First of all in our case the fluctuations of the fluorescence anisotropy within the cell cycle are smaller. Furthermore
and more relevant, in our experiments the behavior of DPH fluorescence anisotropy shows a minimum between G_2 and M phases while a
maximum value is reported by these authors just in the same phases.
We must point out that these results are presented in terms of
"microviscosity" while we prefer to report the corresponding fluorescence anisotropy values. This discrepancy can be due to the use of
a different cell line and to the different growth and synchronization
conditions. The K562 cells grow in suspension while neuroblastoma
cells grow attached to a plastic or glass surface. As a consequence
our measurements were always carried out in suspension while the
"microviscosity" measurements were done on a glass slide except for
the cells in mitosis which were labeled and measured in suspension.
Nevertheless many different data reported in the literature reasonably agree with our results.

294

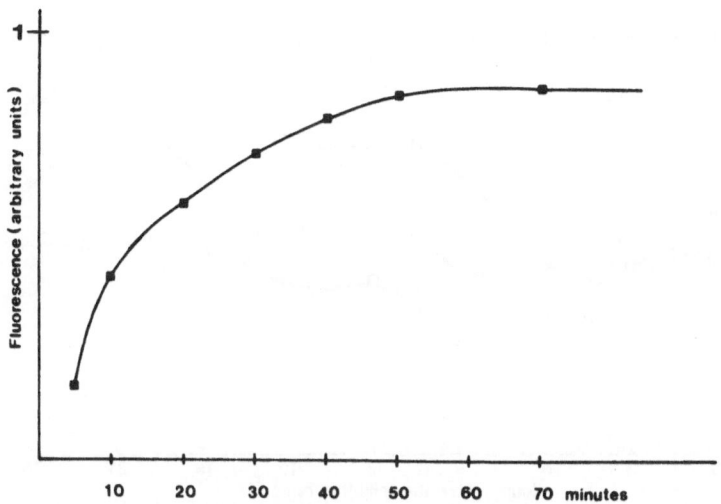

Fig. 3. Time course of DPH fluorescence intensity during incorporation in K 562 cells.

It is known that during the cell cycle a progressive disappearance of the cell cytoskeleton starting from the S phase to the M phase occurs, with a consequent possible decrease of the membrane rigidity[2]. Membranes from regenerating liver cells exhibited a significantly lower DPH fluorescence polarization than observed in quiescent liver[3], suggesting that cell cycling results in significant increases in membrane fluidity. Upon the transformation processes a similar lipid fluidity increase has been reported by several authors in different cellular systems[13]. Inbar and Shinitzky[14] attributed this to a lower cholesterol/phospholipid ratio in the transformed plasma membranes. On the contrary Johnson et al.[15] using a more precise cholesterol determination method have failed to show such a variation. Similar results were obtained by Maldonado and Blough using normal chicken embryo or Rous Sarcoma virus transformed cells[5]. The fact that the mitotic cells are more agglutinable by lectins than interphase cells, is in agreement with an increase of the bilayer mobility reflected on the decreased probe fluorescence polarization[16]. The activity of Na^+/K^+ ATPase, as well as other membrane enzymes, is a function of the bilayer mobility and it has its maximum just in G_2-M phases[17]. Moreover this enzyme could play a regulatory effect on the cell division.

In our conclusion, our preliminary results show clearly that fluorescence is a powerful tool for revealing some unknown and contrasting aspects of the plasma membrane modification during the cell cycle. However more experiments with different cell lines and different probes are needed.

295

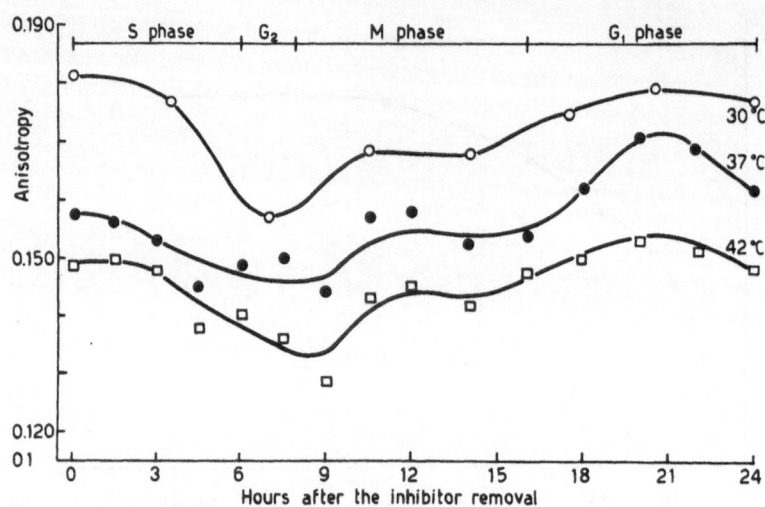

Fig. 4. DPH fluorescence anisotropy at 30°, 37°, 42°C during the cell cycle of K 562 cells.

REFERENCES

1. G. L. Nicolson, Biochim.Biophys.Acta, 457:57 (1976).
2. G. L. Nicolson, G. Poste, and T. H. Ji, in: "Cell Surface Reviews", G. Poste and G. L. Nicolson, eds., North Holland Publishing Co., Amsterdam, vol.3 (1977).
3. S. Cheng and D. Levy, Arch.Biochem.Biophys., 196:424 (1979).
4. S. W. De Laat, P. T. Van der Saag, E. L. Elson, and J. Schlessinger, Biochim.Biophys.Acta, 558:247 (1979).
5. R. L. Maldonado and H. A. Blough, Virology, 102:62 (1980).
6. C. B. Lozzio and B. B. Lozzio, Blood, 45:321 (1975).
7. L. C. Andersson, M. Jokinen, and C. G. Gahmbery, Nature (London), 278:364 (1979).
8. M. Fukuda, H. P. Koeffler, and J. Minowada, Proc.Nat.Acad.Sci. USA, 78:6299 (1981).
9. T. Ooka, D. N. Weatley, A. M. Badger, and S. R. Cooperband, Meth.Cell Biol., XIV, 287, 298, 314 (1976).
10. O. S. Frankfurt, J.Cell Physiol., 107:115 (1981).
11. W. G. Thilly, Meth.Cell Biol., XIV, 273 (1976).
12. S. W. De Laat, P. T. Van der Saag, and M. Shinitzky, Proc.Nat. Acad.Sci.USA, 74:4458 (1977).
13. M. Shinitzky and M. Inbar, J.Mol.Biol., 85:603 (1974).
14. M. Inbar, M. Shinitzsky, and L. Sachs, FEBS Lett., 38:268 (1974).
15. S. M. Johnson and R. Robinson, Biochim.Biophys.Acta, 558:282 (1979).
16. J. Garrido, Exp.Cell Res., 94:159 (1975).
17. C. L. Mummery, J. Booustra, P. Van der Saag, and S. W. De Laat, J.Cell Physiol., 107:1 (1981).

STABLE BIOMEMBRANE SURFACES FORMED BY PHOSPHOLIPID POLYMERS

Otto Albrecht, David S. Johnston, Carmen Villaverde and
Dennis Chapman

Department of Biochemistry and Chemistry
Royal Free Hospital School of Medicine
8 Hunter Street
London WC1N 1BP

INTRODUCTION

Many studies have now been made of the structure and dynamics
of biomembranes and it is generally accepted that a lipid bilayer
provides a matrix in or upon which proteins and glycoproteins are
located[1]. It is significant that within some cell membranes the
lipids are asymmetrically distributed. All the phosphatidylserine
and phosphatidylethanolamine of blood plasma erythrocyte[2] and platelet
cells[3] lie in the inner lamella of the bilayer and all the phosphatidylcholine and sphingomyelin in the outer layer[2,3]. It has been
suggested that this asymmetric phospholipid distribution may serve a
biological purpose by helping to maintain the delicate balance between
haemostasis and thrombosis. Recent results indicate that the outer
surface of blood cells is devoid of phospholipids which are active
in blood coagulation processes[4].

For a variety of purposes it would be useful to have stable
surfaces which contain these polar groupings. For example, studies
could be made of the blood coagulation process or the adsorption
characteristics of proteins (i.e. prothrombin) on surfaces of different ionic character.

We have recently shown that phospholipid molecules which contain
diacetylene groups in the acyl chains can be synthesized and that
these phospholipid molecules form cross-linked polymers upon irradiation with ultraviolet light, Fig. 1. The polymer chain which is
made up of conjugated multiple bonds, absorbs in the visible region
of the spectrum and the polymers are strongly coloured[5].

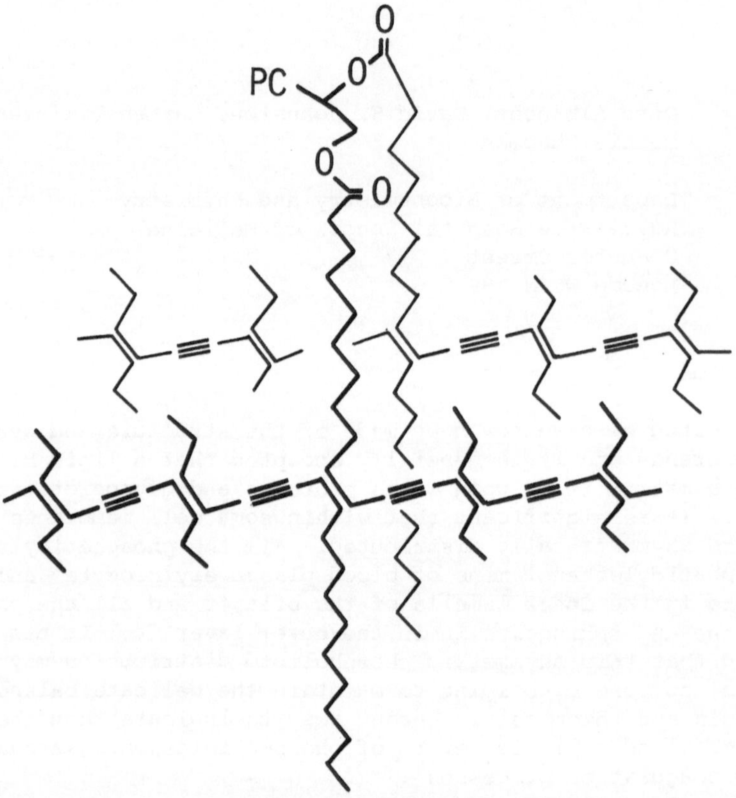

Fig. 1. Schematic diagram of conjugated phospholipid polymer.

Recently we have described the production of Langmuir-Blodgett type multilayers of these types of phospholipid and the characteristics of the layers after polymerisation. Our aim has been to produce stable layers with a hydrophilic outer surface, i.e. with phosphatidylcholine or sphingomyelin molecules to produce a surface which should be nearly identical to the exterior surface of erythrocyte and platelet biomembranes (i.e. of the lipid region).

Using the same technique with diacetylenic phosphatidylethanolamine and phosphatidylserine molecules, stable polymeric surfaces which are similar to the inner lipid surfaces of biomembranes may also be formed.

MATERIALS

The synthesis of phospholipids of general structure

$$CH_3(CH_2)_n-C \equiv C-C \equiv C-(CH_2)_8-\overset{\overset{\text{O}}{\|}}{C}-OCH_2$$

$$CH_3(CH_2)_n-C \equiv C-C \equiv C-(CH_2)_8-\overset{\overset{\text{O}}{\|}}{C}-OCH$$

$$(CH_3)_3\overset{+}{N}CH_2CH_2O-\overset{\overset{\text{O}}{\|}}{\underset{\underset{\text{O}^-}{|}}{P}}-OCH_2$$

has been reported earlier[5].

Standardised procedures were adopted for preparing the various substrates, glass, quartz, perspex, steel and teflon slides. The slide is immersed in a hot detergent solution (5% RBS 35) for an hour and a half and then thoroughly rinsed with pure water. Hydrophilic slides are dried with a hairdryer and coated as soon as possible. Occasionally, the surfaces of these slides have been smoothed with five layers of palmitic or arachidic acid before deposition of the phospholipid. Teflon could also be cleaned satisfactorily by wiping and rinsing with diethyl ether.

An electronically controlled film lift is used to pass the slides through the surface of the monolayer covered subphase. Up to 18 layers (9 down and 9 up trips) can be deposited completely automatically. The speed of upward and downward movements is independently adjustable and can be varied between 0.3 and 14 cm/min.

Phospholipid multilayers are polymerised by placing them in front of a high intensity ultraviolet light lamp (Mineralight R-52, Ultraviolet Products Inc., CA) for 30 s.

The surface characteristics of a multilayer depend on the orientation of molecules in the final layer. In the case of phospholipids, phosphatidylcholine groups uppermost will render the surface hydrophilic, with acyl chains uppermost, the surface will be hydrophobic. When the substrate is below the surface of the film, phospholipid molecules in the outermost layer will be orientated so that the phosphatidylcholine groups are in contact with water, i.e. the surface will be hydrophilic. When the substrate is raised either another layer will deposit or, if there is no film on the subphase, the top layer will peel off. In either case the new surface will be composed of phospholipid molecules with their acyl chains outermost and is expected to be hydrophobic. Attempts to obtain hydrophilic surfaces were made in two ways. In the first method, the multilayer was polymerised while it was still submerged after the deposition of the final layer. The remainder of the phospholipid film was stripped from the subphase before the slide was raised.

In the second method, the subphase is stripped of diacetylenic phospholipid and covered with a non-polymerisable amphiphile (DPPC, palmitic acid) before the slide was raised. After the slide is raised and irradiated, the non-polymeric material is removed by washing.

The hydrophobic or hydrophilic character of the coated surfaces is checked qualitatively by estimating the contact angle which water makes with the surface.

Deposition starts on hydrophobic surfaces on the first pass through the monolayer. It does not start on hydrophilic surfaces until the second pass, i.e. the first time the substrate is raised. In this respect, teflon behaved as a hydrophobic surface, glass, quartz and steel as hydrophilic surfaces. Unexpectedly, perspex also behaves as a hydrophilic surface, deposition not occurring on the first down trip. This holds for the deposition of fatty acids as well as phospholipids.

The conjugated triple bonds of the phospholipid absorb strongly in the ultraviolet region of the spectrum enabling a measure of the consistency of deposition on transparent substrates to be gained from plots of maximum absorbance versus number of layers. A linear relationship was found between these two properties when deposition did not occur on the down trip and measurements were made on the lower 2 cm of the slide.

MULTILAYER PROPERTIES

Diacetylenic phospholipid polymer in multilayers are pink in

300

colour although there needs to be about 6 layers on a slide before
colour is apparent to the eye. The layers become yellow when heated.
Such colour changes have been observed before with other polymeric
diacetylenes[6]. Fig. 2 shows the visible spectrum of a multilayer
at various irradiation times. There is an increase in absorption in
both visible and ultraviolet regions as polymerisation proceeds.

Phospholipid multilayers have strongly hydrophobic surfaces if
they are polymerised after withdrawal from the subphase. Such
surfaces emerge from water absolutely dry and a drop of water rolls
freely across them with a contact angle of almost 180°. However,
layers irradiated under the subphase or under a coating of non-
polymerisable amphiphile have surfaces which are hydrophilic. After
immersion in water, a water film adheres to them for several seconds.
A drop of water applied to an inclined surface rolls across it slowly
and the contact angle lies between 70 and 90°. While these surfaces
must be classified as hydrophilic when compared with the strongly
hydrophobic ones described previously, they are not totally wettable
like clean glass or metal oxides.

As anticipated, the mechanical and chemical stability of phospholipid films is markedly increased by the cross-linking process. Films
were intact after a day's immersion in strong acid or alkali. Ten
minutes treatment with a hot, strong detergent solution (5% RBS 35)
is needed before films start to peel off the substrate. However, it
was noticed that if strong pressure is applied to hydrophilic surfaces
they rapidly become hydrophobic.

CONCLUSION

We have shown that it is possible to coat many materials (glass,
quartz, perspex, teflon and steel) with ordered layers of diacetylene
containing phosphatidylcholine molecules and that upon irradiation
these molecules polymerise. Furthermore, layers can be produced in
such a way that the polar groups of the lipid form the outer coated
surface. The layers after polymerisation are quite stable in
aggressive media and can also, with some precautions, be handled
without damage. It is expected that the same technique can be used
to deposit other phospholipids and glycolipids so that particular
types of charged and zwitterion phospholipid polar groups form the
outer surfaces. We are presently synthesizing other types of
phospholipids such as phosphatidylethanolamine and phosphatidylserine
containing the diacetylene group. In this way, stable polymerised
surfaces consisting of the charged polar groups which make up the
inner surface of erythrocytes and platelets can be modelled. Stable
polymeric surfaces consisting of the carbohydrate groups of certain
cell membranes may be modelled by the biosynthesis of various
glycolipid molecules containing these diacetylene groups.

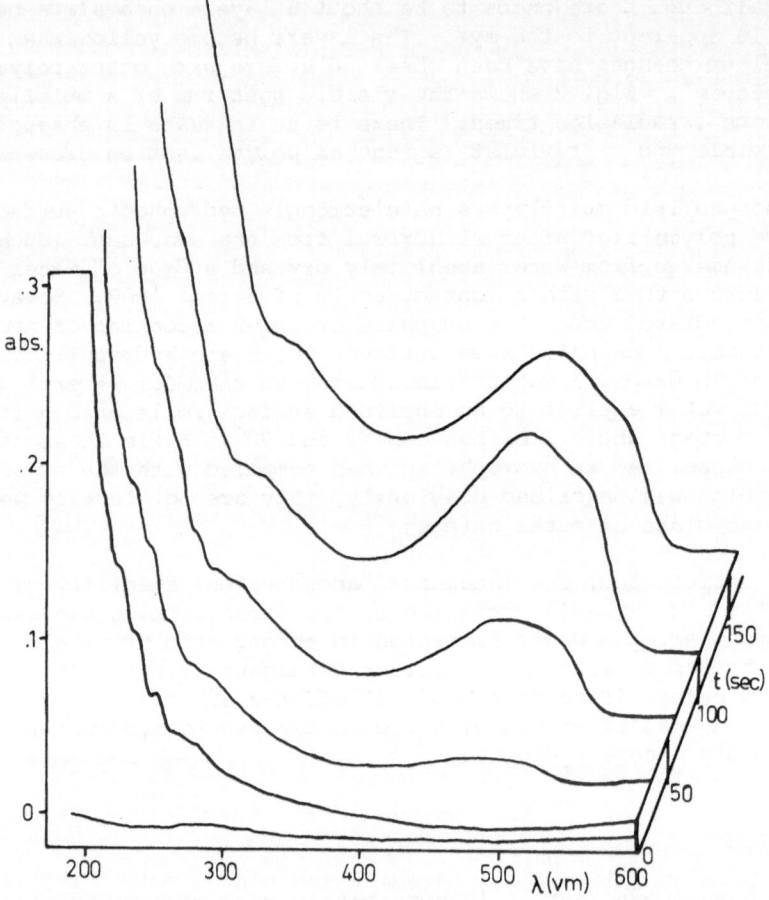

Fig. 2.   Visible spectra of 86 layers of diacetylenic phosphatidyl-
          choline at various irradiation times.   (43 layers on each
          side of a quartz slide.)

These various stable surfaces may be useful for studies of blood coagulation processes, the adsorption of various types of protein such as fibrinogen, and also for cell-cell contact investigations. The ability to coat a surface such as glass or metal to produce a stable surface having the polar characteristics akin to that of the outer lipid layer of erythrocyte membranes may make such surfaces useful in certain biomedical applications such as the production of bio-compatible surfaces.

We shall report our further studies of such stable phospholipid and glycolipid surfaces in future publications.

ACKNOWLEDGEMENTS

We wish to acknowledge support for these studies from the Wellcome Trust, Science Research Council, Deutsche Forschung Gemeinenschaft and Leverhulme Trust.

REFERENCES

1. P.J. Quinn and D. Chapman, (1980) in: "Critical Reviews in Biochemistry", G. Fasman, ed., CRC Press, Cleveland, pp. 1-117.
2. M.S. Bretscher (1973), Science 181, 622-629.
3. H.J. Chap, R.F.A. Zwaal and L.L.M. van Deenen (1977), Biochim. Biophys. Acta 467, 146-164.
4. R.F.A. Zwaal, P. Comfurius and L.L.M. van Deenen (1977), Nature 268, 358-360.
5. D.S. Johnston, S. Sanghera, M. Pons and D. Chapman(1980), Biochim. Biophys. Acta 602, 57-69.
6. G. Lieser, B. Tieke and G. Wegner (1980), Thin Solid Films 68, 77-90.

303

UTILIZING MAST CELLS AND BASOPHILS TO STUDY LIPOXYGENASE INVOLVEMENT IN MEMBRANE ACTIVATION

Albert M. Magro

Kidney Disease Institute, Center for Laboratories and
Research, New York State Department of Health, Albany
New York   12201, U.S.A.

GENERAL ASPECTS

An integral part of the immune response is the ability of
membrane-bound immunoglobulin to initiate cellular events upon
interaction with antigen. The ability of the membrane to translate
messages of antigen presence is an important aspect of cellular
activation and is dependent upon the enzymatic processes which are
initiated. In exploring the enzymatic processes of cellular activation, one should select a system where the response of the cell
can take place, on interaction of antigen with antibody, without
the necessity of other fluid phase extracellular macromolecules.
In addition, stimulation should lead to an immediate response for
which there are efficient quantitative assays. A system which has
these attributes is the in vitro release of histamine from mast
cells or basophils. In this system the interaction of cells and
inducing agent initiates the exocytosis of histamine within minutes,
and this activation can take place in vitro totally independent of
extracellular macromolecular species. The rapid time-course allows
experiments to be completed within convenient time periods, and the
response in terms of the histamine released can be accurately
measured with facility.

It is becoming increasingly clear that lipoxygenases are
essential components of the enzymatic systems which are initiated
when membranes translate activation signals to the cells. The
intermediate lipoxygenase products and their involvement in the
activation process remain to be elucidated. Reasons for selecting
the in vitro release of histamine from mast cells and basophils as a
system in which to study the involvement of lipoxygenases in cellular
activation phenomenon include: 1) A vast amount of knowledge is

305

available on mast cells and basophils which can serve as a foundation
for the study; 2) In vitro techniques for mast cells and basophils
have been established and are reliable; 3) Mast cells can be obtaine
relatively pure and in reasonable quantities; 4) Very high cell
populations can be obtained by using basophilic leukemias and
mastocytomas which are readily available; 5) Electron microscopic
structural aspects of mast cells and basophils have been extensively
studied; 6) The antigen-specific activating antibody, IgE, has been
well-defined; 7) The membrane receptor for IgE has been well
characterized; 8) Antibodies directed against IgE and the receptor
for IgE are available; 9) A number of IgE producing plasmacytomas
exists; 10) There are many non-antigenic inducing agents which
have been characterized and which vary in their mechanisms of
activation; 11) The degree of activation by antigenic and non-
antigenic means can be closely dose controlled; 12) The energy
producing metabolic processes required to secrete have been
extensively studied; 13) Large numbers of inhibitors of histamine
release are available, many of which have well-defined mechanisms;
14) The activation-secretion process demonstrates a temperature
dependence where the kinetics of the enzymatic processes can be slowe
by temperature reduction; 15) Secretion of histamine from mast cells
and basophils requires an intact lipoxygenase system.

It is an objective of this report to provide background and
information which will demonstrate the feasibility of utilizing mast
cells and basophils in vitro to investigate lipoxygenase involvement
and other aspects of cellular activation.

THE DEVELOPMENT OF IN VITRO SYSTEMS

General anaphylaxis is a systemic manifestation resulting from
the in vivo interaction at cellular sites of antigen with specific
cytotropic antibody. The shock-like state results from the immediate
release of vasoactive mediators from cells following the appropriate
antigen-cytotropic antibody reaction. Although general anaphylaxis
is an immediate and definitely visible reaction, it is one which is
difficult to quantitate.

Local anaphylaxis may occur in any organ which is sensitized
with specific cytotropic antibody. A local anaphylactic skin reactio
due to passive sensitization with cytotropic antibody is referred to
as passive cutaneous anaphylaxis (PCA). PCA occurs experimentally
when the skin of an animal is injected with cytotropic antibody and,
after an appropriate latent period, antigen is given intravenously
and an immediate local anaphylactic reaction results in the skin at
the site of antibody injection. PCA enables a quantitation of the
anaphylactic reaction particularly if the antigen is given intravenously along with a dye such as Evans blue. When the antigen
interacts with the fixed antibody in the skin, vasoactive mediators
are released which increase permeability, producing a blue spot in

the skin at the site of antibody injection. The diameter of the spot is a measure of the degree of the local anaphylactic reaction (Ovary, 1958).

Although PCA has the advantage of quantitation, it has the disadvantages of generalized anaphylaxis in being an in vivo phenomenon in which the release of mediators and the biological effects of the mediators cannot be separated from the activation phase of the reaction. In addition, being an in vivo reaction it is not possible to adjust the environment of the releasing cells. The realization that one of the predominate vasoactive mediators of anaphylaxis is histamine led to the development of in vitro systems in which the degree of the reaction is measured by the quantity of histamine released (Schild, 1936). One of the first reliable in vitro systems was developed and utilized by Mongar and Schild (1955). In this system, isolated chopped lungs from guinea pigs which have been previously immunized and suspended in vitro are challenged with antigen and the released histamine is quantitated (Mongar and Schild, 1956). With this in vitro system the advantages of being able to separate the release phase and the activation phase of the reaction and being able to adjust the environment of the releasing cells were utilized. The effects of temperature (Mongar and Schild, 1957b), pH and Ca^{2+} (Mongar and Schild, 1958), and metabolic inhibitors (Mongar and Schild, 1957a) and many other parameters (Mongar and Schild, 1962) were effectively investigated in the in vitro guinea pig lung system.

There were early suggestions of a relationship between histamine and mast cells (Cazal, 1955) which promoted the understanding that degranulation of mast cells is associated with systemic anaphylaxis (Riley and West, 1952, 1953; Fawcett, 1954; Jasmin, 1956) and led investigators to focus on the isolated mast cell as a cell type in which to investigate cellular activation phenomenon. Peritoneal rat mast cells, which can be easily harvested by a saline wash of the peritoneal cavity, offer an in vitro histamine release system in which cells are totally suspended in the reaction medium free of any tissue (Uvnäs and Thon, 1959; Mota and Dias, 1960). Like the chopped lung system, histamine release from isolated peritoneal rat mast cells has produced quantitative information by enabling an in vitro approach to the study of cellular activation phenomenon (Uvnäs, 1964). The development of techniques to produce high reagenic antibody in the rat by nematode infection (Ogilvie, 1967) has rendered the peritoneal rat mast cell system useful in studying antibody mediated cellular activation.

The chopped lung system and the rat peritoneal wash system both utilize tissue mast cells. Additional in vitro systems make use of the isolated blood basophil. The relationship of the basophil to blood histamine was studied by Graham et al. (1952) and Valentine et al. (1955). Shelly and Juhlin (1961, 1962) recognized the human blood basophil as a cell type in which anaphylactic sensitivity could be detected. Shelly (1963) showed how basophil degranulation could be used as an in vitro test for allergies to penicillin and other drugs. Lichtenstein and Osler, 1964, combined the elements of a number of in vitro systems and were able to develop a human basophil in vitro system utilizing washed leukocytes isolated from whole blood. This system quantitates the degree of the response by utilizing the fluorometric assay for released histamine as first developed by Shore et al. (1959). This histamine analyzation method has been automated which adds to the convenience of the method (Siraganian, 1974). The in vitro human basophil histamine release system has been utilized to advantage particularly in revealing the presence of β-adrenergic receptors (Lichtenstein and Margolis, 1968) and other membrane receptors (Plaut et al., 1980) capable of elevating cellular levels of cyclic AMP which inhibits the histamine release process. An in vitro blood basophil system utilizing the basophils of rabbits has also been developed (Siraganian and Osler, 1970).

Neoplastic histamine containing cells such as rat basophilic leukemic cells (Eccleston et al., 1973) and murine mastocytoma cells (Furth et al., 1957; Dunn and Potter, 1957) can also be utilized in vitro to study cellular activation phenomenon. The utilization of cultured neoplastic cells can be particularly useful when long-term incubation is required, in the characterization and investi-gation of membrane receptors, and in the isolating and identifying of cellular activation products where a high-cell population is required to obtain sufficient products for analyzation.

In spite of common elements, there are differences in the various in vitro histamine release systems. The systems are dissimilar in terms of species (guinea pig, rat, rabbit, man), effector cells (mast cells, blood basophils), and the location from which the cells are harvested (lung, peritoneal cavity, peripheral blood leukocytes). Extending observations made in one cell type to that of another cell type can be misleading. For example, it has been known for sometime that β-adrenoreceptor agonists inhibit histamine release much more effectively in lung mast cells (Schild, 1936; Assem and Schild, 1969) than they do in basophils (Lichenstein and Margolis, 1968). Rat peritoneal mast cells are particularly refractory to agents which elevate cyclic AMP, and there is a question of how large a role cyclic nucleotides have in the regu-lation of anaphylactic histamine release in isolated mast cells (Sydbom et al., 1981). Likewise, chromones inhibit histamine release effectively in lung mast cells (Assem and Mongar, 1970) somewhat effectively in peritoneal mast cells, but are totally ineffective in basophils (Conroy and Blancuzzi, 1979). In addition to their varied response to inhibitors (Mielens and Magro, 1982), lung mast cells, peritoneal mast cells, and peripheral blood basophils differ in their kinetics of release, exogenous ion requirements, response to potentiating agents, and in the density of adrenergic and other types of surface receptors. The cells also differ in their response to anoxia and may differ in aspects of their glucose metabolisms. However, as far as is known, all histamine releasing cells have a requirement for an intact lip-oxygenase system. Depending on what aspects are to be investigated, the diversity of the cell types of the various in vitro systems can be utilized to advantage.

IgE AND ITS RECEPTOR ARE WELL-DEFINED

Ever since Prausnitz and Küstner (1921) demonstrated that serum from an individual who was hypersensitive to fish protein could sensitize a non-hypersensitive individual, it has been realized that the sensitizing agent of immediate hypersensitivity is a noncellular soluble component of the blood. Prior to its identification, the soluble sensitizing agent of serum had been referred to as reagin. Ever since the classic experiment of Prausnitz and Küstner, it was suspected that reagin was an immunoglobulin. By the early 1960's, considerable effort had been made in attempting to define the particular immunoglobulin class to which reaginic antibody belonged. By then, there were a number of myeloma proteins which were identi-fied as portions of immunoglobulins or immunoglobulins of a particular class. This enabled the preparation of purified immuno-globulins and also allowed the preparation of highly specific anti-bodies directed against determinants of the known classes of immunoglobulins. There were a number of reports in support of the concept that reagin was an antibody of the IgA class (Müller-Eberhard, 1960; Ishizaka et al., 1963; Yagi et al., 1963; Fireman et al., 1963; Vaerman et al., 1964). However, Juhlin and Ohman (1964) could find no significant relationship between the blood basophil and immunoglobulin classes, G, A, and M even though they were aware that basophil degranulation was a phenomenon associated with anaphylactic sensitivity. By precipitating immunoglobulins from the serum of atopic individuals with antibodies specific to A, G, D, and M, Ishizaka et al. (1966a) became convinced that reaginic antibody was not of the IgA class but rather an unique immunoglobulin which they designated γE-globulin. Their careful work enabled them to raise antibodies to IgE which they utilized to indicate that IgE was associated with reaginic activity in the sera of atopic patients (Ishizaka et al., 1966b).

Paralleling the realization that reagin was not a known class of immunoglobulin, Johansson and Bennich (1967) investigated the properties of a myeloma protein which they designated ND and identified as an immunoglobulin not of a known class. There were suggestions that the myeloma immunoglobulin ND was a reaginic immunoglobulin whereupon Stanworth et al. (1967) cleverly showed that myeloma Ig(ND) could inhibit Prausnitz-Küstner reactions by competing with reaginic antibody. Subsequently, immunological evidence was obtained that IgE and Ig(ND) shared antigenic determinants and were structurally related (Bennich et al., 1969).

The realization that the myeloma protein Ig(ND) was an immunoglobulin of the IgE class greatly facilitated investigations of the structure and function of IgE. Bennich and Johansson (1971) were able to make rapid progress in characterizing the carbohydrate content, amino acid profile, molecular weight, cystein residues, antigenic determinants and the number of domains in the Fc and Fab portions of IgE. Although rare as a type, there have been examples of myelomas subsequent to IgE(ND) which have been identified as immunoglobulins of class E (Ogawa et al., 1969).

The understandings that mast cells and basophils are cell types which release mediators of anaphylaxis, and that IgE is the antigen-specific activator, converged to the realization that IgE must bind to the surface membrane of these cells. Stanworth et al. (1968) were the first to present evidence that the Fc portion of IgE is the membrane binding portion of the molecule leaving the Fab portion free to react with antigen. The idea that IgE is bound by its Fc portion to a specific receptor promoted investigations into the properties of the receptor, particularly as to the number of receptor sites per cell and the binding affinity of the receptor for IgE. The appearance of the rat basophilic leukemia (RBL) (Eccleston et al., 1973) and of rat IgE secreting plasmacytomas (Bazin et al., 1974) were a great aid to the studies of the properties of IgE membrane receptors. To study the properties of IgE receptors, Ishizaka et al. (1973) worked with normal human basophils and human myeloma IgE, while Kulcyzki and Metzger (1974) worked with RBL cells and normal rat and mouse IgE, and Conrad et al. (1975) used normal rat mast cells and rat myeloma IgE. These laboratories roughly agreed that the number of IgE receptors per cell was on the order of 10^5 to 10^6 and that the equilibrium constant for the association reaction of IgE and its receptors is approximately 10^9 to 10^{10} M^{-1}. Kulcyzki et al. (1974) were able to pass the RBL cells in vitro producing cultures which made available large numbers of RBL cells for study. This facilitated the identification of the IgE-receptor as being a glycoprotein (Kulcyzki et al., 1976) with an apparent molecular weight of 62,000 (Conrad and Froese, 1976) and enabled the production of highly specific antibodies against determinants of the receptor (Froese, 1980).

The above studies make it obvious that mast cells and basophils are cell types in which IgE and its receptor are well characterized. The availability of several species of myeloma IgE's, several in vitro systems which utilize normal and neoplastic mast cells and basophils, and of highly specific antibodies directed against IgE or its receptor form the basis of a well-defined system in which to investigate aspects of immunoglobulin mediated cellular activation.

CROSS LINKING OF MEMBRANE BOUND IgE INITIATES CELLULAR ACTIVATION

Landsteiner, one of the giants in the field of immunology, had an intense interest in the configurations of antigens which can successfully initiate immunoglobulin mediated cellular events. In his pioneering hapten work, Landsteiner (1924) recognized the anaphylactic reaction as a convenient and useful example of a biological response initiated by antigens. He utilized active systemic anaphylaxis, in hapten sensitized guinea pigs, to determine that monovalent haptens inhibit rather than evoke anaphylactic type reactions, whereas multivalent haptens readily produce a positive response. Most of the in vivo accumulated evidence which followed corroborated the original work of Landsteiner and also indicated that multivalent haptenic groups bound to a molecule render the molecule active in eliciting hapten systemic or passive cutaneous anaphylaxis (Tillet et al., 1929; Landsteiner and Levine, 1930; Ovary and Karush, 1960; Parker et al., 1962; Levine, 1962).

Utilizing the in vitro histamine release basophil system, as developed by Lichtenstein and Osler (1964), Ishizaka et al. (1969) were able to demonstrate that intact antibodies directed against determinants of myeloma IgE(ND) could induce histamine release whereas the monovalent Fab fragments of the anti-IgE were not active (Ishizaka and Ishizaka, 1969). Using a hapten system in vitro, in which basophils were sensitized with hapten specific membrane bound IgE, it was shown definitively that haptens capable of bridging two-receptor-bound IgE molecules could activate the basophils (Magro and Alexander, 1974a; Siraganian, et al., 1975). The in vitro histamine release system revealed that the interaction of membrane bound IgE with very simple and well-defined antigens enables the IgE to translate messages of antigen presence to the cell. The observations can be summarized as follows: 1) IgE of all specificities is on the majority of mast cells and basophils; 2) IgE on the cell surface is not grouped by specificity; 3) Internal bridging between two Fab pieces of a single IgE molecule is not a basic mechanism by which histamine is released; 4) Divalent haptens readily evoke a positive response whereas monovalent haptens inhibit; 5) Haptens must bridge the membrane-bound IgE to another point away from the IgE molecule for the hapten specific release of histamine to be initiated.

The dose-response curve for the in vitro release of histamine progresses to a maximum and then declines with increasing concentrations of inducing agent. The low doses of antigen which can initiate histamine release infer that only a small number of antigen cell interactions are sufficient to trigger the response. This interference was expressed in quantitative terms (Magro, 1975) by applying the Poisson distribution to the in vitro dose-response curve. It was reported that the geometrical configuration necessary to elicit a response can be fulfilled by a single molecule of antigen. The report did not deal with the number of antigenic sites of the single molecule or the number of clusters that a single molecule of antigen could form. However, Lawson et al. (1975), utilizing mast cells and an immunoferritin electron microscopic

technique, demonstrated that cells can be triggered by IgE clusters
considerably smaller in size than their limits of electron microscopic visualization which was 10 molecules. The number of antigen
induced IgE clusters being less than 10 to trigger a basophil was
corroborated by DeLisi and Siraganian (1979a).

Although the enzymatic sequences of cellular activation are
initiated by the bridging of membrane-bound IgE, there is evidence
that simultaneous deactivation processes may be initiated by
clustering or bridging of membrane-bound IgE. This idea goes back
as far as 1958 when Mongar and Schild demonstrated that calcium
deprived histamine releasing cells, in the presence of antigen, are
deactivated. Evidence for deactivation processes due to bridging,
and the idea that a small amount of IgE clustering can activate
mast cells and basophils led to the investigation of the effects of
a high degree of cross linking membrane-bound IgE. Cell turn off
by excess bridging was proposed (Magro and Alexander, 1974b; Magro,
1974a) where it was possible to show that the descending portion of
the dose-response curve, in vitro, could be the result of cells
being rendered unresponsive by a high degree of bridging of membrane-
bound IgE. The report of Becker et al., 1973, indicated a turn
off by excess bridging phenomenon. Their data showed that a high
degree of bridging is required to cap membrane-bound IgE in human
basophils without the release of histamine. Attempts to quantitate
or mathematically express the proposal that activation and deactivation occur simultaneously and that both are dependent upon the
concentration of IgE receptor cross links have been pursued by a
number of laboratories (Delisi and Siraganian, 1979b; Dembo et al.,
1979a; 1979b).

The in vitro histamine release system has activation-deacti-
vation processes which can be adjusted by the degree of bridging of
membrane-bound IgE. The geometrical configurations of antigen
necessary to initiate an IgE mediated response are simple and
well defined. This enables a close control of the degree of
bridging which can be utilized to advantage when investigating
enzymatic processes which are bridging dependent.

NON-ANTIGENIC INDUCERS OF HISTAMINE RELEASE

The release of histamine from mast cells and basophils by
non-antigenic means has been very useful in delineating aspects of
cellular activation. One of the most frequently utilized non-
antigenic liberators of histamine is compound 48/80. This compound
originated from the studies of Baltzly et al., (1949) who in
analyzing the depressor action of certain isoquinaline derivatives,
found that the depressor action was due to contamination with
phenylethylamine condensation products. Several such compounds were
prepared, and one of them, compound 48/80 was found to lower blood
pressure at very low doses. Feldberg and Paton (1951) upon

pharmacological examination of compound 48/80, used the substance
to demonstrate that its depressor action was due to the ability
of the compound to release histamine.

Compound 48/80 is a non-toxic (Diamant, 1967) energy requiring
(Johnson and Moran, 1969) inducer of histamine release which may
not require the presence of millimolar quantities of exogenous Ca^{2+}
and which has been utilized extensively to study induction mechanisms
in rat mast cells (Uvnäs, 1964). Other agents which induce a Ca^{2+}
dependent, energy requiring, non-toxic release of histamine from
rat mast cells include: d-tubocurarine, polymyxin B, stilbamidine
(Alam et al., 1939; Bushby and Green, 1955; Ellis et al., 1970),
ATP and ADP (Diamant and Krüger, 1967) and dextran (Voorhees et al.,
1951). Some rat strains are resistant to dextran (West, 1974) and
substantial release of histamine by dextran from the mast cells of
most rat strains requires the addition of exogenous phosphatidylserine (Goth and Adams, 1970; Goth et al., 1971). Antigen and
dextran induced histamine release from rat mast cells are enhanced
by phosphatidylserine in all rat strains except the fawn-hooded,
a unique rat strain which does not respond to phosphatidylserine
potentiation (Magro, 1981). An additional inducing agent which has
all of the characteristics of antigen induced histamine release
in rat mast cells is the desferrioxamine B chelate of ferric ion
(Magro and Brai, 1982). This inducing agent is of interest
because its mode of action appears to involve free radicals of
oxygen which promote lipoxygenase activity.

An anaphylatoxin formed from antigen-antibody complexes combined
with serum was first reported by Friedberger (1909) who speculated
that anaphylatoxin was formed from the complement system. Mota
(1959), reported that serum generated anaphylatoxin could induce
histamine release from guinea pig mast cells. Osler et al. (1959),
showed that anaphylatoxin could not be generated after complement
was inactivated. Later Cochrane and Müller-Eberhard, 1968, were
able to show that the relationship of anaphylatoxin to the
complement system involved the third and fifth components of
complement and identified anaphylatoxin as the C3a and C5a.
Vallota and Müller-Eberhard (1973) and Johnson et al. (1975) showed
by directly using purified C3a and C5a that these agents are
anaphylatoxic, with C5a being 1000 times more active than C3a on a
molar basis. Grant et al. (1975), produced evidence, using antibodies against C3 and C5, that C3 and C5 induce histamine release
from human basophils. Antigen induced histamine release has an
activation phase which is cyclic AMP sensitive (Lichtenstein and
DeBernardo, 1971). Complement mediated histamine release shares
with antigen intracellular mechanisms for induction in that there
is an activation stage in both which can be inhibited by elevated
levels of cyclic AMP (Grant et al., 1976). However, there are
differences between complement and antigen induced histamine
release. The kinetics for complement is much faster and the cells
cannot be desensitized and rendered unresponsive to C3a and C5a by

incubating cells with antigen prior to the addition of C3a and C5a
(Petersson et al., 1975). Calcium ionophore A23187 (Reed and Lardy,
1972) can induce histamine release (Foreman et al., 1973) but its
mechanism of induction appears to bypass the calcium independent
cyclic AMP sensitive activation stage (Lichtenstein, 1975). Studies
on the ionophorous influx to exogenous Ca^{2+} have been particularly
informative especially since Ca^{2+} plays a central role in coupling
the cells' stimulatory and secretory phases (Douglas et al., 1967).

Many types of histamine releasing compounds have been catalogued
(Rothschild, 1966) but two other agents worthy of mention are the
formyl methionine containing peptides (Siraganian and Hook, 1977)
and the cationic proteins (Seegers and Janoff, 1966; Keller, 1968;
Ranadive and Cochrane, 1968). Histamine release by the formyl
methionine containing peptides, like antigen induced histamine
release, requires exogenous Ca^{2+} and is inhibited by agents which
increase intracellular cyclic AMP. In other aspects, the peptides
resemble C5a induced histamine release particularly in its kinetics
of release and its response to pharmacologic agents. Cationic
protein, isolated from neutrophil liposomes, induces histamine
release from mast cells by mechanisms in which the sequence of biochemical events leading to the release of histamine is similar to
antigen (Ranadive and Cochrane, 1971).

Most of the non-antigenic inducers of histamine release
mentioned are non-IgE mediated. However, there are a number of
non-antigenic inducers of histamine release which initiate cellular
activation by cross linking IgE. Two such agents are anti-IgE and
concanavalin A. Anti-IgE induces histamine release by cross linking
determinants of the Fc portion of membrane-bound IgE (Ishizaka et
al., 1969). Concanavalin A induces histamine release (Keller,
1973) by cross linking sugar moieties of the Fc portion of membrane-
bound IgE(Magro, 1974b; Magro and Bennich, 1977). Concanavalin A
induced histamine release has characteristics similar to IgE
mediated release (Siraganian and Siraganian, 1974).

Antibodies directed against the membrane receptor for IgE can
also induce histamine release (Ishizaka et al., 1977; Isersky
et al., 1978). The cross linking of the IgE receptors by divalent
anti-receptor antibodies even when not occupied by IgE molecules
initiates histamine release (Ishizaka and Ishizaka, 1978). Anti-
receptor antibodies have the potential of being a very useful
bridging method to initiate cellular activation particularly since
the activation can take place with the use of non-sensitized cells.

Postulations that phospholipase A is an enzymatic mechanism
for the release of histamine have been long standing (Feldberg
et al., 1938; Högberg and Uvnäs, 1958). Högberg and Uvnäs (1960)
were also able to show that various metabolic inhibitors and Zn^{2+}
blocked compound 48/80, antigen and phospholipase A induced histamine release similarly. Whelan (1978) reported that phospholipase

A initiation of histamine release has a requirement for extracellular phospholipids. Lytic polypeptides such as melittin, (Sessa et al., 1969) which can activate phospholipase A (Mollay et al., 1976) can also induce histamine release (Jaques, 1965; Jasani and Stanworth, 1972). Phospholipase C is an additional lipase which has the ability to induce histamine release in an energy dependent non-toxic manner (Strandberg et al., 1974).

Mast cells and basophils can be triggered by a wide variety of well-defined inducing agents, most of which have different triggering mechanisms. These agents have been extensively studied and consequently more is known about the mechanisms of triggering mast cells and basophils than any other cell type.

THE ENERGY REQUIREMENTS OF HISTAMINE RELEASE

The energy producing and utilizing mechanisms during exocytosis of histamine-containing granules have been an interest of long standing. Parrot (1942) made the observation that anaphylactic histamine release from sensitized guinea pig lung is inhibited by anoxia, indicating that oxygen metabolism may be an essential component of anaphylaxis. The inhibitory effects of anoxia upon histamine release from sensitized guinea pig lung was subsequently confirmed (Mongar and Schild, 1957; Chakravarty, 1960a). However, Diamant (1962) reported that in chopped guinea pig and rat lung, anoxia inhibits histamine release only in the absence of glucose. There have also been disagreements as to whether oxygen consumption is increased during the anaphylactic reaction in the guinea pig. Mongar and Schild (1962) obtained a slight increase in oxygen consumption of isolated guinea pig lung during histamine release, but Chakravarty (1962a) obtained no increase in oxygen consumption in the same system. The difference in these early reports on the effect of anoxia and oxygen consumption may be due to difficulties in obtaining total oxygen depletion and use of different techniques for measuring oxygen consumption. The disagreement among the early investigators in measuring actual consumption of oxygen raised some doubt as to whether oxygen metabolism was an essential component of histamine release. However, information obtained by exogenous agents which are known inhibitors of oxygen metabolism has been consistent and reproducible. Moussatché and Provost-Dannon (1958) showed that uncouplers of oxidative phosphorylation such as 2-4 dinitrophenol have the ability to inhibit histamine release, suggesting that ATP formation may be an essential feature of anaphylactic histamine release. This idea was promoted further by their work showing the effects of inhibitors of the respiratory chain, such as antimycin A and carbon monoxide upon histamine release (Moussatché and Prouvost-Dannon, 1962). There has been fairly good agreement for the involvement of glycolysis during histamine release from guinea pig lung mast cells. Glycolysis was implicated by the ability of glucose to reverse the inhibitory effects of anoxia (Diamant, 1962) and by the blocking effects of 2-deoxyglucose upon histamine release (Chakravarty, 1962b).

Measurements of oxygen uptake also have produced disagreement as to whether increased oxygen consumption is associated with histamine release from rat mast cells. Mongar and Perera (1965) showed no increase in the rate of oxygen consumption during histamine release from mast cells. However, Chakravarty (1968), using pure rat peritoneal mast cells, reported that a transient stimulation of respiration occurs immediately after contact with antigen. 2-deoxyglucose can inhibit antigen induced histamine release from peritoneal mast cells but fairly high concentrations are required (Chakravarty, 1967). The ineffectiveness of 2-deoxyglucose in the peritoneal rat mast cell system as compared to guinea pig lung mast cells or human basophils indicate that there are differences in their dependence upon glycolysis. The degree of dependence upon aerobic and anaerobic glycolysis is not clear, but Chakravarty (1965) claims that there is a high rate of both anaerobic and aerobic glycolysis in the peritoneal mast cell during histamine release, both of which can produce ATP. The presence of creatine phosphokinase in peritoneal rat mast cells (Magro 1980) suggests an additional method of ATP production and may be related to the ineffectiveness of 2-deoxyglucose in these cells.

Studies have directly promoted the idea that the energy requirements for histamine release are supplied by ATP generation and utilization (Johansen and Chakravarty, 1972; 1975; Diamant et al., 1974). The importance of ATPases in making the chemical energy of ATP available prompted the demonstration of Ca^{2+}-Mg^{2+} activated ATPases in rat mast cells (Cooper and Stanworth, 1976; (Magro, 1977a; Chakravarty and Echetebu, 1978). There are indications that a portion of the divalent cation ATPase may be an actomyosin type. The presence of actin in rat mast cells (Röhlich, 1975) and the presence of creatine phosphokinase (Magro, 1980) promote the opinion that actomyosin-like contractile microfilaments may supply motive power for granule extrusion. Reports illustrating the fact that agents which are known to inhibit ATPases similarly inhibit histamine release could also be considered supportive of ATPase involvement in histamine release (Fewtrell and Gomperts, 1977; Magro, 1977b). In addition, a good rank order of effectiveness has been observed in the ability of analogs to inhibit ATPase and histamine release (Magro et al., 1982). ATPases have also been visualized by a cytochemical/electron-microscope technique in both secreting and non-secreting mast cells (Chakravarty and Nielsen, 1980).

The data as a whole indicate that both glycolysis and oxidative phosphorylation, as ATP producing systems, are capable of supplying energy for histamine release. The ATP can be acted upon by Ca^{2+}-Mg^{2+} activated ATPases and it appears that an intact ATPase system is necessary for the release of histamine from mast cells and basophils.

SRS AND THE LIPOXYGENASE PATHWAY

Bioassays which measure the rate and force of contraction of smooth muscle have been instrumental in the discovery of many biologically important molecules. Shultz (1910) and Dale (1913) demonstrated that a contraction of guinea pig intestinal smooth muscle could be produced during anaphylaxis. The Shultz-Dale reaction evolved into an assay system capable of measuring exogenous agents, such as histamine, which could produce a contraction of a strip of guinea pig ileum, suspended in vitro, in an oxygenated organ bath. Feldberg and Kellaway (1938), in exploring the action of cobra venom upon perfused guinea pig lung, noticed that the guinea pig lung released a slow reacting substance (SRS) distinct from histamine, which caused a delayed and slow contraction of the ileum smooth muscle. Kellaway and Trethewie (1940) subsequently showed that a slow reacting substance is similarly liberated from guinea pig lung during anaphylaxis. Consequently, a nomenclature has developed where SRS refers to a slow reacting substance which is released by means other than anaphylaxis and SRS-A designates a slow reacting substance which is released during anaphylaxis.

Time and again the goal of a scientific pursuit will be clearly defined and capture the imagination of the scientific community. Such an occurrence took place in the determination of the molecular structure of SRS. The chemical nature of SRS was of interest from the onset when Feldberg and Kellaway (1938) postulated that SRS was a lecithin-like substance. Years later, Vogt (1957) obtained evidence that SRS activity was released from cobra venom treated egg yolk raising the possibility that the lecithinases in the venom were releasing SRS-like fatty acids from the egg yolk lecithin. Further evidence that SRS was an acidic lipid was obtained in early studies of the physicochemical properties of SRS(Chakravarty, 1960b; Änggård et al., 1963). Interest in the chemical nature of SRS-A became more acute when it was realized that histamine and SRS-A can be released by antigen in asthmatic (Brocklehurst, 1960) and passively sensitized human lung (Parish, 1967; Sheard et al., 1967). The clinical relevance of SRS-A to bronchoconstrictive diseases was made clear by Brocklehurst (1962) in the demonstration that antihistamines are ineffective against antigen-induced human bronchoconstriction. The huge success of cromoglycate (Altounyan, 1967) as an anti-bronchoconstrictive drug added further impetus to the study of SRS-A particularly by the pharmaceutical companies.

By 1971, Strandberg and Uvnäs reported that SRS activity is dependent upon the molecule having intact hydroxylic and carboxylic groups as well as unsaturated bonds. Orange, one of the stalwart

317

investigators of the properties of SRS, in collaboration with
Murphy and Austen, reported that SRS-A could be inactivated by
high concentrations of arylsulfatases (Orange et al., 1974).
While incorrectly focusing on the importance of the sulfate moiety,
the report did call attention to the sulfur atom which, in the end,
was an important component of SRS. The importance of sulfur
containing thiol groups was made evident when it was reported
that thiols, particularly cysteine, greatly enhance the onset of
SRS-A (Orange and Chang, 1975; Orange and Moore, 1976). The
importance of cysteine was corroborated by Parker et al. (1979) who
reported that ^{35}S cysteine could be incorporated into SRS.

The reports that calcium ionophore A23187 can release SRS from
isolated rat peritoneal mast cells (Bach and Brashler, 1974), and
that A23187 can release SRS from neoplastic cells, such as rat
basophil leukemia cells (Jakschik et al., 1977a) were instrumental
in the utilization of excellent systems in which to generate SRS.
Another contributing factor was the evidence that arachidonic acid
via the lipoxygenase pathway is a precursor molecule for SRS
production (Jákschik et al., 1977b; Bach et al., 1977). Hamberg
(1976) reported that in guinea pig spleen indomethacin while
blocking the cyclooxygenase pathway redirects arachidonic acid
metabolism via the lipoxygenase pathway. The suggestion of
lipoxygenase involvement in SRS production was supported by the
observation that indomethacin in inhibiting cyclooxygenase activity
enhances SRS release (Engineer et al., 1978). Other developments
important to the resolution of the structure of SRS were the
implementation of HPLC as a purification tool, and the realization
that optical absorbance properties are a means of detection of SRS
(Morris et al., 1978).

The efforts to determine the nature of SRS resulted in the
report of Murphy, Hammarström and Samuelsson (1979). Murphy who
had collaborated with Orange and Austen (Orange and Austen, 1969)
had extensive experience in the production and properties of SRS.
While on a sabbatical stay, Murphy combined his experience with the
vast experience of Hammarström and Samuelsson in identifying
arachidonic acid metabolites. The combination produced a break
through in the elucidation of the structure of SRS. Treating
murine mastocytoma cells with calcium ionophore A23187 they were
able to produce a cysteine augmented lipoxygenase derived SRS.
Utilizing high performance liquid chromatography coupled with
gas-liquid chromatography/mass spectrometry they identified the
murine mastocytoma SRS as a cysteine-containing derivative of
5-hydroxy 7, 9, 11, 14 eicosatetraenoic acid. Although this
report did not identify the cysteine containing amino acid portion,
it noted that the cysteine was attached in thioether linkage at the
C-6 position of the molecule. The amino acid portion of the murine
mastocytoma SRS was immediately identified (Hammarström et al.,
1980) as the tripeptide glutathione producing the molecule 5-
hydroxy-6-cysteinyl   glycinyl glutamyl 7, 9, 11, 14 eicosatetraenoic

acid. Because of the previous description of 5, 6 epoxy 7, 9, 11, 14 eicosatetraenoic acid, as leukotriene A and 5, 12 dihydroxy 7, 9, 11, 14 eicosatetraenoic acid as leukotriene B, (Borgeat and Samuelsson 1979b) the mastocytoma SRS was described as leukotriene C. The leukotrienes A and B were originally isolated as lipoxy-genase products generated by leukocytes and have the characteristic of being trienes (Borgeat and Samuelsson 1979a). Depending upon its source and method of generation, the amino acid portion of an SRS molecule can vary but all have the characteristic optically active triene structure and consequently the nomenclature, leukotriene, has been readily adopted (Samuelsson et al., 1979). SRS-A has been identified as the dipeptide, 5-hydroxy-6-cysteinyl glycinyl 7, 9, 11, 14 eicosatetraenoic acid (Morris et al., 1980a) and this has been referred to as leukotriene D. The leukotriene D form has been identified as a major component of the SRS of RBL cells (Morris et al., 1980b) and other cell types (Bach et al., 1980). Another analog, leukotriene E, is the unipeptide 5-dydroxy-6-cysteinyl 7, 9, 11, 14 eicosatetraenoic acid. There have been reports demonstrating the conversion of leukotriene C and D to leukotriene E in human plasma (Parker et al., 1980a) and RBL cells (Parker et al., 1980b).

The major aspects of the chemical nature of various SRS's and SRS-A have been resolved. It appears that the predominant SRS's are trienes generated via the lipoxygenase pathway in combination with enzymes capable of producing the thioether linkage. The background and resolution of SRS have been intertwined with anaphylaxis and histamine release indicating that lipoxygenases are important enzymatic pathways in mast cells and basophils.

MAST CELLS AND BASOPHILS REQUIRE AN INTACT LIPOXYGENASE SYSTEM TO RELEASE HISTAMINE

Generation of arachidonic acid by cellular phospholipases acting on membrane phospholipids appears to be an integral part of cellular activation phenomenon. The arachidonic acid produced can be metabolized by cyclooxygenases, prostaglandin synthetases, isomerases, thromboxane synthetases and lipoxygenases to yield short and long-lived metabolites. Many aspects of the pathways of arachidonic acid metabolism have been elegantly determined. The enzymatic production of prostaglandins was demonstrated over 20 years ago in the laboratory of Bergström et al. (1962). Prior to the production of prostaglandins, by prostaglandin synthetases, cyclooxygenase acting upon arachidonic acid can yield transient endoperoxide intermediates. Although Samuelsson (1965) had reported unstable endoperoxide intermediates in the formation of PGE_1, it was not until the arachidonic acid metabolite endoperoxides were put into the context of platelet function was there a full appreciation of their importance (Hamberg and Samuelsson, 1973; 1974; Nugteren and Hazelhof, 1973; Hamberg et al., 1974a; 1974b).

Adhering to the lettering system that existed for prostaglandins the endoperoxide intermediates were named PGG_2 and PGH_2. These intermediates can be isomerized to PGD_2, PGE_2, or $PGF_{2\alpha}$ (Nugteren and Hazelhof, 1973; Hamberg and Samuelsson, 1973). The intermediate endoperoxide PGH_2 can also be transformed into other metabolites such as prostacyclin, PGI_2, (Johnson et al., 1976) and other metabolites such as thromboxane A_2, TxA_2, and hydroxyheptadecatrienoic acid, HHT, possessing non-prostanoate acid structures (Hamberg and Samuelsson, 1974; Hamberg et al., 1975).

Platelet aggregation and secretion studies have shown that arachidonic acid metabolites generated by the cell can also regulate the cells' response to stimuli and this was an important aspect in dilineating the biological importance of the arachidonate cyclooxy-genase products. Coincident with the isolation of the endoperoxides, PGG_2 and PGH_2, Willis and Kuhn (1973) found an unidentified labile substance formed by platelets, which could stimulate the aggregation of platelets. They found that the synthesis of this labile substance by the platelets could be blocked by aspirin and Vane (1971) had brought about the realization that aspiran's anti-inflammatory action was due to its ability to block cyclooxygenases. Hamberg et al., (1974a) showed that the platelet aggregating factor of Willis and Kuhn (1973) was a mixture of PGG_2 and PGH_2. Moncada et al. (1976) found an enzyme isolated from arteries which can transform prosta-glandin endoperoxides into an unstable substance that inhibits platelet aggregation. The structure of the platelet disaggregation factor was determined and referred to as prostacyclin or PGI_2 (Johnson et al., 1976). Smooth muscle contraction was also an important biological phenomenon that helped detect new cyclooxygenase products. Piper and Vane (1969) found a rabbit aorta contracting substance the production of which could be blocked by aspirin. The molecule was subsequently identified as thromboxane A_2 (Hamberg et al., 1975).

The number of biologically active cyclooxygenase pathway product cast a shadow upon the lipoxygenase pathway, and its importance was initially somewhat ignored. However, since the isolation from platelets of 12-hydroxyeicosatetraenoic acid and its corresponding hydroperoxide, it was suspected that lipoxygenases would be of importance (Hamberg and Samuelsson, 1974). Chemotaxis and other interesting biological properties of hydroxyeicosatetraenoic acid (HETE) and corresponding peroxides (HPETE) were reported (Turner et al., 1975; Goetzl et al., 1977). Other biological properties of products of the lipoxygenase pathway include plasma leakage in addition to smooth muscle contraction (Dahlén et al., 1981). The laboratory of Samuelsson, realizing the biological importance of cyclooxygenase pathway intermediates, and having the experience in their isolation, took a similar approach in the structural analysis of the hydroxylated arachidonate lipoxygenase intermediates, (Borgeat and Samuelsson, 1979a; 1979b; 1979c; 1979d). In addition to the awareness of biologically active products, there has been a realization that these products can be regulatory to the cells ability to respond to stimuli. Gryglewski et al. (1975) noticed that cyclooxygenase inhibition does not block but enhances histamine release. The effect of cyclooxygenase inhibition upon histamine release is similar to its effect upon SRS production. Since the investigation of SRS and histamine release were intertwined, the importance of the lipoxygenase pathway to SRS production acted as a beacon in revealing the importance of lipoxygenases in mast cell and basophil activation. The investigations that exogenous arachidonic acid can potentiate histamine release (Sullivan and Parker, 1979; Marone et al., 1979) followed reports that exogenous arachidonic acid can potentiate SRS production (Jakschik et al., 1977b; Bach et al., 1977). Likewise, the reports that the lipoxygenase products, 5-HETE and 5-HPETE can augment IgE mediated histamine release (Stenson et al., 1980; Peters et al., 1981) grew out of the knowledge that these lipoxygenase products are an important aspect of the structure of SRS (Murphy et al., 1979), and when 5-HPETE is added exogenously it can be incorporated into and augment SRS production (Parker et al., 1980c; Falkenhein et al., 1980). The same was true for the inhibitor of histamine release, 5, 8, 11, 14·eicosatetraynoic acid, which was known previously to inhibit SRS production. Studies dealing directly with the effect of inhibitors of lipoxygenase activity upon histamine release (Marone et al., 1980, Magro, 1982) showed that lipoxygenase activity is essential to mast cell and basophil activation. Marone et al., 1980, reported that the acetylenic competitive inhibitor of arachidonic acid, 5, 8, 11 eicosatriynoic acid, which inhibits lipoxygenase more effectively than cyclooxygenase, blocks histamine release. The report of Magro (1982) tested a large number of inhibitors of lipoxygenase, cyclooxygenases, prostaglandin isomerases and thromboxane synthetases for their ability to block histamine release. Agents inhibitory to the activity of lipoxy- genases were effective blockers of IgE and non-IgE mediated release while agents antagonistic to cyclooxygenases, isomerases, and thromboxane synthetases, were not. In addition, there was a fairly good rank order of effectiveness between the concentration ranges at which compounds are antagonistic to lipoxygenase activity and to histamine release.

Virtually every tissue studied has some capacity to metabolize arachidonic acid. Histamine releasing cells can generate both prostaglandin and non-prostaglandin metabolites. By the cyclooxy- genase pathway, resting mast cells can produce significant quanti- ties of PGE_2 (Talone et al., 1978). Stimulated cells can produce thromboxanes and prostaglandins with PGD_2 being the major prostaglandin metabolite (Roberts et al., 1978). Although histamine- releasing cells have the capacity to generate arachidonic acid metabolites by pathways other than lipoxygenase, these pathways do not appear to be necessary for the functioning of the cellular activation and secretory processes.

CONCLUSIONS AND SPECULATIONS

The role of arachidonic acid metabolism in the control of phospholipid metabolism is an important aspect of mast cell and basophil activation. Electron micrographs demonstrate that during histamine secretion, the perigranular and plasma membrane fuse, whereupon the intragranular contents of the cell are expelled and the cytosol contents are retained (Röhlich et al., 1971; Lagunoff, 1973; Lawson et al., 1977; Dvorak et al., 1980). The fusion of membranes following stimulation is closely linked to changes in arachidonic acid metabolism and phospholipid turnover. Hokin and Hokin (1953) first reported that phospholipid metabolism can be increased in cholinergic stimulated cells of the pancreas. This was later articulated as the "inositol effect" where increased phosphatidylinositol metabolism is associated with physiological stimuli directed toward plasma membranes (Michell, 1975). The increased phospholipid metabolism is detected by the incorporation of $^{32}PO_4$ into the phospholipids following stimulation. The current explanation is that cell stimuli initiate the breakdown of pre-existing phospholipids to diacylglycerols which are then rephosphorylated during phospholipid resynthesis by the cell. This would implicate phospholipase C as an enzyme closely associated with the ability of cells to respond to stimuli. Increased phosphatidylinositol turnover has been observed in IgE mediated and non-IgE mediated stimulation of mast cells (Kennerly et al., 1979a; Cockcroft and Gomperts, 1979). Kennerly et al. (1979b) have suggested that diacylgycerol metabolism has a role in membrane fusion in activated mast cells. It has also been reported that stimulated phospholipid metabolism in mast cells can be modulated by pharmacologic agents that increase cyclic 3',5' adenosine monophosphate levels (Kennerly et al., 1979c). In addition, the stimulation of arachidonic acid production following mast cell activation may feedback and regulate the degree of phosphatidyl-inositol breakdown in mast cells (Marquardt et al., 1981). This implies that following mast cell stimulation there is an interconnection between phospholipase C, which generates diacylglycerols, and phospholipase A which generates arachidonic acid. Evidence that the mechanism of arachidonic acid release involves two enzymes, a phosphatidylinositol-specific phospholipase C and an arachidonic acid liberating diglyceride lipase, has been presented (Bell et al., 1979). Phospholipase A_1 and phospholipase A_2 catalyze the hydrolysis of the fatty ester bonds at the 1 and 2 positions respectively, of 1,2 diacyl Sn-phosphoglycerides. How phospholipase C, diglyceride lipase, phospholipase A_1, and phospholipase A_2 all interrelate to intermediates of the lipoxygenase pathway and how arachidonate lipoxygenase products stimulate cellular lipases in mast cells during cellular activation are particularly interesting questions. How mast cell lipases, which catalyze the hydrolysis of fatty acid ester bonds of the 1,2 diacyl Sn-phosphoglycerides, are affected by lipoxygenases hydroxylating the fatty acids prior to their hydrolysis is not known. The sequence of phospholipase C, diglyceride lipase

322

phospholipase A_1, phospholipase A_2 and lipoxygenase enzymatic activities and how they interrelate and feedback on one another is one of the essential areas in the investigation of mast cell activation which remains to be elucidated.

Phospholipid methylation is another cellular enzymatic system which may initiate physical changes in membranes which allow cells to respond to stimuli (Hirata and Axelrod, 1980). Phospholipid methyltransferase activity, capable of converting phosphatidyl-ethanolamine to phosphatidylcholine, has been demonstrated in antigen stimulated mast cells and basophils (Hirata et al., 1979; Crews et al., 1980). The phospholipid methylation initiates immediately post activation of the mast cells, slightly precedes the Ca^{2+} influx and histamine release, while being in close association with arachidonic acid release. Utilizing the in vitro human basophil system Morita et al. (1981) reported that methylation reactions play an important role in IgE-mediated activation, but histamine release induced by calcium ionophore A23187 bypasses the methylation step. Since inhibitors of lipoxygenase activity block both IgE-mediated and A23187 induced histamine release, a resolution of whether phospholipid methylation and lipoxygenase activity are interdependent may be of interest.

Systems which can affect the generation of arachidonic acid will be regulatory to arachidonic acid dependent cellular activation phenomenon. Following Vane's (1971) explanation of the cyclooxy-genase inhibitory action of aspirin-like drugs, there were a number of inquiries as to whether anti-inflammatory corticosteroids had a similiar mechanism. Initial studies recognized that corticosteroids and aspirin-like anti-inflammatories have different mechanisms (Vane, 1971; Flower et al., 1972; Greaves and McDonald-Gibson, 1972) especially since corticosteroids do not prevent the production of cyclooxygenase products from free arachidonic acid. However, it was soon demonstrated, in a number of systems known to produce prostaglandins, that corticosteroids inhibit prostaglandin release or production (Kantrowitz et al., 1975; Tashjian et al., 1975; Lewis and Piper, 1975). Gryglewski et al. (1975) noted, that in examining the effluent of perfused antigen challenged guinea pig lung, corticosteroids inhibited histamine release and prostaglandin production, whereas indomethacin enhanced histamine release while inhibiting prostaglandin production. This report also noted that the inhibitory properties of corticosteroids could be reversed by arachidonic acid. They suggested that corticosteroids cause an impairment in the availability of arachidonic acid as a substrate for enzymes which generate arachidonic acid metabolites. Subse-quently, investigators focused on the ability of corticosteroids to block arachidonic acid generation postulating that corticosteroids may be acting as inhibitors of phospholipase A (Nijkamp et al., 1976; Hong and Levine, 1976; Tam et al., 1977). Blackwell et al. (1978) utilizing a phospholipase A_2 assay in guinea pig lung showed directly that corticosteroids inhibit phospholipase activity.

Previous investigations into the mechanisms of corticosteroids proved useful in guiding the studies of how corticosteroids inhibit phospholipase A. It has been known for some time that inhibitors of RNA and protein synthesis, such as actinomycin D, puromycin and cycloheximide, can block certain actions of corticosteroids (Fain, 1967a; 1967b). It was also known that corticosteroid-induced increases of cellular enzymatic levels, such as aminotransferase, required RNA synthesis (Peterkofsky and Tomkins, 1967; Granner et al., 1968). It was further shown that the mechanism of induction was dependent upon cytoplasmic corticosteroid hormone receptors (Baxter and Tomkins, 1971). Saeed et al. (1977) reported that mammalian serum and plasma contain an inhibitor of prostaglandin production and that the levels of the inhibitor were dependent upon corticosteroid action. The corticosteroid-induced inhibitor being a protein, was made likely by the report of Danon and Assouline (1978) which showed that inhibition of prostaglandin biosynthesis by corticosteroids requires RNA and protein synthesis. Futhermore, the steroid induced protein synthesis which inhibits prostaglandin production is mediated through corticosteroid cellular receptors (Russo-Marie et al., 1979). Utilizing a phospholipase A_2 assay Flower and Blackwell (1979) demonstrated that protein biosynthesis is required for the production of the phospholipase A_2 inhibitor by anti-inflammatory steroids. The involvement of a protein causing the anti-phospholipase A effect of corticosteroids was substantiated in the studies of Blackwell et al. (1980) and Hirata et al. (1980). Blackwell et al. (1980) have referred to the phospholipase A inhibitor as a polypeptide which they call macrocortin. Hirata et al. (1980) claim the inhibitor is a protein(s) of approximately 40,000 daltons. Carnuccio et al. (1980) demonstrated that rat peritoneal cells incubated with hydrocortisone release a protein which inhibits prostaglandin generation by having an anti-phospholipase effect. Carnuccio et al. (1981) subsequently showed that the rat peritoneal cells store and secrete the phospholipase inhibitor. It was not clear from these reports whether the macrocortin was being produced by mast cells or other cell types in the peritoneal cell population. Gryglewski et al. (1975) demonstrated that corticosteroids inhibit histamine release from guinea pig lung. However, whether the steroids inhibit the histamine release by producing macrocortin directly in the mast cell or by producing macrocortin in other cell types which subsequently affects the mast cell is not clear. The recent report of Schleimer et al. (1981), which showed that anti-inflammatory steroids inhibit histamine release from human basophils in vitro also does not resolve whether the steroid is acting on the basophils directly. A long-term incubation of the cells and steroid was required and the system they utilized consisted of a mixed cell population. Whether mast cells and basophils do have corticosteroid receptors and whether macrocortin endogenous to mast cells is regulatory to phospholipase activity remains to be investigated.

It is obvious that phospholipase inhibitors generated by the cell can have a profound effect on the ability of the cell to generate arachidonic acid. Since arachidonic acid is a lipoxygenase substrate, it appears that the interrelationship of macrocortin and lipoxygenases is of great importance in membrane associated events and is an area of cellular activation that requires investigation.

Reactive forms of molecular oxygen may be associated with arachidonic acid metabolism in activated mast cells and basophils. Superoxide anion (O_2^-) generation has been observed to occur concomitantly with mediator release in stimulated mast cells and basophils (Henderson and Kaliner, 1978). The generation of free radical oxygen appears to be an essential component of lipid peroxidation. Cellular lipoxygenases are a class of enzymes which have a requirement for ferric iron (Aharony et al., 1980; Greenwald et al., 1980). Iron-chelates can have stimulatory effects in systems which utilize molecular oxygen to initiate lipid peroxidation. This was the rationale which initiated the report of Magro and Brai (1982) which demonstrated that a ferric iron-desferrioxamine B chelate effectively induces histamine release from rat peritoneal mast cells. The chelated-iron induced histamine release is a non-toxic energy requiring process and in many aspects, parallels IgE-mediated release. Lipoxygenase inhibitors effectively blocked the chelated-iron induced histamine release indicating involvement of fatty acid metabolism via the lipoxygenase pathway. Superoxide dismutase and benzoate, which scavenge the O_2^- free radical and the hydroxyl free radical ($\cdot OH$) respectively, inhibited the chelated-iron induced histamine release. Overall, the data raised the possibility that endogenous cellular iron may be involved in the generation of free radicals and lipid peroxidation, and these may be early events in the IgE-mediated release of histamine.

In many biological systems, the participation of O_2^- in lipid peroxidation is assumed to be due to the formation of $\cdot OH$. The reaction, $H_2O_2 + O_2^- \rightarrow \cdot OH + OH^- + O_2$, (Haber and Weiss, 1934) has been suggested as a source of this activity. Because of the slow rate of the Haber-Weiss reaction, its biological importance as a mechanism of $\cdot OH$ generation is a matter of controversy, except in the presence of Fe^{3+}-chelates which greatly accelerate the reaction (Czapski and Ilan, 1978). Whether mast cells and basophils can generate $\cdot OH$ by means of a Haber-Weiss-like reaction is not yet clear. Mast cells and basophils have other mechanisms to generate free radical oxygen, which could promote lipid peroxidation. The location and types of nicotinamide adenine dinucleotide (NADPH) dependent cytochrome-flavoprotein reductases, which may be essential to cellular activation and which can generate O_2^- by catalyzing the reaction ($2O_2 + NADPH \rightarrow 2O_2^- + NADP^+ + H^+$), are not well delineated in histamine releasing cells. Various dehydrogenases have been cytochemically displayed in histamine releasing cells (Quaglino and Hayhoe, 1960; Balogh and Cohen, 1961), but the relationship to cellular activation is not resolved.

However, there are various types of NADPH-utilizing cytochrome reductases, which can initiate lipid peroxidation and are amplified by chelated iron (Hochstein et al., 1964; Pederson and Aust, 1975; Svingen et al., 1978). In many NADPH cytochrome reductase systems, the initiation of lipid peroxidation is dependent upon the production of a hydroperoxide generating perferryl ion complex (chelate-Fe^{3+}-O_2^-). Perferryl ions can initiate the formation of hydroperoxides. There is evidence that the initiation and propagation of lipid peroxidation takes place via different mechanisms. It has been proposed that propagation of membrane lipid peroxidation arises from a breakdown of hydroperoxides which generate alkoxyl radicals ($RO\cdot$), and it is the alkoxyl radical which actually amplifies and propagates lipid peroxidation (Pederson and Aust, 1975; Svingen et al., 1978). The perferryl ion and other free radicals could possibly be important components in the initiation and propagation of lipid peroxidation by cellular lipoxygenases.

The hydroxylation properties of free radicals such as O_2^- and $\cdot OH$ are considered to be harmful to biological systems. Damaging processes in which free radicals are implicated include radiation (Oberley et al., 1976), inflammatory joint disease (Lund-Oleson and Menander, 1974), DNA inactivation (Van Hemmen and Meuling, 1975), and ageing (Reiss and Gershon, 1976). The production of free radicals has also been considered an important component in the killing of bacteria during phagocytosis (Curnutte and Babior, 1974). However, the report of Magro and Brai (1982) raises the possibility that the generation and usage of free radicals is a normal component of the chemistry of cellular activation. In addition, the report of Smith and Weidemann (1980) suggests that the lipoxygenase pathway may be a significant source of reactive species of oxygen in activated mouse peritoneal macrophages. Whether cellular lipoxygenase and free radical formation are interconnected events in IgE-mediated histamine release, and how lipid peroxidation of the various membranes, i.e., plasma, perigranular and microsomal relate to this process, remain to be elucidated.

The enzymatic sequences which follow the activation of mast cells and basophils are immediate. Part of the function of these enzymes is to convert the activating membrane perturbation, which is a small change in energy, into the larger energy changes which are required for the secretion of histamine. Consequently, amplification systems must exist which, in part, are enzymatic proteins catalyzing changes in lipids causing alterations in the fluidity, water content and the protein-lipid structure of the membranes. An understanding of these enzymatic systems and how they alter the state of the membranes is an understanding of the essence of cellular activation. An essential component of this process is the oxygenation of unsaturated fatty acids by cellular lipoxygenases. It was an object of this report to illustrate that lipoxygenase activity, during cellular activation, is a study of emerging interest and that basophils and mast cells offer ideal cell types in which to investigate the phenomenon.

REFERENCES

Aharony, D., Smith, J. B., and Silver, M. S., 1980, Human platelet lipoxygenase requires ferric iron, Fed. Proc., 39:424.

Alam, M., Anrep, G. V., Barsoum, G. S., Talaat, M., and Weininger, E., 1939, Liberation of histamine from the skeletal muscle by curare, J. Physiol. (Lond), 95:148.

Altounyan, R. E. C., 1967, Inhibition of experimental asthma by a new compound disodium cromoglycate "intal", Acta Allergy, 22: 487.

Änggård, E., Bergzvist, U., Högberg, B., Johansson, K., Thon, I. L., Uvnäs, B., 1963, Biologically active principles occurring on histamine release from cat paw, guinea pig lung and isolated rat mast cells, Acta Physiol Scand., 59:97.

Assem, E. S. K., and Schild, H. O., 1969, Inhibition by sympatho-mimetic amines of histamine release induced by antigen in passively sensitized human lung, Nature, 224:1028.

Assem, E. S. K., and Mongar, J. L., 1970, Inhibition of allergic reactions in man and other species by cromoglycate, Int. Arch. Allergy Appl. Immunol., 38:68.

Bach, M. K., and Brashler, J. R., 1974, In vivo and in vitro production of slow reacting substance in the rat upon treatment with calcium ionophores, J. Immunol., 113:2040.

Bach, M. K., Brashler, J. R., and Gorman, R. R., 1977, On the structure of slow reacting substance of anaphylaxis: evidence of biosynthesis from arachidonic acid, Prostaglandins, 14:21.

Bach, M. K., Brashler, J. R., Hammarström, S., and Samuelsson B., 1980, Identification of a component of rat mononuclear cell SRS as leukotriene D, Biochem. Biophys. Res. Commun., 93:1121.

Balogh, K., Jr., and Cohen, B., 1961, Histochemical demonstration of diaphorases and dehydrogenases in normal human leukocytes and platelets, Blood, 17:491.

Baltzly, R., Buck, J. S., de Beer, E. J., and Webb, F. J., 1949, A family of long-acting depressors, J. Amer. Chem. Soc., 71: 1301.

Baxter, J. D., and Tomkins, G. M., 1971, Specific cytoplasmic glucocorticoid hormone receptors in hepatoma tissue culture cells, Proc. Natl. Acad. Sci. USA, 68:932.

Bazin, H., Querinjean, P., Beckers, A., Heremans, J. I., and Dessy, F., 1974, Transplantable immunoglobulin secreting tumors in rats. IV. Sixty-three IgE-secreting immunocytoma tumors, Immunology, 26:713.

Becker, K. E., Ishizaka, T., Metzger, H., Ishizaka, K., and Grimly, P. M., 1973, Surface IgE on human basophils during histamine release, J. Exp. Med., 138:394.

Bell, R. L., Kennerly, D. A., Stanford, N., and Majerus, P. W., 1979, Diglyceride lipase: a pathway for arachidonate release from human platelets, Proc. Natl. Acad. Sci. USA, 76:3238.

Bennich, H., Ishizaka, K., Ishizaka, T., and Johansson, S. G. O., 1969, A comparative antigenic study of γE-globulin and myeloma-IgND, J. Immunol., 102:826.

Bennich, H., and Johansson, S. G. O., 1971, Structure and function of human immunoglobulin E, Adv. Immunol., 13:1.

Bergström, S., Ryhage, R., Samuelsson, B., and Sjövall, J., 1962, The structure of prostaglandin E, F_1 and F_2, Acta Chem. Scand., 16:501.

Blackwell, G. J., Flower, R. J., Nijkamp, F. P., and Vane, J. R., 1978, Phospholipase A_2 activity of guinea-pig isolated perfused lungs: stimulation and inhibition by anti-inflammatory steroids, Br. J. Pharmacol., 62:79.

Blackwell, G. J., Carnuccio, R., Di Rosa, M., Flower, R. J., Parente, L., and Perisco, P., 1980, Macrocortin: a polypeptide causing the anti-phospholipase effect of glucocorticoids, Nature, 287:147.

Borgeat, P., and Samuelsson, B., 1979a, Arachidonic acid metabolism in polymorphonuclear leukocytes. Unstable intermediate in the formation of dihydroxyacids, Proc. Natl. Acad. Sci. USA, 76:3213.

Borgeat, P., and Samuelsson, B., 1979b, Transformation of arachidonate acid by rabbit polymorphonuclear leukocytes. Formation of a novel dihydroxy eicosatetraenoic acid, J. Biol. Chem., 254:2643.

Borgeat, P., and Samuelsson, B., 1979c, Arachidonic acid metabolism in polymorphonuclear leukocytes: effects of ionophore A23187, Proc. Natl. Acad. Sci. USA, 76:2148.

Borgeat, P., and Samuelsson, B., 1979d, Metabolism of arachidonic acid in polymorphonuclear leukocytes. Structural analysis of novel hydroxylated compounds, J. Biol. Chem., 254:7865.

Brocklehurst, W. E., 1960, The release of histamine and the formation of a slow reacting substance (SRS-A) during anaphylactic shock, J. Physiol. (Lond), 151:416.

Brocklehurst, W. E., 1962, Slow reacting substance and related compounds, Prog. Allergy, 6:539.

Bushby, S. R. M., and Green, A. F., 1955, The release of histamine by polymyxin B and polymyxin E, Br. J. Pharmacol., 10: 215.

Carnuccio, R., Di Rosa, M., and Persico, P., 1980, Hydrocortisone-induced inhibitor of prostaglandin biosynthesis in rat leukocytes, Br. J. Pharmacol., 68:14.

Carnuccio, R., Di Rosa, M., Flower, R. J., Pinto, A., 1981, The inhibition by hydrocortisone of prostaglandin biosynthesis in rat peritoneal leukocytes is correlated with intracellular macrocortin levels, Br. J. Pharmacol., 74:322.

Cazal, P., 1955, Mastocytose médullaire et aplasie, Rev. Belge Pathol., 24:107.

Chakravarty, N. K., 1960a, The mechanism of histamine release in anaphylactic reaction in guinea pig and rat, Acta Physiol. Scand., 48:146.

Chakravarty, N., 1960b, The occurrence of a lipid-soluble smooth muscle stimulating principle (SRS) in anaphylactic reaction, Acta Physiol. Scand., 48:167.

Chakravarty, N. K., 1962a, Aerobic metabolism in anaphylactic reaction in vitro, Am. J. Physiol., 203:1193.

Chakravarty, N., 1962b, Inhibition of anaphylactic histamine release by 2-deoxyglucose, Nature, 194:1182.

Chakravarty, N., 1965, Glycolysis in rat peritoneal mast cells, J. Biol. Chem., 25:123.

Chakravarty, N., 1967, Further observations on the inhibition of histamine release by 2-deoxyglucose, Acta Physiol. Scand., 72:425.

Chakravarty, N., 1968, Respiration of rat peritoneal mast cells during histamine release induced by antigen-antibody reaction, Exp. Cell. Res., 49:160.

Chakravarty, N., and Echetebu, Z., 1978, Plasma membrane adenosine triphosphatase in rat peritoneal mast cells and macrophages: the relation of mast cell enzyme to histamine release, Biochem. Pharmacol., 27:1561.

Chakravarty, N., and Nielsen, E. H., 1980, Ca^{2+}-Mg^{2+} activated adenosine triphosphatase in plasma and granule membranes in non-secreting and secreting mast cells, Exp. Cell. Res., 130: 175.

Cochrane, C. G., and Müller-Eberhard, H. J., 1968, The derivation of two distinct anaphylatoxin activities from the third and fifth components of human complement, J. Exp. Med., 127:371.

Cockcroft, S., and Gomperts, B. D., 1979, Evidence for a role of phosphatidylinositol turnover in stimulus secretion coupling. Biochem. J., 178:681.

Conrad, D. H., Bazin, H., Sehon, A. H., and Froese, A., 1975, Binding parameters of the interaction between rat IgE and rat mast cell receptors, J. Immunol., 114:1688.

Conrad, D. H., and Froese, A., 1976, Characterization of the target cell receptors for IgE. II. Polyacrylamide gel analysis of the surface IgE receptor from normal rat mast cells and rat basophilic leukemia cells, J. Immunol., 116:319.

Conroy, M. C., and Blancuzzi, V., 1979, Differential ability of rat mast cells and human leukocytes to detect inhibitors of histamine release, Monogr. Allergy, 14:307.

Cooper, P. H., and Stanworth, D. R., 1976, Characterization of calcium ion-activated adenosine triphosphatase in the plasma membrane of rat mast cells, Biochem. J., 156:691.

Crews, F. T., Morita, Y., Hirata, F., Axelrod, J., and Siraganian, R. P., 1980, Phospholipid methylation affects immunoglobulin E-mediated histamine and arachidonic acid release in rat leukemic cells, Biochem. Biophys. Res. Commun., 93:42.

Curnutte, J. T., and Babior, B. M., 1974, Biological defense mechanisms. The effect of bacteria and serum on superoxide production by granulocytes, J. Clin. Invest., 53:1662.

Czapski, G., and Ilan, Y. A., 1978, On the generation of the hydroxylation agent from superoxide radical. Can the Haber-Weiss reaction be the source of ·OH radicals? Photochem. Photobiol., 28:651.

Dahlén, S., Björk, J., Hedqvist, P., Arfors, K., Hammarström, S., Lindgren, J., and Samuelsson, B., 1981, Leukotrienes promote plasma leakage and leukocyte adhesion in postcapillary venules: in vivo effects with relevance to the acute inflammatory response, Proc. Natl. Acad. Sci. USA, 78:3887.

Dale, H. H., 1913, The anaphylactic reaction of plain muscle in the guinea pig, J. Pharmacol., 4:167.

Danon, A., and Assouline, G., 1978, Inhibition of prostaglandin biosynthesis by corticosteroids requires RNA and protein synthesis, Nature, 273:552.

De Lisi, C., and Siraganian, R. P., 1979a, Receptor cross linking and histamine release. I. The quantitative dependence of basophil degranulation on the number of receptor doublets, J. Immunol., 122:2286.

Delisi, C., and Siraganian, R. P., 1979b, Receptor cross linking and histamine release. II. Interpretation and analysis of anomalous dose response patterns, J. Immunol., 122:2293.

Dembo, M., Goldstein, B., Sobotka, A. K., and Lichtenstein, L. M., 1979a, Degranulation of human basophils: quantitative analysis of histamine release and desensitization due to a bivalent penicilloyl hapten, J. Immunol., 123:1864.

Dembo, M., Goldstein, B., Sobotka, A. K., and Lichtenstein, L. M., 1979b, Histamine release due to bivalent penicilloyl haptens: the relation of activation and desensitization of basophils to dynamic aspects of ligand binding to cell surface antibody, J. Immunol., 122:518.

Diamant, B., 1962, Further observations on the effect of anoxia on histamine release from guinea pig and rat lung tissue in vitro, Acta Physiol. Scand., 56:1.

Diamant, B., 1967, Observations of some metabolic enzymes of the mast cell and macrophage fraction of rat peritoneal cells, Int. Arch. Allergy Appl. Immunol., 32:236.

Diamant, B., and Krüger, P. G., 1967, Histamine release from isolated rat peritoneal mast cells induced by adenosine-5'-triphosphate, Acta Physiol. Scand., 71:291.

Diamant, B., Norn, S., Felding, P., Olsen, N., Ziebell, A., and Nissen, J., 1974, ATP level and CO_2 production of mast cells in anaphylaxis, Int. Arch. Allergy Appl. Immunol., 47:894.

Douglas, W. W., Kanno, T., and Sampson, S. R., 1967, Influence of the ionic environment on the membrane potential of adrenal chromaffin cells and on the depolarizing effect of acetyl-choline, J. Physiol. (Lond), 191:107.

Dunn, T. B., and Potter, M., 1957, A transplantable mast cell neoplasm in the mouse, J. Natl. Cancer Inst., 18:587.

Dvorak, A. M., Newball, H. H., Dvorak, H. F., and Lichtenstein, L. M., 1980, Antigen-induced IgE-mediated degranulation of human basophils, Lab. Invest., 43:126.

Eccleston, E., Leonard, B. J., Lowe, J. S., and Welford, H. J., 1973, Basophilic leukemia in the albino rat and a demonstration of the basoprotein, Nature (New Biol.), 244:73.

Ellis, H. V., Johnson, A. R., and Moran, N. C., 1970, Selective release of histamine from rat mast cells by several drugs, J. Pharmacol. Exp. Ther., 175:267.

Engineer, D. M., Niederhauser, U., Piper, P. J., and Sirois, P., 1978, Release of mediators of anaphylaxis: inhibition of prostaglandin synthesis and the modification of release of slow reacting substance of anaphylaxis and histamine, Br. J. Pharmacol., 62:61.

Fain, J. N., 1967a, Studies on the role of RNA and protein synthesis in the lipolytic action of growth hormone in isolated fat cells, Adv. Enzyme Regul., 5:39.

Fain, J. N., 1967b, Inhibition of lipolytic action of growth hormone and glycocorticoid by ultraviolet and X-radiation, Science, 157:1062.

Falkenhein, S. F., MacDonald, H., Huber, M. M., Koch, D., and Parker, C. W., 1980, Effect of the 5-hydroperoxide of eicosatetraenoic acid and inhibitors of the lipoxygenase pathway on the formation of slow reacting substance by rat basophilic leukemia cells; direct evidence that slow reacting substance is a product of the lipoxygenase pathway, J. Immunol., 125:163.

Fawcett, D. W., 1954, Cytological and pharmacological observations on the release of histamine by mast cells, J. Exp. Med., 100:217.

Feldberg, W., and Kellaway, C. H., 1938, Liberation of histamine and formation of a lecithin-like substance by Cobra venom, J. Physiol. (Lond), 94:187.

Feldberg, W., Holden, H. F., and Kellaway, C. H., 1938, Formation of lysocithin and of muscle-stimulating substance by snake venom, J. Physiol. (Lond), 94:232.

Feldberg, W., and Paton, W. D. M., 1951, Release of histamine from skin and muscle in cat by opium alkaloids and other histamine liberators, J. Physiol. (Lond), 114:490.

Fewtrell, C. M. S., and Gomperts, B. D., 1977, Effect of flavone inhibitors of transport ATPases on histamine secretion from rat mast cells, Nature, 265:635.

Fireman, P., Vannier, W. E., and Goodman, H. C., 1963, The association of skin-sensitizing antibody with the β_2A-globulins in sera from ragweed-sensitive patients, J. Exp. Med., 117:603.

Flower, R., Gryglewski, R., Herbaczyńska-Cedro, K., and Vane, J. R., 1972, Effects of anti-inflammatory drugs on prostaglandin biosynthesis, Nature, 238:104.

Flower, R. J., and Blackwell, G. J., 1979, Anti-inflammatory steroids induce biosynthesis of a phospholipase A_2 inhibitor which prevents prostaglandin generation, Nature, 278:456.

Foreman, J. C., Mongar, J. L., and Gomperts, D., 1973, Calcium ionophores and movement of calcium ions following the physiological stimulus to a secretory process, Nature, 245:249.

Friedberger, E., 1909, Weitere untersuchungen über eiweissanaphylaxie, Ztschr. Immunitätsforsh., 4:636.

Froese, A., 1980, Structure and function of the receptor for IgE, CRC Crit. Rev. Immunol., 1:79.

Furth, J., Hagen, P., and Hirsh, E., 1957, Transplantable mastocytoma in the mouse containing histamine heparin and 5-hydroxytryptamine, Proc. Soc. Exp. Biol. Med., 95:824.

Goetzl, E. J., Woods, J. M., and Gorman, R. R., 1977, Stimulation of human eosinophil and neutrophil polymorphonuclear leukocyte chemotaxis and random migration by 12-L-hydroxy-5,8,10,14-eicosatetraenoic acid, J. Clin. Invest., 59:179.

Goth, A., and Adams, H. R., 1970, Selective effect of phosphatidylserine on macromolecular histamine release in the rat, Fed. Proc., 69:2087.

Goth, A., Adams, H. R., and Knoohuizen, M., 1971, Phosphatidylserine selective enhancer of histamine release, Science, 173:1034.

Graham, H. T., Wheelwright, R., Parish, H. H., Jr., Marks, A. R., and Lowry, O. H., 1952, Distribution of histamine among blood elements, Fed. Proc., 11:350.

Granner, D. K., Hayashi, S., Thompson, E. B., and Tomkins, G. M., 1968, Stimulation of tyrosine aminotransferase synthesis by dexmethasone phosphate in cell culture, J. Mol. Biol., 35:291.

Grant, J. A., Dupree, E., Goldman, A. S., Schultz, D. R., and Jackson, A. L., 1975, Complement-mediated release of histamine from human leukocytes, J. Immunol., 114:1101.

Grant, J. A., Settle, L., Whorton, E. B., and Dupree, E., 1976, Complement-mediated release of histamine from human basophils. II. Biochemical characterization of the reaction, J. Immunol., 117:450.

Greaves, M. W., and McDonald-Gibson, W., 1972, Inhibition of prostaglandin biosynthesis by corticosteroids, Br. Med. J., 2:83.

Greenwald, J. E., Alexander, M. S., Fertel, R. H., Beach, C. A., Wong, L. K., and Bianchine, J. R., 1980, Role of ferric iron in platelet lipoxygenase activity, Biochem. Biophys. Res. Commun., 96:817.

Gryglewski, R. J., Panczenko, B., Korbut, R., Grodzińska, L., and Ocetkiewicz, A., 1975, Corticosteroids inhibit prostaglandin release from perfused mesenteric blood vessels of rabbit and from perfused lungs of sensitized guinea pig, Prostaglandins, 10:343.

Haber, F., and Weiss, J., 1934, The catalytic decomposition of hydrogen peroxide by iron salts, Proc. R. Soc. A., 147:332.

Hamberg, M., 1976, On the formation of thromboxane B_2 and 12L-hydroxy-5,8,10,14-eicosatetraenoic acid (12 No. 20:4) in tissue from the guinea pig, Biochim. Biophys. Acta, 431:651.

Hamberg, M., and Samuelsson, B., 1973, Detection and isolation of an endoperoxide intermediate in prostaglandin biosynthesis, Proc. Natl. Acad. Sci. USA, 70:899.

Hamberg, M., and Samuelsson, B., 1974, Prostaglandin endoperoxides. Novel transformations of arachidonic acid in human platelets, Proc. Natl. Acad. Sci. USA, 71:3400.

Hamberg, M., Svensson, J., Wakabayashi, T., and Samuelsson, B., 1974a, Isolation and structure of two prostaglandin endoperoxides that cause platelet aggregation, <u>Proc. Natl. Acad. Sci.</u>, 71:345.

Hamberg, M., Svensson, J., and Samuelsson, B., 1974b, Prostaglandin endoperoxides. A new concept concerning the mode of action and release of prostaglandins, <u>Proc. Natl. Acad. Sci.</u>, 71:3824.

Hamberg, M., Svensson, J., and Samuelsson, B., 1975, Thromboxanes: a new group of biologically active compounds derived from prostaglandin endoperoxides, <u>Proc. Natl. Acad. Sci.</u>, 72:2994.

Hammarström, S., Murphy, R. C., Samuelsson, B., Clark, D. A., Misokowski, C., and Corey, E. J., 1980, The structure of leukotriene C identification of the amino acid part, <u>Biochem. Biophys. Res. Commun.</u>, 91:1266.

Henderson, W. R., and Kaliner, M., 1978, Immunologic and nonimmunologic generation of superoxide from mast cells and basophils, <u>J. Clin. Invest.</u>, 61:187.

Hirata, F., Axelrod, J., and Crews, F. T., 1979, Concanavalin A stimulates phospholipid methylation and phosphatidylserine decarboxylation in rat mast cells, <u>Proc. Natl. Acad. Sci.</u>, 76:4813.

Hirata, F., and Axelrod, J., 1980, Phospholipid methylation and biological signal transmission, <u>Science</u>, 209:1082.

Hirata, F., Schiffman, E., Venkatasubramanian, K., Salomon, J., and Axelrod, J., 1980, A phospholipase A2 inhibitory protein in rabbit neutrophils induced by glucocorticoids, <u>Proc. Natl. Acad. Sci.</u>, 77:2533.

Hochstein, P., Nordenbrand, K., and Ernster, L., 1964, Evidence for the involvement of iron in the ADP-activated peroxidation of lipids in microsomes and mitochondria, <u>Biochem. Biophys. Res. Commun.</u>, 14:323.

Högberg, B., and Uvnäs, B., 1958, Inhibitory action of allicin on degranulation of mast cells produced by compound 48/80, histamine liberator from ascaris, lecithinase A and antigen, <u>Acta Physiol. Scand.</u>, 44:157.

Högberg, B., and Uvnäs, B., 1960, Further observations on the disruption of rat mesentery mast cells caused by compound 48/80 antigen-antibody reaction, lecithinase A and decylamine, <u>Acta Physiol. Scand.</u>, 48:133.

Hokin, M. R., and Hokin, L. E., 1953, Enzyme secretion and incorporation of P^{32} into phospholipides of pancreas slices, <u>J. Biol. Chem.</u>, 203:967.

Hong, S. L., and Levine, L., 1976, Inhibition of arachidonic acid release from cells as the biochemical action of anti-inflammatory corticosteroids, <u>Proc. Natl. Acad. Sci.</u>, 73:1730.

Isersky, C., Taurog, J. T., Poy, G., and Metzger, H., 1978, Triggering of cultured neoplastic cells by antibodies to the receptor for IgE, <u>J. Immunol.</u>, 121:549.

Ishizaka, K., Ishizaka, T., and Hornbrook, M. M., 1963, Blocking of Prausnitz-Küstner sensitization with reagin by normal human beta-2A globulin, <u>J. Allergy</u>, 34:395.

Ishizaka, K., Ishizaka, T., and Hornbrook, M. M., 1966a, Physico-chemical properties of human reaginic antibody. IV. Presence

of a unique immunoglobulin as a carrier of reaginic antibody,
J. Immunol., 97:75.

Ishizaka, K., Ishizaka, T., and Hornbrook, M. M., 1966b, Physico-
chemical properties of reaginic antibody. V. Correlation of
reaginic antibody with γ-E globulin antibody, J. Immunol.,
97:840.

Ishizaka, K., and Ishizaka, T., 1969, Immune mechanisms of reversed
type reagenic hypersensitivity, J. Immunol., 103:588.

Ishizaka, T., Ishizaka, K., Johansson, S. G. O., and Bennich, H.,
1969, Histamine release from human leukocytes by anti-γE
antibodies, J. Immunol., 102:884.

Ishizaka, T., Soto, C. S., and Ishizaka, K., 1973, Mechanisms of
passive sensitization. III. Number of IgE molecules and
its receptor sites on human basophil granulocytes, J. Immunol.,
111:500.

Ishizaka, T., Chang, T. H., Taggart, M., and Ishizaka, K., 1977,
Histamine release from mast cells by antibodies against rat
basophilic leukemia cell membrane, J. Immunol., 119:1589.

Ishizaka, T., and Ishizaka, K., 1978, Triggering of histamine
release from rat mast cells by divalent antibodies against
IgE-receptors, J. Immunol., 120:800.

Jakschik, B. A., Kulzycki, A., MacDonald, H. H., and Parker, C. W.,
1977a, Release of slow reacting substance (SRS) from rat
basophil leukemia (RBL-1) cells, J. Immunol., 119:618.

Jakschik, B. A., Falkenhein, S., and Parker, C. W., 1977b, Precursor
role of arachidonic acid in release of slow-reacting substance
from rat basophilic leukemia cells, Proc. Natl. Acad. Sci.,
74:4577.

Jaques, R., 1965, Non-specific effects of synthetic corticotropin
polypeptides, Int. Arch. Allergy Appl. Immunol., 28:16.

Jasani, B., and Stanworth, D. R., 1972, Studies on the mast cell
triggering action of certain artificial histamine liberators,
Int. Arch. Allergy Appl. Immunol., 45:74.

Jasmin, G., 1956, Etude de l'inflammation anaphylactoide, Rev.
Can. Biol., 15:107.

Johansen, T., and Chakravarty, N., 1972, Dependence of histamine
release from rat mast cells on adenosine triphosphate,
Naunyn-Schmiedebergs Arch. Pharmacol., 275:457.

Johansen, T., and Chakravarty, N., 1975, The utilization of adenosine
triphosphate in rat mast cells during histamine release induced
by anaphylactic reaction and compound 48/80, Naunyn-
Schmiedebergs Arch. Pharmacol., 288:243.

Johansson, S. G. O., and Bennich, H., 1967, Immunological studies of
an atypical (meyeloma) immunoglobulin, Immunology, 13:381.

Johnson, A. R., and Moran, N. C., 1969, Selective release of histamine
from the mast cell by compound 48/80 and antigen, Am. J.
Physiol., 216:453.

Johnson, A. R., Hugli, T. E., and Müller-Eberhard, H. J., 1975,
Release of histamine from rat mast cells by the complement
peptides C3a and C5a, Immunology, 28:1067.

Johnson, R. A., Morton, D. R., Kinner, J. H., Gorman, R. R.,

McGuire, J. C., Sun, F. F., Whittaker, N., Bunting, S., Salmon, J., Moncada, S., and Vane, J. R., 1976, The chemical structure of prostaglandin X (prostacyclin), Prostaglandins, 12:915.

Juhlin, L., and Ohman, S., 1964, Basophil and eosinophil leukocytes in cantharidin blisters of patients with various dermatoses, Acta Derm. Venereol., 44:303.

Kantrowitz, F., Robinson, D. R., McGuire, M. B., and Levine, L., 1975, Corticosteroids inhibit prostaglandin production by rheumatoid synovia, Nature, 258:737.

Kellaway, C. H., and Trethewie, E. R., 1940, The liberation of slow-reacting smooth-muscle stimulating substance in anaphylaxis, Q. J. Exp. Physiol., 30:121.

Keller, R., 1968, Interrelations between different types of cells. II. Histamine-release from the mast cells of various species by cationic polypeptides of polymorphonuclear luekocyte lysosomes and other cationic compounds, Int. Arch. Allergy Appl. Immunol., 34:139.

Keller, R., 1973, Concanavalin A, a model "antigen" for the in vitro detection of cell-bound reaginic antibody in the rat, Clin. Exp. Immunol., 13:139.

Kennerly, D. A., Sullivan, T. J., and Parker, C. W., 1979a, Activation of phospholipid metabolism during mediator release from stimulated rat mast cells, J. Immunol., 122:152.

Kennerly, D. A., Sullivan, T. J., Sylwester, P., and Parker, C. W., 1979b, Diacyglycerol metabolism in mast cells: a potential role in membrane fusion and arachidonic acid release, J. Exp. Med., 150:1039.

Kennerly, D. A., Secosan, C. J., Parker, C. W., and Sullivan, T. J., 1979c, Modulation of stimulated phospholipid metabolism in mast cells by pharmacologic agents that increase cyclic 3',5' adenosine monophosphate levels, J. Immunol., 123:1519.

Kulczcki, A., McNearney, T. A., and Parker, C. W., 1976, The rat basophilic leukemia cell receptor for IgE. I. Characterization as a glycoprotein, J. Immunol., 117:661.

Kulczycki, A., Jr., Isersky, C., and Metzger, H., 1974, The interaction of IgE with rat basophilic leukemia cells. I. Evidence for specific binding of IgE, J. Exp. Med., 139:600.

Kulczycki, A., Jr., and Metzger, H., 1974, The interaction of IgE with rat basophilic leukemia cells. II. Quantitative aspects of the binding reaction, J. Exp. Med., 140:1676.

Lagunoff, D., 1973, Membrane fusion during mast cell secretion, J. Cell. Biol., 57:252.

Landsteiner, K., 1924, Experiments on anaphylaxis to azoproteins, J. Exp. Med., 39:631.

Landsteiner, K., and Levine, P., 1930, Experiments on anaphylaxis to azoproteins. Third paper, J. Exp. Med., 52:347.

Lawson, D., Fewtrell, C., Gomperts, B., and Raff, M. C., 1975, Anti-immunoglobulin-induced histamine secretion by rat peritoneal mast cells studied by immunoferritin electron microscopy, J. Exp. Med., 142:391.

Lawson, D., Raff, M. C., Gomperts, B., Fewtrell, C., and Gilula, N. D., 1977, Molecular events during membrane fusion. A study of exocytosis in rat peritoneal cells, J. Cell. Biol., 72:242.

Levine, B. B., 1962, N (α-D-Penicilloyl) amines as univalent hapten inhibitors of antibody-dependent allergic reactions to penicillin, J. Med. Chem., 5:1025.

Lewis, G. P., and Piper, P. J., 1975, Inhibition of release of prostaglandins as an explanation of some of the actions of anti-inflammatory corticosteroids, Nature, 254:308.

Lichtenstein, L. M., 1975, The mechanism of basophil histamine relea induced by antigen and by the calcium ionophore A23187, J. Immunol., 114:1692.

Lichtenstein, L. M., and Osler, A. G., 1964, Studies on the mechanis of hypersensitivity phenomena: histamine release from human leukocytes by ragweed pollen antigen, J. Exp. Med., 120:507.

Lichtenstein, L. M., and Margolis, S. C., 1968, Histamine release in vitro: inhibition by catecholamines and methylxanthines, Science, 161:902.

Lichtenstein, L. M., and De Bernardo, R., 1971, The immediate allerg response: in vitro action of cyclic AMP active and other drugs on the two stages of histamine release, J. Immunol., 107:1131.

Lund-Oleson, K., and Menander, K. B., 1974, Orgotein: a new anti-inflammatory metalloprotein drug: preliminary evaluation of clinical efficacy and safety in degenerative joint disease, Curr. Ther. Res., 16:706.

Magro, A. M., 1974a, In vitro studies of concanavalin-A-induced histamine release from human basophils: excess bridging in th inhibitory region of the dose-response curve, Int. Arch. Allergy Appl. Immunol., 47:433.

Magro, A. M., 1974b, Evidence for IgE involvement in Con A induced histamine release from human basophils in vitro, Nature, 249: 512.

Magro, A. M., 1975, Evidence for one hit activation for the in vitro release of histamine from human basophils, Immunochem., 12:389

Magro, A. M., 1977a, Ethacrynic acid inhibitable Ca^{2+} and Mg^{2+} activated membrane adenosine triphosphatase in rat mast cells, Clin. Exp. Immunol., 30:160.

Magro, A. M., 1977b, Blocking of histamine release from human basophils in vitro by the ATPase inhibitor, ethacrynic acid, Clin. Exp. Immunol., 29:436.

Magro, A. M., 1980, Creatine phosphokinase in rat mast cells, Immunology, 39:323.

Magro, A. M., 1981, Histamine release from fawn-hooded rat mast cells is not potentiated by phosphatidylserine, Immunology, 44:1.

Magro, A. M., 1982, Effect of inhibitors of arachidonic acid metabolism upon IgE and non-IgE mediated histamine release, Int. J. Immunopharmacol., 4:15.

Magro, A. M., and Alexander, A., 1974a, In vitro studies of histamin release from rabbit leukocytes by divalent haptens, J. Immunol., 112:1757.

336

Magro, A. M., and Alexander, A., 1974b, Histamine release: in vitro studies of the inhibitory region of the dose-response curve, J. Immunol., 112:1762.

Magro, A. M., and Bennich, H., 1977, Concanavalin A induced histamine release from human basophils in vitro, Immunology, 33:51.

Magro, A. M., and Brai, M., 1982, Evidence for lipoxygenase activity in induction of histamine release from rat peritoneal mast cells by chelated iron, Immunology, in press.

Magro, A. M., Cragoe, E. J., Jr., and Hurtado, I., 1982, Effect of sulfhydryl-reactive ATPase inhibitors upon mast cell and basophil activation,(personal observation; to be submitted).

Marone, G., Kagey-Sobotka, A., and Lichtenstein, L., 1979, Effects of arachidonic acid and its metabolites on antigen-induced histamine release from human basophils in vitro, J. Immunol., 123:1669.

Marone, G., Hammarström, S., and Lichtenstein, L. M., 1980, An inhibitor of lipoxygenase inhibits histamine release from human basophils, Clin. Immunol. Immunopathol., 17:117.

Marquardt, D. L., Nicolotti, R. A., Kennerly, D. A., and Sullivan, T. J., 1981, Lipid metabolism during mediator release from mast cells: studies of the role of arachidonic acid metabolism in the control of phospholipid metabolism, J. Immunol., 127: 845.

Michell, R. H., 1975, Inositol phospholipids and cell surface receptor function, Biochim. Biophys. Acta, 415:81.

Mielens, Z. E., and Magro, A. M., 1982, Comparison of effects of investigational compounds upon allergic phenomenon in rodents in vivo and human basophils in vitro, Methods Find. Exp. Clin. Pharmacol., 4:111.

Mollay, C., Kreil, G., and Berger, H., 1976, Action of phospholipases on the cytoplasmic membrane of escherichia coli. Stimulation by melettin, Biochim. Biophys. Acta, 426:317.

Moncada, S., Gryglewski, R., Bunting, S., and Vane, J. R., 1976, An enzyme isolated from arteries transforms prostaglandin endoperoxides to an unstable substance that inhibits platelet aggregation, Nature, 263:663.

Mongar, J. L., and Schild, H. O., 1955, Inhibition of histamine release in anaphylaxis, Nature, 176:163.

Mongar, J. L., and Schild, H. O., 1956, Effect of antigen and organic bases on intra-cellular histamine in guinea-pig lungs, J. Physiol. (Lond), 131:207.

Mongar, J. L., and Schild, H. O., 1957a, Inhibition of the anaphylactic reaction, J. Physiol. (Lond), 135:301.

Mongar, J. L., and Schild, H. O., 1957b, Effect of temperature on the anaphylactic reaction, J. Physiol. (Lond), 135:320.

Mongar, J. L., and Schild, H. O., 1958, The effect of calcium and pH on the anaphylactic reaction, J. Physiol. (Lond), 140:272.

Mongar, J. L., and Schild, H. O., 1962, Cellular mechanisms in anaphylaxis, Physiol. Rev., 42:226.

Mongar, J. L., and Perera, B. A. V., 1965, Oxygen consumption during

337

histamine release by antigen and compound 48/80, Immunology, 8:511.

Morita, Y., Chiang, P. K., and Siraganian, R. P., 1981, Effect of inhibitors of transmethylation on histamine release from human basophils, Biochem. Pharmacol., 30:785.

Morris, H. R., Taylor, G. W., Piper, P. J., Sirois, P., and Tippins, J. R., 1978, Slow-reacting substance of anaphylaxis: purification and characterization, FEBSC (letters), 87:203.

Morris, H. R., Taylor, G. W., Piper, P. J., and Tippins, J. R., 1980a, Structure of slow-reacting substance of anaphylaxis from guinea pig lung, Nature, 285:104.

Morris, H. R., Taylor, G. W., Piper, P. J., Samhoun, M. N., and Tippins, J. R., 1980b, Slow reacting substance (SRSs): the structure identification of SRSs from rat basophil leukaemia (RBL-1) cells, Prostaglandins, 19:185.

Mota, I., 1959, The mechanism of action of anaphylatoxin. Its effect on guinea pig mast cell, Immunology, 2:403.

Mota, I., and Dias Da Silva, W., 1960, Antigen induced damage to isolated sensitized mast cells, Nature, 186:245.

Moussatché, H., and Prouvost-Dannon, A., 1958, Influence of oxidative phosphorylation inhibitors on the histamine release in the anaphylactic reaction in vitro, Experientia, 14:414.

Moussatché, H., and Prouvost-Dannon, A., 1962, Influence of inhibitor of the respiratory chain on the release of histamine during the anaphylactic reaction in vitro. Action of antimycin A and carbon monoxide, Biochem. Pharmacol., 11:603.

Müller-Eberhard, H. J., 1960, A new supporting medium for preparative electrophoresis, Scand. J. Clin. Lab. Invest., 12:33.

Murphy, R. C., Hammarström, S., and Samuelsson, B., 1979, Leukotriene C: a slow-reacting substance from murine mastocytoma cells, Proc. Natl. Acad. Sci., 76:4275.

Nugteren, D. H., and Hazelhof, E., 1973, Isolation and properties of intermediates in prostaglandin biosynthesis, Biochim. Biophys. Acta, 326:448.

Nÿkamp, F. P., Flower, R. J., Moncada, S., and Vane, J. R., 1976, Partial purification of rabbit aorta contracting substance-releasing factor and inhibition of its activity by anti-inflammatory steroids, Nature, 263:479.

Oberley, L. W., Lindgreen, S. A., and Stevens, R. H., 1976, Superoxide ion as the cause of the oxygen effect, Radiat. Res., 68:320.

Ogawa, M., Kochwa, S., Smith, C., Ishizaka, K., and McIntyre, O. R., 1969, Clinical aspects of IgE myeloma, N. Engl. J. Med., 281:1217.

Ogilvie, B. M., 1967, Reagin-like antibodies in animals immune to helminth parasites, Nature, 204:91.

Orange, R. P., and Austen, K. F., 1969, Slow reacting substance of anaphylaxis, Adv. Immunol., 10:106.

Orange, R. P., Murphy, R. C., and Austen, K. F., 1974, Inactivation of slow reacting substance of anaphylaxis (SRS-A) by arylsulfatases, J. Immunol., 113:316.

Orange, R. P., and Chang, P. L., 1975, The effect of thiols on
immunologic release of slow reacting substance of anaphylaxis.
I. Human lung, J. Immunol., 115:1072.

Orange, R. P., and Moore, E. G., 1976, The effect of thiols on the
immunologic release of slow reacting substance of anaphylaxis.
II. Other in vitro and in vivo models, J. Immunol., 116:392.

Osler, A. G., Randall, H. G., Hill, B. M., and Ovary, Z., 1959,
Studies on the mechanisms of hypersensitivity phenomena.
III. The participation of complement in the formation of
anaphylatoxin, J. Exp. Med., 110:311.

Ovary, Z., 1958, Immediate reactions in the skin of experimental
animals provoked by antigen-antibody interaction, Prog.
Allergy, 5:460.

Ovary, Z., and Karush, F., 1960, Studies on the immunologic mechanism
of a anaphylaxis. I. Antibody-hapten interactions studied by
passive cutaneous anaphylaxis in the guinea pig, J. Immunol.,
84:409.

Parish, W. E., 1967, Release of histamine and slow reacting substance
with mast cell changes after challenge of human lung sensitized
with reagin in vitro, Nature, 215:738.

Parker, C. W., Kern, M., and Eisen, H. N., 1962, Polyfunctional
dinitrophenyl haptens as reagents for elicitation of immediate
type allergic skin responses, J. Exp. Med., 115:789.

Parker, C. W., Huber, M. T., and Falkenhein, S., 1979, Incorporation
of ^{35}S into slow reacting substance (SRS), Fed. Proc., 38:1167.

Parker, C. W., Koch, D., Huber, M. M., and Falkenhein, S. F., 1980a,
Formation of the cysteinyl form of slow reacting substance
(leukotriene E_4) in human plasma, Biochem. Biophys. Res.
Commun., 97:1038.

Parker, C. W., Falkenhein, S. F., and Huber, M. M., 1980b, Sequential
conversion of the glutathionyl side chain of slow reacting
substance (SRS) to cysteinyl-glycine and cysteine in rat
basophilic leukemia cells stimulated with A-23187,
Prostaglandins, 20:863.

Parker, C. W., Koch, D., Huber, M. M., and Falkenhein, S. F., 1980c,
Incorporation of radiolable from [1-14C] 5-hydroperoxy-
eicosatetraenoic acid into slow reacting substance, Biochem.
Biophys. Res. Commun., 94:1037.

Parrot, J. L., 1942, Sur la reaction cellulaire de l'anaphylaxie.
Son charatere aérobie, C.R. Soc. Biol. (Paris), 136:361.

Pederson, C. T., and Aust, S. D., 1975, The mechanism of liver
microsomal lipid peroxidation, Biochim. Biophys. Acta, 385:
232.

Peterkofsky, B., and Tomkins, G. M., 1967, Effect of inhibitors of
nucleic acid synthesis on steroid-mediated induction of
tyrosine aminotransferase in hepatoma cell cultures, J. Mol.
Biol., 30:49.

Peters, S. P., Siegel, M. I., Kagey-Sobotka, A., and Lichtenstein,
L. M., 1981, Lipoxygenase products modulate histamine release
in human basophils, Nature, 292:455.

Petersson, B., Nilsson, A., and Stålenheim, G., 1975, Induction of

histamine release and desensitization in human leukocytes,
J. Immunol., 144:1581.

Piper, P. J., and Vane, J. R., 1969, Release of additional factors
in anaphylaxis and its antagonism by anti-inflammatory drugs,
Nature, 223:29.

Plaut, P., Marone, G., Thomas, L. L., and Lichtenstein, L. M., 1980,
Cyclic-nucleotides in immune responses and allergy, Adv.
Cyclic Nucleotide Res., 12:161.

Prausnitz, C., and Küstner, H., 1921, Ustudien über allergie,
Zentr. Bakteriol. Parasitenk. Abt. I Orig., 86:160.

Quaglino, D., and Hayhoe, F. G., 1960, Acetone fixation for the
cytochemical demonstration of dehydrogenases in blood and
bone marrow cells, Nature, 187:85.

Ranadive, N. S., and Cochrane, C. G., 1968, Isolation and characterization of permeability factors from rabbit neutrophils, J.
Exp. Med., 128:605.

Ranadive, N. S., and Cochrane, C. G., 1971, Mechanism of histamine
release from mast cells by cationic protein (band 2) from
neutrophil lysosomes, J. Immunol., 106:506.

Reed, P. W., and Lardy, H. A., 1972, A23187: a divalent cation
ionophore, J. Biol. Chem., 247:6970.

Reiss, U., and Gershon, D., 1976, Rat liver superoxide dismutase.
Purification and age-related modifications, Eur. J. Biochem.,
63:617.

Riley, J. F., and West, G. B., 1952, Histamine in tissue mast cells,
J. Physiol.(Lond), 117:72.

Riley, J. F., and West, G. B., 1953, The presence of histamine
release in tissue mast cells, J. Physiol. (Lond), 120:528.

Roberts, L. J., II, Lewis, R. A., Lawson, J. A., Sweetman, B. J.,
Austen, K. F., and Oates, J. A., 1978, Arachidonic acid
metabolism by rat mast cells, Prostaglandins, 15:717.

Röhlich, P., 1975, Membrane associated actin filaments in the cortical
cytoplasm of the rat mast cell, Exp. Cell. Res., 93:293.

Röhlich, P., Anderson, P., and Uvnäs, B., 1971, Electron Collier
microscope observations on compound 48/80-induced degranulation
in rat mast cells, J. Cell. Biol., 51:465.

Rothschild, A. M., 1966, Histamine release by basic compounds, in:
"Handbook of Experimental Pharmacology," O. Eichler and
A. Farah, ed., Vol. 18, Histamine and anti-histaminics, Part 1,
M. Rocha e Silva, ed., pp. 386-430, Springer-Verlag,
New York.

Russo-Marie, F., Paing, M., and Duval, D., 1979, Involvement of
glucocorticoid receptors in steroid-induced inhibition of
prostaglandin secretion, J. Biol. Chem., 254:8498.

Saeed, S. A., McDonald-Gibson, W. J., Cuthbert, J., Copas, J. L.,
Schneider, C., Gardiner, P. J., Butt, N. M., and
H. O. J., 1977, Endogenous inhibitor of prostaglandin
synthetase, Nature, 270:32.

Samuelsson, B., 1965, On the incorporation of oxygen in the conversion
of 8,11,14-eicosatrienoic acid to prostaglandin E, J. Am. Chem.
Soc., 87:3011.

Samuelsson, B., Borgeat, P., Hammarström, S., and Murphy, R. C.,
    1979, Introduction of a nomenclature: leukotrienes,
    Prostaglandins, 17:785.
Schild, H. O., 1936, Histamine release and anaphylactic shock in
    isolated lungs of guinea pigs, Q. J. Exp. Physiol., 26:165.
Schleimer, R. P., Lichtenstein, L. M., and Gillespie, E., 1981,
    Inhibition of basophil histamine release by anti-inflammatory
    steroids, Nature, 292:454.
Schultz, W. H., 1910, Physiological studies in anaphylaxis. I.
    The reaction of smooth muscle of the guinea pig sensitized
    with horse serum, J. Pharmacol., 1:549.
Seegers, W., and Janoff, A., 1966, Mediators of inflammation in
    leukocyte lysosomes. VI. Partial purification and
    characterization of a mast cell-rupturing component, J. Exp.
    Med., 124:833.
Sessa, G., Freer, J. H., Colacicco, G., and Wiessmann, G., 1969,
    Interaction of a lytic polypeptide melittin with lipid
    membrane systems, J. Biol. Chem., 244:3575.
Sheard, P., Killingback, P. G., and Blair, A. M. J. N., 1967,
    Antigen induced release of histamine and SRS-A from human
    lung passively sensitized with reaginic serum, Nature, 216:
    283.
Shelley, W. B., 1963, Indirect basophil degranulation test for
    allergy to penicillin and other drugs, J. Am. Med. Assoc.,
    184:171.
Shelley, W. B., and Juhlin, L., 1961, A new test for detecting
    anaphylactic sensitivity. The basophil reaction, Nature,
    191:1056.
Shelley, W. B., and Juhlin, L., 1962, In vitro effect of lecithinase
    A on the cytology of the human basophil, J. Lab. Clin. Med.,
    60:589.
Shore, P. A., Burkhalter, A., and Cohn, V. H., Jr., 1959, A method
    for the fluorometric assay of histamine in tissues, J.
    Pharmacol. Exp. Ther., 127:182.
Siraganian, P. A., and Siraganian, R. P., 1974, Basophil activation
    by concanavalin A: characteristics of the reaction, J. Immunol.,
    112:2117.
Siraganian, R. P., 1974, An automated continuous-flow system for the
    extraction and fluorometric analysis of histamine, Anal.
    Biochem., 57:383.
Siraganian, R. P., and Osler, A. G., 1970, Antigenic release of
    histamine from rabbit leukocytes, J. Immunol., 104:1340.
Siraganian, R. P., Hook, W. A., and Levine, B. B., 1975, Specific
    in vitro histamine release from basophils by bivalent haptens:
    evidence for activation by simple bridging of membrane bound
    antibodies, Immunochemistry, 12:149.
Siraganian, R. P., and Hook, W. A., 1977, Mechanism of histamine
    release by formyl methionine containing peptides, J. Immunol.,
    119:2078.
Smith, R. L., and Weidemann, M. J., 1980, Reactive oxygen production
    associated with arachidonic acid metabolism by peritoneal

macrophages, Biochem. Biophys. Res. Commun., 97:973.

Stanworth, D. R., Humphrey, J. H., Bennich, H., and Johansson, S. G. O., 1967, Specific inhibition of the prausnitz-küstner reaction by an atypical human myeloma protein, Lancet, 2:330.

Stanworth, D. R., Humphrey, J. H., Bennich, H., and Johansson, S. G. O., 1968, Inhibition of prausnitz-küstner reaction by proteolytic-cleavage fragments of a human myeloma protein of immunoglobulin class E, Lancet, 2:17.

Stenson, W. F., Parker, C. W., and Sullivan, T. J., 1980, Augmentation of IgE-mediated release of histamine by 5-hydroxyeicosatet-raenoic acid, Biochem. Biophys. Res. Commun., 96:1045.

Strandberg, K., and Uvnäs, B., 1971, Purifications and properties of slow reacting substance formed from cat paw perfused with compound 48/80, Acta Physiol. Scand., 82:358.

Strandberg, K., Möolby, R., and Wadström, T., 1974, Histamine release from mast cells by highly purified phospholipase C (alpha-toxin) and thetatoxin from clostridium prefringens, Toxicon, 12:199.

Sullivan, T., and Parker, C. W., 1979, Possible role arachidonic acid and its metabolites in mediator release from rat mast cells, J. Immunol., 122:431.

Svingen, B. A., O'Neal, F. O., and Aust, S. D., 1978, The role of superoxide and singlet oxygen in lipid peroxidation, Photochem. Photobiol., 28:803.

Sydbom, A., Fredholm, B., and Uvnäs, B., 1981, Evidence against a role of cyclic nucleotides in the regulation of anaphylactic histamine release in isolated mast cells, Acta Physiol. Scand., 112:47.

Tam, S., Hong, S. L., and Levine, L., 1977, Relationships among the steroids of anti-inflammatory properties and inhibition of prostaglandin production and arachidonic acid release by transformed mouse fibroblasts, J. Pharmacol. Exp. Ther., 203:162.

Tashjian, A. H., Jr., Voelkel, E. F., McDonough, J., and Levine, L., 1975, Hydrocortisone inhibits prostaglandin production by mouse fibrosarcoma cells, Nature, 258:739.

Tillett, W. S., Avery, O. T., and Geobel, W. F., 1929, Chemoimmuno-logical studies on conjugated carbohydrate-proteins. III. Active and passive anaphylaxis with synthetic sugar-proteins, J. Exp. Med., 50:551.

Tolone, G., Bonasera, L., and Tolone, C., 1976, Biosynthesis and release of prostaglandins by mast cells, Br. J. Exp. Pathol., 59:105.

Turner, S. R., Tainer, J. A., and Lynn, W. W., 1975, Biogenesis of chemotactic molecules by arachidonate lipoxygenase system of platelets, Nature, 257:680.

Uvnäs, B., 1964, Release processes in mast cells and their activation by injury, Ann. NY Acad. Sci., 116:880.

Uvnäs, B., and Thon, I. L., 1959, Isolation of biologically intact "mast cells". Exp. Cell. Res., 18:512.

Vaerman, J. P., Epstein, W., Fudenberg, H., and Ishizaka, K., 1964, Direct demonstration of reagin activity in purified γlA-globulin, Nature, 203:1046.

Valentine, W. N., Lawrence, J. S., Pearse, M. L., and Beck, W. S., 1955, Relationship of basophil to blood histamine in man, Blood, 10:154.

Vallota, E. H., and Müller-Eberhard, H. J., 1973, Formation of C3a and C5a anaphylatoxins in whole human serum after inhibition of the anaphylatoxin inactivator, J. Exp. Med., 137:1109.

Vane, J. R., 1971, Inhibition of prostaglandin synthesis as a mechanism of action for aspirin like drugs, Nature, 231:232.

Van Hemmen, J. J., and Meuling, W. J. A., 1975, Inactivation of biologically active DNA by gamma induced superoxide radicals and their dismutation products singlet molecular oxygen and hydrogen peroxide, Biochim. Biophys. Acta, 402:133.

Vogt, W., 1957, Pharmacologically active substance formed in egg yolk by Cobra venom, J. Physiol. (Lond), 136:131.

Voorhees, A. B., Baker, H. J., and Pulaski, E. J., 1951, Reactions of albino rats to injections of dextran, Proc. Soc. Exp. Biol. Med., 76:254.

West, C. B., 1974, Further analysis of the resistance of a colony of rats to dextran, Int. Arch. Allergy Appl. Immunol., 47:296.

Whelan, C. J., 1978, Histamine release from rat peritoneal mast cells by phospholipase A. The "activation" of phospholipase A by phospholipids, Biochem. Pharmacol., 27:2115.

Willis, A. L., and Kuhn, D. C., 1973, A new potential mediator of arterial thrombosis whose biosynthesis is inhibited by aspirin, Prostaglandins, 4:127.

Yagi, Y., Maier, P., Pressman, D., Arbesman, C., and Reisman, R. E., 1963, The presence of the ragweed-binding antibodies in the β2A-B2M and γ-globulins of the sensitive individuals, J. Immunol., 91:83.

IMMUNOGLOBULIN SURFACE ADSORPTION STUDIED BY TOTAL INTERNAL

REFLECTION WITH FLUORESCENCE CORRELATION SPECTROSCOPY

Nancy L. Thompson   Daniel Axelrod

Chemistry Department   Biophysics Research Division
Stanford University   University of Michigan
Stanford, Cal. 94305   Ann Arbor, Mich. 48109

ABSTRACT

Total internal reflection with fluorescence correlation spectroscopy (TIR/FCS) is a new technique for studying binding and
surface diffusion of fluorescent molecules in equilibrium at a
surface. We apply TIR/FCS to the rapidly reversible nonspecific
adsorption of rhodamine-labeled immunoglobulin to fused silica coated
with bovine serum albumin. The measured characteristic time is 6.3ms
and is limited by the rate of diffusion in solution. TIR/FCS is
shown to be a feasible technique for measuring nonspecific binding
kinetic rates at equilibrium and should prove useful for measuring
specific solute/surface site kinetic rates in systems where nonspecific binding is low.

INTRODUCTION

Association/dissociation reactions between biological molecules
in solution and sites on a surface are involved in many important
processes, including serum protein adsorption, antibody detection,
biochemical manufacture and purification, and the attachment of
viruses or antigens to immunological cells. Total internal reflection with fluorescence correlation spectroscopy (TIR/FCS)[1-4] is a
technique for measuring the rate parameters critical to such processes, i.e., the binding and unbinding rates and the surface diffusion coefficient.

In TIR/FCS, fluorescent-labeled molecules are in chemical equilibrium between a solution and a surface to which they reversibly
adsorb. A laser beam internally reflects at the surface/solution
interface, illuminating a very thin layer ($\sim$1000 A) of solution

immediately adjacent to the surface[5], thus selectively exciting only
surface-adsorbed molecules. The fluorescence arising from a small
well-defined surface area fluctuates in time as molecules bind and
unbind within the observation area or surface diffuse across it. The
fluctuations are analyzed by on-line minicomputer autocorrelation.
The decay time of the autocorrelation function depends on the rates
of binding/unbinding and surface diffusion. The time zero value is
inversely proportional to the average number of fluorescent molecules
in the observation area and provides a measure of molar adsorbate
concentration. We have applied TIR/FCS to rhodamine-labeled immunoglobulin or insulin adsorbing in equilibrium to a fused silica microscope slide coated with irreversibly adsorbed bovine serum albumin.

EXPERIMENTAL APPARATUS

    As shown in Fig. 1(a), the 514.5 nm beam of an argon ion laser
(Lexel 95-3) passes through a focusing lens into a $(1.5 \text{ cm})^3$ fused
silica prism (Precision Cells) in optical contact with a Suprasil
fused silica slide (Amersil). The beam is internally reflected at
the interface of the slide with a solution of fluorescent protein.
A thin teflon spacer separates the slide from the bottom of the
sample holder.

    As shown in Fig. 1(b), the sample holder is mounted on the stage
of an inverted microscope (Leitz Diavert). Light is collected by an
objective (40X, 1.3 numerical aperture, Leitz), transmitted through a
filter which blocks excitation-wavelength light, transmitted through
a diaphragm at a microscope image plane which defines the observation

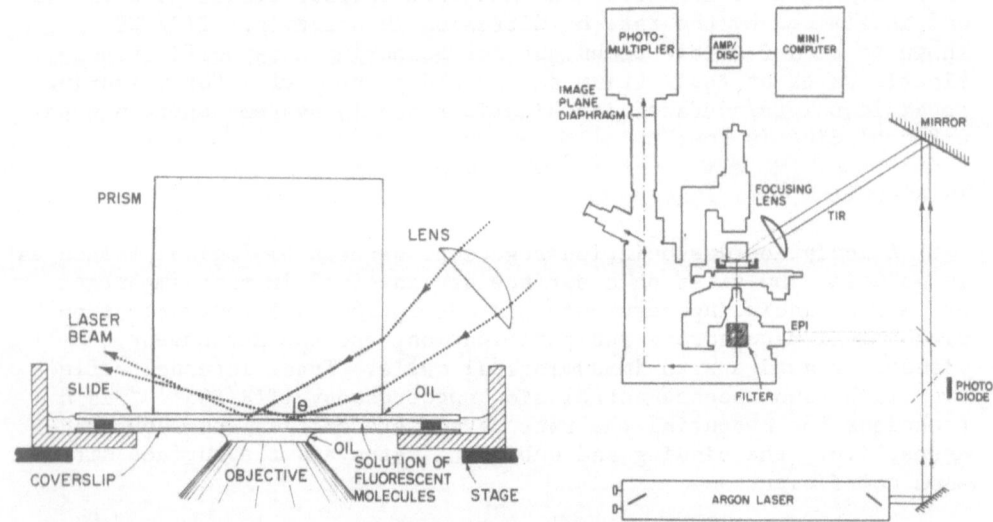

Figure 1. Optical Apparatus. (a) Sample holder. (b) Laser,
inverted microscope, minicomputer, and accessories.

area, and detected by a thermoelectrically cooled photomultiplier
(RCA C31034A). The photomultiplier output passes through an ampli-
fier/discriminator to a minicomputer (NOVA 3/12), which counts the
number of photoelectron pulses occurring during consecutive sample
times of user-specified duration. The data, counted in 16-bit words,
is on-line autocorrelated by assembly language software.

RESULTS

IgG (Cappel) and insulin (Sigma) are labeled with tetramethyl-
rhodamine isothiocyanate (BBL) so that the final molar ratio of label
to protein is .16 (IgG) or $\leq$ 1 (insulin).[6,7]   All adsorption experiments are performed in .1 M NaPO$_4$ .15 M NaCl .02% NaN$_3$ pH 7.0 (PBS).
Slides are cleaned first with chromic acid and then with an argon
gas plasma cleaner (Harrick), incubated with 3 mg/ml BSA/PBS, rinsed
with PBS, coated with the fluorescent protein solution, and mounted
in the sample holder on the microscope stage.

Surface concentration fluctuations can be seen through the
microscope eyepiece as randomly flashing fluorescence fluctuations.
Fig. 2 displays the autocorrelation function $G(\tau)$ of .023 $\pm$ .005
mg/ml R-IgG adsorbing to BSA-glass, observed over (2.1 μm)^2. $G(\tau)$
has been corrected for slow drifts in average fluorescence intensity, for background intensity, and for systematic fluctuations, as
described elsewhere.[3]   No difference in the shape of $G(\tau)$ is observed
for a 3-fold decrease in incident laser intensity.

A plot of fluorescence (proportional to surface concentration)
vs. bulk concentration indicates that surface sites are far from
saturated. No difference in the shape of $G(\tau)$ is observed for a

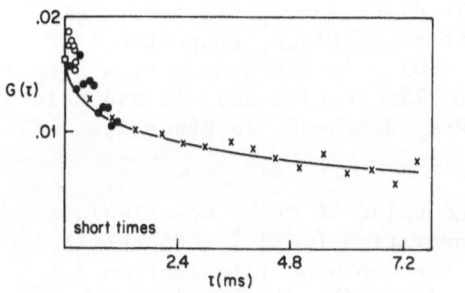
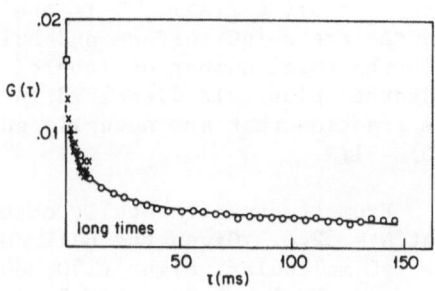

Figure 2.   Autocorrelation of R-IgG Adsorbing to BSA-glass. Data
points are composited from sample times of .03 ms (o), .1 ms (●),
.5 ms (×), and 5 ms (o). The laser intensity is .02 mW/μm^2 (5 ms
and .5 ms), .06 mW/μm^2 (.1 ms), or .18 mW/μm^2 (.03 ms). The line
represents the best fit as described in the text. The deviation
of the best fit from the data points at early times represents
correlations arising from R-IgG diffusing through the small layer
of illuminated solution.

347

3-fold decrease in concentration. $G(\tau)$ also does not change with a 10-fold decrease in observation area; this means that surface diffusion across an area of .65 μm^2 does not occur during the average surface residency time of an R-IgG.

The experimental data points shown in Fig. 2 are a minicomputer estimate of the theoretical autocorrelation function defined as

$$G(\tau) = <\delta F(t+\tau)\,\delta F(t)>/<F>^2$$

where $F(t)$ is the fluorescence at time t, brackets <> indicate an ensemble average, and δF is the deviation of F from its average value <F>. Theoretical expressions for $G(\tau)$ have been derived.[1-3] If $G(\tau)$ does not depend on observation area (as in our experiments), its shape is determined by two rates: one depends on the surface binding desorption rate and the degree of surface site saturation; the other depends on the ratio of surface to bulk concentration (a function of the equilibrium constant), the diffusion coefficient in solution, and the degree of surface site saturation.

The data in Fig. 2 is curvefit to the known analytical forms of observation area-independent $G(\tau)$, in the limit far from surface binding site saturation, plus an arbitrary constant. Only times τ greater than 1 ms are included in the curvefitting, since the experimental values at earlier times may be partially determined by diffusion through the thin layer of illumination. Because of the rapidity of its decay, $G(\tau)$ is almost certainly in the "bulk diffusion limit", a case for which

$$G(\tau) = \alpha\,\exp(R\tau)\,\mathrm{erfc}(\sqrt{R\tau})\,/\,N \quad \text{(applied here for } \tau > 1 \text{ ms)}$$

where $R = D/(<C>/<A>)^2$; D is the bulk diffusion coefficient; <C> and <A> are R-IgG surface and bulk concentrations, respectively; N is the total number of labeled molecules in the observation area (adsorbed plus bulk dissolved in the illumination layer); and α is the fraction that are actually surface adsorbed. In general, $G(0) = 1/N$.

From the experimentally observed value of $G(0)$, we calculate that N = 52.6. Given the bulk concentration (.023 $\pm$ .005 mg/ml), the IgG molecular weight (150,000), the depth of illumination (.1 μm), and the labeled to total IgG ratio (.16), we calculate that an average of only 6.5 labeled IgG within the observation area are not actually adsorbed, but instead are diffusing in solution within the illumination layer.

The best fit to the data of Fig. 2 is for $R = (6.3 \text{ ms})^{-1}$ and an additive arbitrary constant $\ll G(0)$. Ratio <C>/<A> can be calculated from R and a $D = 5 \times 10^{-7}$ cm^2/sec. Thus, for an estimated surface concentration of free binding sites $<B> = 10^4/\mu m^2$, equilibrium constant <C>/<A><B> equals 3.3×10^4 M^{-1}.

We have also obtained $G(\tau)$ for R-insulin at a bulk concentration of .002 mg/ml reversibly adsorbing to BSA-glass. The average time for decay of $G(\tau)$ to the midpoint of its values at $\tau = 5$ ms and $\tau = 295$ ms is somewhat longer for R-insulin (40 ± 1 ms) than for R-IgG (30 ± 1 ms).

DISCUSSION

These experiments demonstrate the capability of TIR/FCS to measure, in equilibrium, rapidly reversible binding rates and absolute concentrations of adsorbed fluorescent molecules at a surface. In contrast to other kinetics techniques, TIR/FCS requires no macroscopic perturbation from equilibrium and no spectroscopic or thermodynamic change between the dissociated and complexed states of the reaction. As a surface chemistry technique, TIR/FCS should be particularly useful where both surface adsorption and surface diffusion occur. TIR/FCS should also prove useful for measuring specific reversible solute/surface kinetic rates at equilibrium in systems where nonspecific binding is low.

ACKNOWLEDGEMENTS

The authors thank Thomas Burghardt and Doran Smith for their invaluable help with the optics and electronics, respectively. This project was supported by USPHS NIH grants NS14565 and HL24039.

REFERENCES

1. N.L. Thompson, T.P. Burghardt, and D. Axelrod, Measuring surface dynamics of biomolecules by total internal reflection fluorescence with photobleaching recovery or correlation spectroscopy, Biophys. J. 33:435 (1981).
2. N.L. Thompson, Surface binding rates of nonfluorescent molecules may be obtained by total internal reflection with fluorescence correlation spectroscopy, Biophys. J. 38:327 (1982).
3. N.L. Thompson, Ph.D. Thesis, University of Michigan (1982).
4. T. Hirschfeld and M.J. Block, Virometer: real time virus detection and identification in biological fluids, Opt. Eng. 16: 406 (1977).
5. N.J. Harrick, "Internal Reflection Spectroscopy", Wiley & Sons, New York (1967).
6. L. Amante, A. Ancora, and L. Forni, The conjugation of immunoglobulins with TMR-ITC: a comparison between the amorphous & the crystalline fluorochrome, J. Immunol. Meth. 1:289 (1972).
7. Y. Shechter, J. Schlessinger, S. Jacobs, K. Chang, and P. Cuatrecasas, Fluorescent labeling of hormone receptors in viable cells: prep. & properties of highly fluorescent derivatives of EGF and insulin, Proc. Natl. Acad. Sci. USA 75:2135 (1978).

ENTHALPY STATE OF DIFFERENT LIPOSOME PREPARATIONS OF DIMYRISTOYL PHOSPHATIDYLCHOLINE AND THEIR STABILITY IN THE PRESENCE OF A COMPLEX FORMING PROTEIN

Ignace Hanssens, Willy Herreman, Jean-Claude Van Ceunebroeck, and Frans Van Cauwelaert

Interdisciplinair Research Centrum, Katholieke Universiteit Leuven, campus Kortrijk, B-8500 Kortrijk

Bovine α-lactalbumin was used to study the influence of the protein conformation on its interaction with phosphatidylcholine liposomes. The isoionic point of α-lactalbumin is at pH 5, and it has been demonstrated that between pH 4 and 2, the protein undergoes a conformational change with a concommitant increase in α-helicity.[1] Thé α-helices in α-lactalbumin are amphipatic in the same way as in apolipoproteins.[2]

In a first series of experiments small unilamellar vesicles (SUV) of dimyristoylphosphatidylcholine (DMPC) obtained by sonication above the transition temperature, for 25 min with a MSE 150 disintegrator at maximum output, were used in the interaction study.[3,4]

From measurements with a batch microcalorimeter (LKB, type 2107) it has been outlined that at pH 7.4 (Tris buffer 0.01 M and NaCl 0.1 M) the heat released on mixing α-lactalbumin and the vesicles decreases when the reagents were at higher temperature. However at pH 4 and 4.7 (acetate buffer 0.01 M and NaCl 0.1 M) the released heat is more important and reaches a maximum value at about 23° C. It indicates that in the acidic medium the interaction is strongly favoured at the transition temperature of the phospholipid i.e. the temperature at which the apolar phase is more accessible. Phosphatidylcholine-α-lactalbumin mixtures, respectively at pH 7.4 and pH 4, were negatively stained with an 0.5% uranylacetate solution. On the electron micrographs at pH 7.4, the original small vesicles are still seen, whereas at pH 4, bar shaped entities are formed. At pH 7.4 the mixture of vesicles and protein is easily separated by gel chromatography on Sepharose 6B. On the contrary at pH 4, complex particles elute between the elution volume of pure SUV's and that of pure α-lactalbumin. From the

351

product analysis a lipid-to-protein molar ratio of 70/1 has been calculated in this complex fraction.

COMPARISON OF THE ENTHALPY STATE OF VESICLES OF DIFFERENT SIZE

Lipid-protein complex particles of the same size and shape and elution volume in gel chromatography are obtained after interaction of α-lactalbumin at pH 4 and 23° C or 25° C with either small (SUV) or large unilamellar vesicles (LUV) or multilamellar vesicles (MLV). Also the temperature scans of the fluorescence polarization of 1,6-diphenyl-1,3,5-hexatriene, used as a lipid embedded probe,

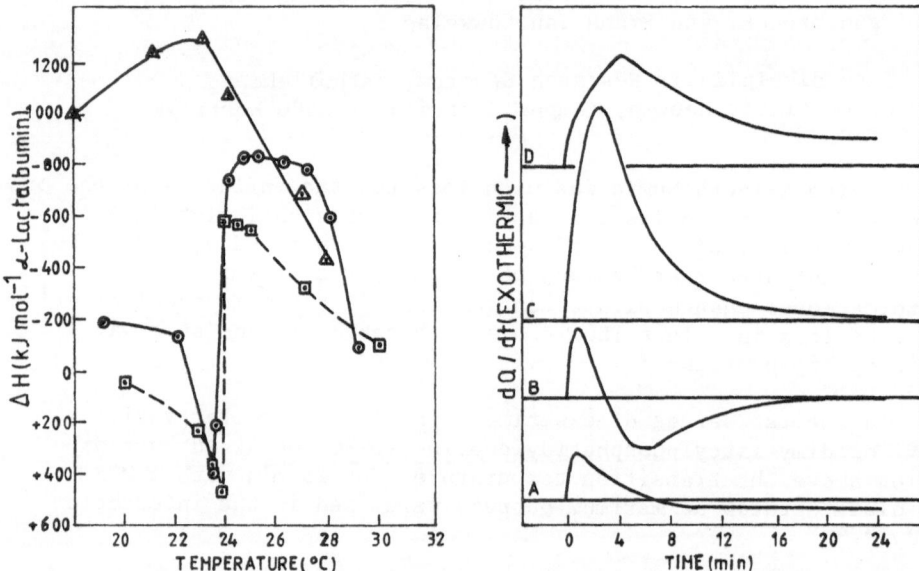

Fig. 1. Enthalpy change upon mixing DMPC and α-lactalbumin at pH 4 and a molar ratio of 100 as a function of the temperature. The phospholipid is added as SUV's (▲), as LUV's (☉) or as MLV's(▣). The enthalpy changes are measured within 30 min after mixing. Only when SUV's or LUV's interact with α-lactalbumin near the transition temperature the interaction is completed in that time.

Fig. 2. Experimental tracing of the heat released from the reaction vessel (dZ/dt) versus time upon mixing large unilamellar vesicles of DMPC with α-lactalbumin at pH 4, a molar ratio of 100 and at different temperatures. A: 22.5° C, B: 23.3° C, C: 24.0° C, D: 27.0° C.

becomes independent of the original vesicle structure after interaction with α-lactalbumin at these temperatures.[5] Therefore it is concluded that the complex formed after incubation, is independent of the vesicle preparation and of the temperature between 23° C and 25°C.

However, as shown in fig. 1, the enthalpy changes of the protein interation with vesicles of different sizes are very different. Whereas the α-lactalbumin interaction with small vesicles is always exothermic, the interaction with large vesicles (unilamellar and multilamellar) changes from a maximal endothermic value to a maximal exothermic one near the transition temperature.

As the lipid interacts with the protein in the complex in a molar ration 70/1, it can be calculated that for large unilamellar vesicles the difference in enthalpy between the maximal exothermic value (850 kJ/mol α-lactalbumin) and the maximal endothermic value (400 kJ/mol α-lactalbumin) correspond to 18 kJ/mol DMPC; this approximates the transition enthalpy of the lipid.

As the three vesicle systems form identical complexes at a given temperature, it can also be stated in a more general way that the different heat exchanges reflect a difference in the enthalpy state of the lipid in the original vesicles at that temperature. As a consequence the dimyristoyl phophatidylcholine in small vesicles has a higher enthalpy content than in both large vesicle types, even above the transition temperature (23.9° C).

As shown in fig. 2 the comparison of the enthalpy state of the different vesicles should be limited to the region of the transition temperature. Indeed, only in that temperature range the heat exchange on mixing the phospholipid and the protein was finished within a reasonable measuring time of 30 min.

STABILITY OF VESICLES IN PRESENCE OF A COMPLEXATING PROTEIN

At pH 4, 23° C and a starting lipid-to-protein molar ratio 200 (and at higher molar ratios), the protein of a SUV-α-lactalbumin mixture elutes at the volume of the small vesicles indicating that no smaller lipid-protein complexes are formed. However when adding the protein in the same molar ratio to LUV's, the small lipid-protein complexes are formed.[6] Therefore the lipid-to-protein molar ratio itself is not determining for the formation of the complex particle. Also from the energetic viewpoint, it should not be expected that small vesicles which have a higher enthalpy state than large vesicles resist better the breakdown of the complexating protein. For the interpretation of that unexpected behavior, the following reaction mechanism is considered. From the calorimetric tracing of the LUV-α-lactalbumin interaction at 23° C (fig. 2) it is obvious that at least two separate steps (or groups of steps) must be present. The first one which is always exothermic must include the adsorption of the protein to the vesicles, the latter must result in a complex of definite size and lipid-to-protein molar ratio. As a consequence a definite number of proteins have to agglomerate with a definite number of lipids. If the latter step occurs intravesicularly, we have to accept that the vesicles only break up if after the random adsorption step(s), sufficient proteins are on one vesicle.

To verify this interpretation we have looked for the number of α-lactalbumin molecules that a small vesicle has to adsorb in order to break down and we have compared it with the number of protein molecules in a complex particle. In a sonicated vesicle there are about 2850 dimyristoyl phosphatidylcholine molecules.[7] At a lipid-to-protein molar ratio between 200 and 150 the vesicles start to disrupt i.e. when a mean of 14 to 19 α-lactalbumin molecules accumulate on it by random adsorption. The mass of the complex particle isolated at 23° C, determined from the sedimentation and diffusion coefficients using the Svedberg equation was 1.05 10^6 g/mol. It can be calculated from this that at a lipid-to-protein molar ratio 70/1, about 17 protein molecules are incorporated in a complex particle. Data in the litterature indicate that the suggested interaction mechanism between vesicles and a complexating protein intervenes in a more general way. Apolipoprotein AI and DMPC vesicles form complex particles with 3 proteins incorporated, whereas the vesicles with less than 3 to 4 adsorbed apoproteins are not broken down.[8] It is also observed that plasma has the capacity to degrade liposomes, resulting in a massive transfer of the liposomal lecithin to plasma high density lipoproteins and after an intravenous injection, large liposomes clear faster from blood than small vesicles.[9]

REFERENCES

1. C.C. Contaxis and C.C. Bigelow, Free energy changes in α-lactalbumin denaturation, Biochemistry 20:1618 (1981).
2. J.P. Segrest and R.L. Jackson, Molecular properties of membrane proteins, in "Membrane proteins and their interaction with lipids", R.A. Capaldi, ed., M. Dekker, New York (1977).
3. I. Hanssens, C. Houthuys, W. Herreman and F.H. Van Cauwelaert, Interaction of α-lactalbumin with dimyristoyl phosphatidylcholine vesicles. I. A microcalorimetric and fluorescence study, Biochim. Biophys. Acta 602:539 (1980).
4. W. Herreman, P. Van Tornout, F.H. Van Cauwelaert and I. Hanssens, Interaction of α-lactalbumin with dimyristoyl phosphatidylcholine vesicles. II. A fluorescence polarization study, Biochim. Biophys. Acta 640:419 (1981).
5. F. Van Cauwelaert, I. Hanssens, W. Herreman, J.C. Van Ceunebroeck, J. Baert and H. Berghmans, Comparison of the enthalpy state of vesicles of different size by their interaction with

α-lactalbumin, submitted for publication in Biochim. Biophys. Acta.

6. I. Hanssens, W. Herreman, J.C. Van Ceunebroeck, H. Dangreau, C. Gielens, G. Preaux and F. Van Cauwelaert, Interaction of α-lactalbumin with dimyristoyl phosphatidylcholine vesicles III. Influence of the temperature and of the lipid-to-protein molar ratio on the complex formation, Biochim. Biophys. Acta, accepted for publication.

7. A. Watts, D. Marsh and P.F. Knowles, Characterization of dimyristoyl phosphatidylcholine vesicles and their dimensional changes through the phase transition: molecular control of membrane morphology, Biochemistry 17:1792 (1978).

8. A. Jonas, S.M. Drengler and B.W. Patterson, Two types of complexes formed by the interaction of apolipoprotein A-I with vesicles of L-α-dimyristoyl phosphatidylcholine, J. Biol. Chem. 255:2183 (1980)

9. G. Scherphof, F. Roerdink, D. Hoekstra, J. Zborowski and E. Wisse, Stability of liposomes in presence of blood constituents: consequences for uptake of liposomal lipid and entrapped compounds by rat liver cells, in "Liposomes in biological systems" G. Gregoriadis and A.C. Allison, ed., John Wiley & Sons Ltd., New York (1980).

FREE-FLOW ELECTROPHORESIS AS A SUITABLE TOOL TO EXPLORE

INTERVESICULAR INTERACTIONS

Marcel De Cuyper and Marcel Joniau

Interdisciplinary Research Center, K.U. Leuven, Campus
Kortrijk, B-8500 Kortrijk, Belgium

INTRODUCTION

Membrane surface characteristics can be modulated by lipid
exchange processes which are either dependent or independent of
proteins. To unravel the underlying mechanism(s) of the spontaneous
way of lipid transfer, we used artificial phospholipid vesicles
which are expected to mimic – at least in part – cell membranes.
In this report, lipid movement between differently charged vesicles is followed by the free-flow electrophoresis technique
which we recently introduced in the liposome field[1].

MATERIALS AND METHODS

DMPC and DPPC (Sigma) were transformed to DMPG and DPPG as
described[2]. $|^3H|$DMPG was synthetized using the same procedure with
$|^3H|$ glycerol (Amersham). $|^3H|$ DMPC was prepared according to
Stockton et al.[3]. DOPG and DOPC were obtained from Serdary. Small
unilamellar vesicles were prepared by sonication and characterized
by gelfiltration and electron microscopy [1]. Free-flow electrophoresis was performed on a Desaga FF-48 apparatus as described [1].
After each run, the fractions were analyzed for light scattering
(400 nm) and occasionally also for radioactivity.

The transfer processes are treated with first-order reaction
mathematics. Either time-dependent changes in the migration dis-

---

* Abbreviations PC, phosphatidylcholine ; PG, phosphatidylglycerol ;
  O, oleoyl ; M, myristoyl ; P, palmitoyl ; $t_{1/2}$, halftime.

tance or - in the case of the immobile electropherograms - time-dependent changes in the amount of radiolabeled phospholipid associated with donor or acceptor vesicles, are measured.

RESULTS AND DISCUSSION

Electrophoresis of zwitterionic phospholipids, cosonicated with increasing amounts of anionic ones, shows that particularly in the low charge range small differences in the content of anionic phospholipids are easily distinguished in the electropherogram. Based on this observation we studied the interaction of DMPC/DMPG (molar ratio 9/1) and DMPC vesicles in 5 mM morpholinoethane-sulfonic acid, 10 mM potassium chloride pH 6.0. Fig. 1 shows some representative electropherograms at different times after incubation. Clearly, the overall changes in migration distances are associated with a net flux of $|^3H|$ DMPG. The transfer rate, deduced from the electrophoretic migration distances is first-order for several halftimes ($t_{1/2}$ = 40 min). The transfer phenomenon is strongly temperature-dependent ; above the gel-to-liquid crystal phase transition temperature (23° C) the activation energy equals 30 kcal/mol.

The fact that first-order kinetics are obtained with equimolar amounts of donor and acceptor vesicles already points to a mechanism of lipid transferring through the aqueous phase. Transfer during collision of vesicles can be further excluded since at a donor/acceptor ratio of 1, the transfer rate per vesicle is independent of total phospholipid concentration. Also, the higher transfer rates found at lower ionic strengths of the incubation medium are in agreement with an enhanced water solubility of the anionic DMPG [4].

Replacing DMPC in donor and/or acceptor vesicles by the more fluidizing DOPC does not considerably change the $t_{1/2}$-value of DMPG-transfer, indicating that the degree of fluidity of "melted" membranes is not a critical factor in the transfer process (Table 1). In contrast, the structure of the apolar moiety of the molecule to be transferred is of major importance. The transfer potency increases in the order DPPG << DMPG << DOPG.

The impact of the polar headgroup is assessed by using the immobile electropherograms obtained for the equimolar mixture of DMPC/DPPG (9/1) and DMPC vesicles. Either the neutral or the negatively charged vesicles were labelled with $|^3H|$DMPC or $|^3H|$ DMPG. From these experiments it appears that the transfer rate of $|^3H|$DMPC from the anionic vesicle population towards the neutral ones ($t_{1/2}$ = 90 min) only slightly differs from its transfer rate in the opposite direction ($t_{1/2}$ = 130 min). Moreover, a curved line is found in the first-order kinetic plot (not shown), indi-

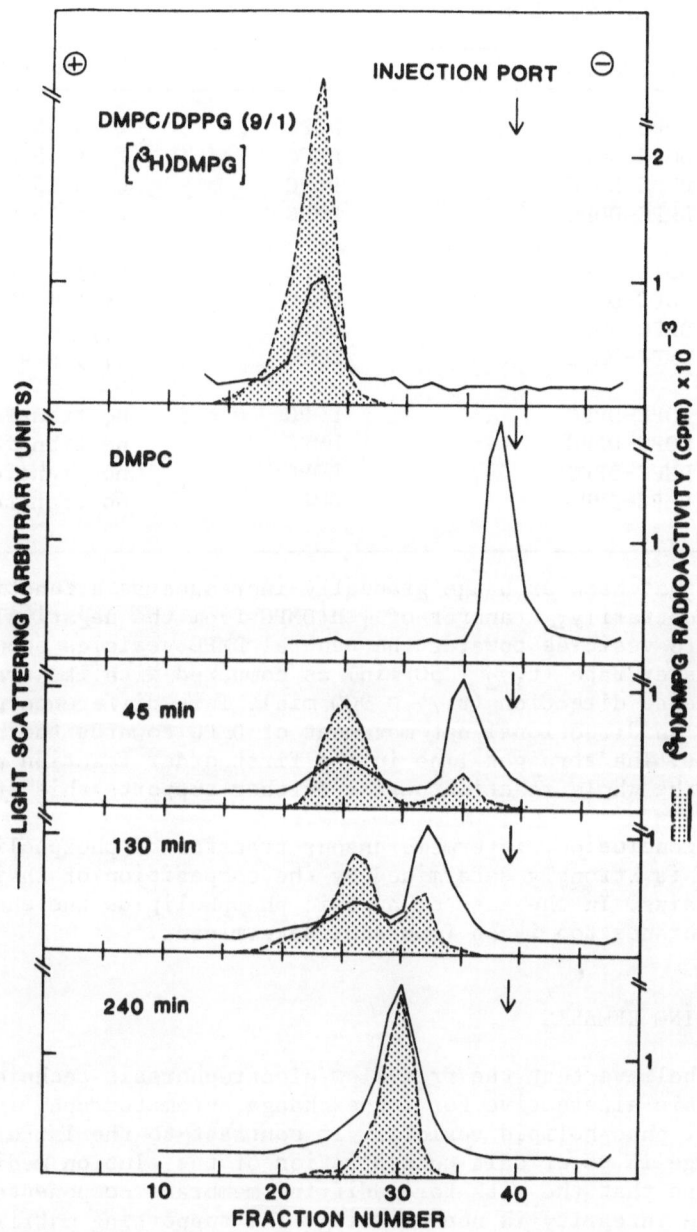

Fig. 1. Kinetics of spontaneous phospholipid transfer between
equimolar amounts of DMPC/DMPG (molar ratio 9/1),
labelled with a trace of |³H|DMPG, and DMPC vesicles
at 33° C, pH 6.0.

Table 1. Summary of kinetic data for the transfer occurring between different donor and acceptor vesicle types at 33° C, pH 6.0.

Donor (PC/PG : 9/1)	Acceptor	Rate
DOPC–DOPG	DOPC	$t_\infty < 15$ min
DOPC–DOPG	DMPC	$t_\infty < 15$ min
DMPC–DOPG	DOPC	$t_\infty < 15$ min
DMPC–DOPG	DMPC	$t_\infty < 15$ min
DOPC–DMPG	DOPC	$t_{1/2} = 86$ min
DOPC–DMPG	DMPC	$t_{1/2} = 86$ min
DMPC–DMPG	DOPC	$t_{1/2} = 50$ min
DMPC–DMPG	DMPC	$t_{1/2} = 41$ min
DOPC–DPPG	DOPC	no transfer
DOPC–DPPG	DMPC	no transfer
DMPC–DPPG	DOPC	no transfer
DMPC–DPPG	DMPC	no transfer

cating that back exchange gradually increases as a function of time. Contrarily, transfer of $|^3H|$DMPG from the negatively charged DMPC/DPPG vesicles towards the neutral DMPC vesicles occurs at a much faster rate ($t_{1/2} = 50$ min) as compared with its transfer in the reverse direction ($t_{1/2} \cong 900$ min). This difference points to a quasi–unidirectional net movement of DMPG towards the neutral vesicles. The straight line in the first-order reaction plot which covers the whole reaction period further supports this statement.

In conclusion, intermembraneous transfer of phospholipid molecules is strongly determined by the composition of their fatty acyl chains. In the case of anionic phospholipids the charge of the membranes too is an important determinant.

CONCLUDING REMARKS

We believe that the free-flow electrophoresis technique is a reasonable alternative for ion-exchange chromatography in the study of phospholipid vesicles. In contrast to the latter technique no change in pH or salt concentration of the elution medium is needed so that the risk for modifying membrane components and/or particle integrity is non-existing. The supporting matrix used in the ion-exchange experiments also adsorbs a considerable amount of the lipid vesicles [5]. By free-flow electrophoresis both the donor and acceptor vesicles injected in the separation chamber are completely recovered.

Our results clearly demonstrate that the important parameters which determine the spontaneous phospholipid transfer phenomenon between differently charged vesicles can be studied in detail by free-flow electrophoresis. Undoubtedly the technique can also be successfully applied to study other aspects of model membranes. In this respect, we recently reported the interaction of pig brain tubulin with artificial phospholipid vesicles [6].

ACKNOWLEDGEMENTS

We acknowledge the help of Mr. W. Noppe and S. Van Cauwenberghe. This investigation was supported by a grant of the Belgian "Fonds voor Geneeskundig Wetenschappelijk Onderzoek".

REFERENCES

1. M. De Cuyper, M. Joniau & H. Dangreau, Spontaneous phospholipid transfer between artificial vesicles followed by free-flow electrophoresis, Biochem. Biophys. Res. Commun. 95:1224 (1980).
2. D. Papahadjopoulos, K.Jacobson, S. Nir & T. Isac, Phase transitions in phospholipid vesicles. Fluorescence polarization and permeability measurements concerning the effect of temperature and cholesterol, Biochim. Biophys. Acta 311:330 (1973).
3. G.W. Stockton, C.F. Polnaszek , L.C. Leitch, A.P. Tulloch & I.C.P. Smith, A study of mobility and order in model membranes using ^{2}H NMR relaxation rates and quadrupole splittings of specifically deuterated lipids, Biochem. Biophys. Res. Commun. 60:844 (1974).
4. C. Tanford, Thermodynamics of micelle formation, Chapter VII, in : "The Hydrophobic Effect", 2nd edition, John Wiley and Sons, New York (1980).
5. A.M.H.P. van den Besselaar, G.M. Helmkamp, Jr. & K.W.A. Wirtz, Kinetic model of the protein-mediated phosphatidylcholine exchange between single bilayer liposomes, Biochemistry 14:1852 (1975).
6. M. Beyls, M. Joniau, M. De Cuyper & H. Dangreau, Interaction of tubulin and model membranes, Arch. internat. Physiol. Biochim. 89:BP17 (1981).

SOLID STATE ^{19}F NMR IN 'MODEL' MEMBRANE SYSTEMS

J. Post*, Y.C.W. van der Leeuw and
H.J.C. Berendsen

Laboratory of Physical Chemistry
University of Groningen
The Netherlands

Solid state NMR techniques can be used to study the behaviour of molecules in membranes and their model systems[1,2,3,4]. We used a 4,4-difluoromyristate-D$_2$O mixture to test the possibilities. The major interactions are the F-F dipolar coupling, the F-H dipolar coupling and the F chemical shift anisotropy. A normal ^{19}F Free Induction Decay will be featureless and will after fourier transformation lead to a broad resonance signal (fig. 1). It can be shown, however, that a Carr-Purcell-Meiboom-Gill pulse sequence selects the F-F dipolar coupling and fourier transformation of the F.I.D. leads to a so-called Pake doublet spectrum[1] (fig. 1). From the splitting the $|S_{FF}|$ order parameter of the fluorine pair can be directly determined, according to

$$\Delta = |S_{FF}|\Delta(o) \qquad\qquad \Delta(o) = 16,0 \text{ kHz}$$

To solve the diagonal terms of the order parameter tensor we need another order parameter. The S_{CF} can be calculated by measuring the ^{19}F chemical shift anisotropy with a MERV pulse experiment which averages out the homonucelar dipolar interactions. We find an anisotropic linewidth of 44.0 ppm ± 2.5 ppm. This leads to the following order parameters:

$$S_{xx} = -0.201 \pm 0.004 = S_{FF}$$
$$S_{yy} = -0.164 \pm 0.022$$
$$S_{zz} = +0.365 \pm 0.023$$

---

* Department of Biological Sciences, Carnegie Mellon Un., Pittsburgh.

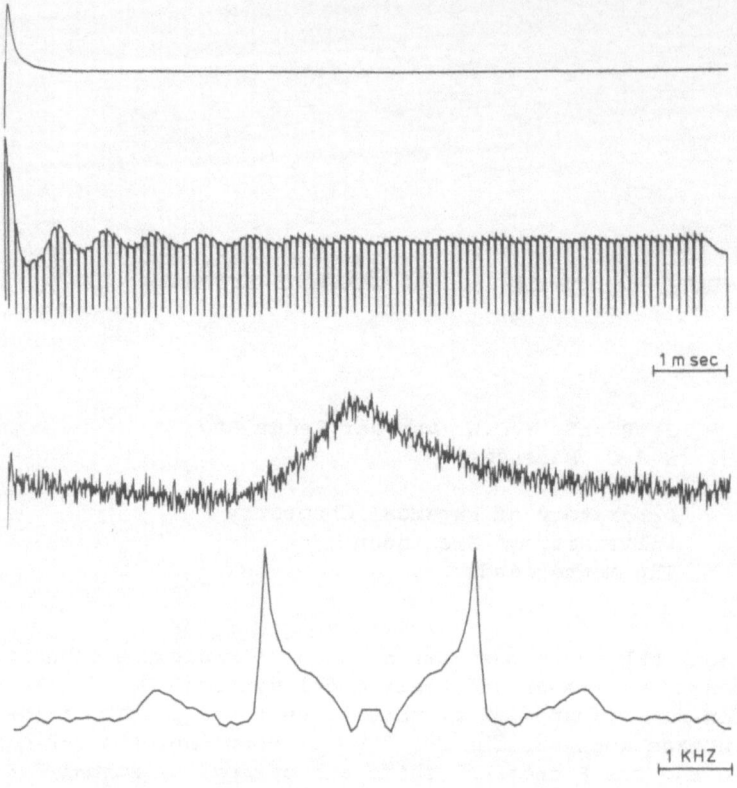

Fig. 1. Above an FID following a single 90° pulse, compared
with the response of a 90°- $(\tau-180°_{90°}-\tau)_N$ pulse sequence.
The fourier transforms are shown below.

where the z-direction is perpendicular to the CF_2 plane. The
asymmetry of this tensor is in good agreement with the results of a
computer simulation studies of a bilayer membrane by the method of
molecular dynamics, carried out in this laboratory[5,6,7]. The main
features of the simulation are the occurrence of a fluctuating
cooperative tilt of the hydrocarbon chains, with a strong
correlation between the molecular plane and the plane of tilt,
resulting in a deviation from axial symmetry around the molecular

z-axis. An F-F pair incorporated in a membrane system can now be used to study the structure and dynamics of the molecules, providing the introduction of such a group does not perturb the system too much.   In the literature[8],[9] there is some uncertainty about these perturbing effects. We calculated the S_{CF} order parameter and compared it with a S_{CD} value from a similar system[10] and found a good agreement:

$$S_{CF} = -0.19 \qquad\qquad S_{CD} = -0.21$$

We used undiluted systems, but due to the high sensitivity of the ^{19}F resonance, it is possible to reduce the perturbing effects by using considerable dilution. The temperature dependence of the S_{FF} gives information about the phases of the system (fig. 2). Below 50°C the splitting decreases gradually until at 37°C a sharp phase transition from L_α to L_β occurs[2]. The gradual decrease is probably due to an increasing tilt angle. From 27°C to 18°C there is a further averaging of the interaction, probably resulting from an extra axis of rotation in a hexagonal or inverted hexagonal phase. To study more biologically relevant systems we used a fluorinated lipid[3],[4], i.e. 1,2 - 4,4 difluoro-myristoyl-SN-glycero-3-phosphocholine (F-DMPC). (Fig. 3).
The difference between the two chains of the molecule gives rise to spectra with two splittings (Fig. 3A). Introduction or a protein leads to a broadening of the signals due to exchange between unperturbed molecules and molecules in a region perturbed by the protein. The effect of addition of cholesterol is clearly an increase of the order parameter.

In conclusion we may say that the use of ^{19}F as a label is a useful method to study complex membrane systems as the perturbing effect does not appear to be strong and ^{19}F NMR is much more sensitive than ^{31}P NMR or ^2H NMR.

ACKNOWLEDGEMENTS

This research was supported by the foundation for Biophysics under the auspices of the Netherlands Organization for Pure Research.

REFERENCES

1. J.F.M. Post, E. James and H.J.C. Berendsen, Miltipulse Fluorine NMR Experiments on Lyotropic Liquid Crystals. I., J. of Magn. Res. 47, 251 (1982).

2. J.F.M. Post, R.L. Kamman and H.J.C. Berendsen, Militpulse Fluorine NMR Experiments on Lyotropic Liquid Crystals. II., J. of Magn. Res. 47, 264 (1982).

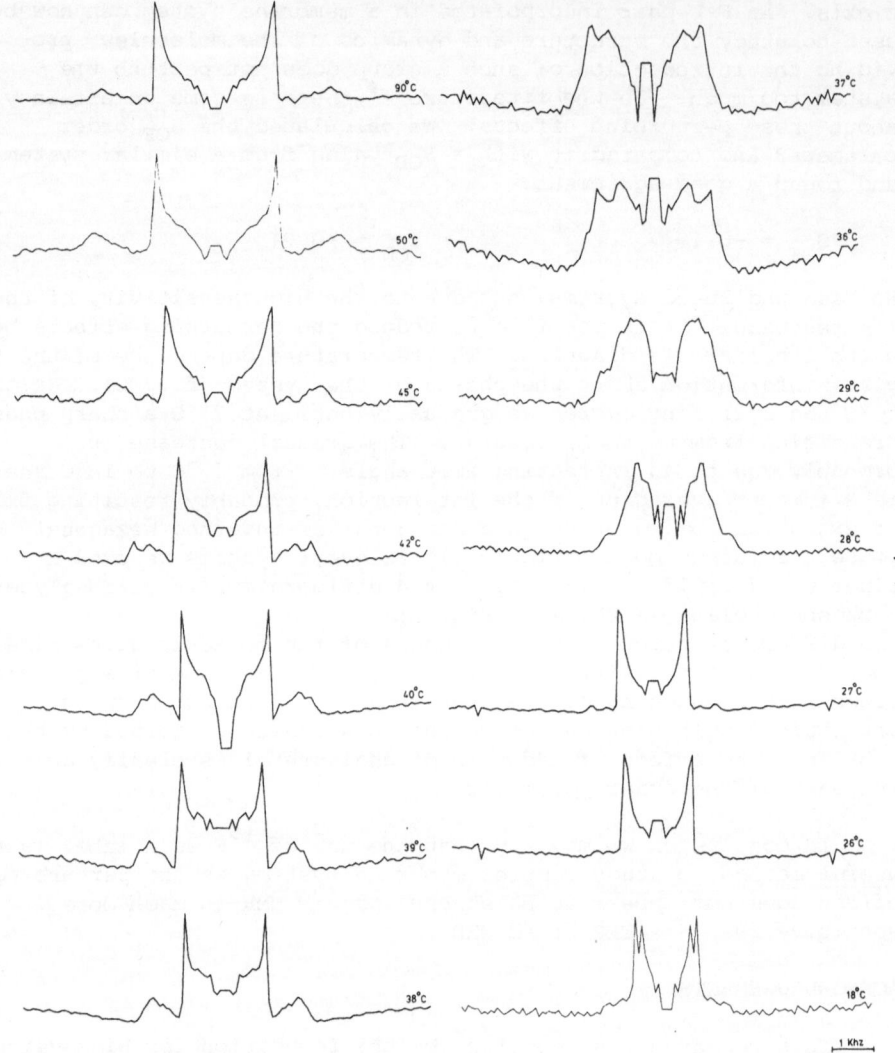

Fig. 2. ^{19}F dipolar spectra, obtained with the $90°-(\tau-180°_{90°}-\tau)_N$ pulse sequence. 200 scans were averaged. Spectral resolution is 100 Hz.

3. J. Post, Ph.D. thesis, University of Groningen, the Netherlands 1981.

4. J.F.M. Post, E.E.J. de Ruiter and H.J.C. Berendsen, A fluorine NMR study of model membranes containing ^{19}F-labeled Phospholipids and an intrinsic membrane protein, Febs. Lett. 132, 2, 257 (1981).

5. P. van der Ploeg, Ph.D. thesis, University of Groningen, the Netherlands, 1982.

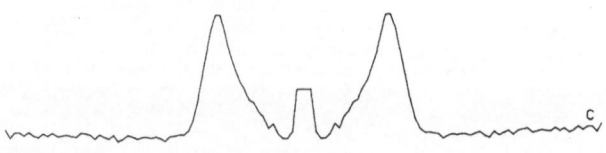

Fig. 3. ^{19}F dipolar spectra obtained at 40°C using the
90°-$(\tau$-180°$_{90°}$-$\tau)_N$ pulse sequence.   A. F-DMPC/D$_2$O.
B. F-DMPC/Cholesterol/D$_2$O.   C. F-DMPC/Lipophilin/D$_2$O.
Spectra are plotted as mirror images around the central
frequency. The central signal has no physical significance
but is due to baseline distortions which contribute in
the fourier transformation.

[6]. P. van der Ploeg and H.J.C. Berendsen, Molecular Dynamics
     Simulation of a bilayer membrane, J. of Chem. Phys. 76,
     3271 (1982).
[7]. P. van der Ploeg and H.J.C. Berendsen, Molecular Dynamics
     of a bilayer membrane, submitted to Mol. Phys.
[8]. E. Oldfield, R.W.K. Lee, M. Meadows, S.R. Dowd and C. Ho
     Deuterium NMR of specifically deuterated fluorine spin probes.
     J. Biol. Chem., 255, 11652 (1980).
[9]. J.M. Sturlevant, C. Ho and A. Reimann, Thermotropic behaviour
     of some fluorodimyristoylphophaticlylcholines, Proc. Natl.Acad.
     Sci. USA 76, 2239 (1979).
[10]. B. Mely, J. Charvolin and P. Keller, Chem. Phys. Lip. 15,
      161 (1975).

ORGANIZED CULTURES OF NERVE TISSUE: A NOVEL MODEL SYSTEM
FOR STUDIES OF LIPID PROTEIN INTERACTION ON THE FUNCTIONAL
LEVEL

Michael Giesing

Nattermann Research Laboratories

Nattermannallee 1

D-5000 Köln 30, G.F.R.

ABSTRACT

Organized fragments of grey matter of cerebral cortex
(CC) maintained in vitro during cell maturation in the
explant culture assembly are presented as a novel tool
for studies on the functional impact of membrane phospholipids (PL). PL have been found to affect two transport
systems and one receptor system. The specificity of
the PL effects has been examined by analyzing the
kinetics of proteins that bind glycine - an irrelevant
amino acid neurotransmitter in CC - and γ-aminobutyric
acid (GABA), a major inhibitory neurotransmitter in the
tissue. The composition of membrane constituents has
been changed either through introduction of exogenous
PL into the lipid matrix of viable cells or through
specific degradation of asymmetrically distributed
components. The following findings that were made are
discussed:

1) The glycine carrier is a transversal protein
that is governed by phosphatidylcholine (PC) but not
by phosphatidylethanolamine (PE) in a fatty acid specific
fashion. PC binds the ligand as well as it affects the
protein. The carrier does not undergo lateral motion
between ordered and fluid lipid domains. On the whole
the glycine carrier is a protein with relatively little
functional relation to the compositional mosaicism of
the plasma membrane. 2) The GABA carrier is a mobile
protein that seems to be localized to a major extent
in the outer leaflet. Asymmetrically distributed PL

regulate the activity of the protein in a fatty acid specific manner. The carrier can be inhibited as well as stimulated. 3) The GABA receptor is not a lipoprotein in nature. PL affect the receptor activity either through a single transversal modulator protein or through two asymmetrically distributed modulator proteins. PL that are components of the outer leaflet such as phosphatidyl-N-dimethylethanolamine (PDE) activate the inhibitory capacity of the modulator leading to a state of desensitization of the receptor that is characterized by an increase in the strength of ligand binding. Cytoplasmic PL, among them mainly PE and phosphatidyl-N-monoethanolamine (PME), induce a stimulation of GABA receptor binding through activation of the inner part of the transversal modulator or of the cytoplasmic modulator , thus creating a state of supersensitivity. Cholinergic excess stimulation of the networ of cultured neurons by bath application of carbachol increases PDE formation via the methylation pathway. This is accompanied by desensitization of the GABA receptor. The state of desensitization should be terminated through PE. The findings illustrate the functional specificity of PL in nerve tissue. It seems to be essential that viable cells are used since results stemming from artificial membranes of membrane preparations are not always the same.

INTRODUCTION

Studies on structural and functional impacts of lipids in membranes can be carried out in viable nervous tissues providing the maintenance of the typical cellular heterogeneity. The results obtained may be different from those in artificial bilayer systems or even in brain membrane preparations. The viability of nerve cells is maintained in a culture system under appropriate conditions. Most in vitro techniques, however, induce abnormal lipid composition[33]. Previous work in our laboratory has shown that nervous tissue cultured in the form of organized explants exerts a distribution of phospholipids and fatty acids that is very close to in vivo conditions[14]. The culture system may, therefore, be used as a model for biological and also physico-chemical approaches. The advantages emerging from studies in explants are obvious:
1. Exposure of viable cells to lipids dispersed in a buffer by sonication leads to changes of the cellular lipid composition as shown first in our laboratory with the aid of fluorescent labeled phospholipids[18].

2. Incubation of viable cells with lipases leads to degradation of the hydrophobic moiety facing the external leaflet of the plasma membrane. Addition of the enzymes to postnuclear homogenates results in hydrolysis of lipids on both the external and cytoplasmic leaflet. This would allow to differentiate between asymmetrically distributed lipids since it is unlikely that trans-bilayer movement occurs during the procedure[10].
3. Stimulation of the cellular biosynthetic machinery under living conditions affects the membrane lipid composition. This can be achieved by activation of neurotransmitter receptors resulting either in degradation and reformation of phosphatidylinositol[21], in methylation of phosphatidylethanolamine[15,16,24] or in de novo biosynthesis of zwitterionic phospholipids[16]. Also ingrowing synapses affect the metabolism of glycerolipids and fatty acids in target cells[19].

Structural and functional impacts of phospholipids, acidic galactolipids and sialoglycolipids have been demonstrated for a variety of neurotransmitters and neurotransmitter-receptor systems among them acetylcholine[2,9], dopamine[1], gamma-aminobutyric acid (GABA)[37], opiate-like peptides[34], serotonin[35,39], substance P[27] and taurine[4]. Our main interest was focussed on the specificity of phospholipid action on nerve cell membrane functioning. We, therefore, compared the activity of the receptor for GABA in tissue specimens of grey matter of rat cerebral cortex (CC) with the carrier and transport systems for the amino acid. GABA is a major inhibitory neurotransmitter in CC[31]. In addition, GABA transport was compared with glycine transport. The latter amino acid acts as inhibitory neurotransmitter in the spinal cord but not in CC[3]. For these studies cultures were used prior to glia maturation in vitro[16]. Highly differentiated synapses, however, were then already established[18]. It will be shown in this paper that the receptor and carrier systems for the two amino acids are specifically affected through phospholipids depicting a characteristic distribution and mobility of the proteins.

RESULTS AND DISCUSSION

Carrier-Mediated Transport of GABA and Glycine

Fig. 1 shows that the transport systems for both amino acids have similar kinetic values in CC explants although they are of different biological relevance.

Maintenance of tissue specimens in lipid-free medium[25] does not allow the cells to incorporate lipids from the nutrient. Under these circumstances the transport was greatly reduced. Two carrier populations of higher affinity appeared.

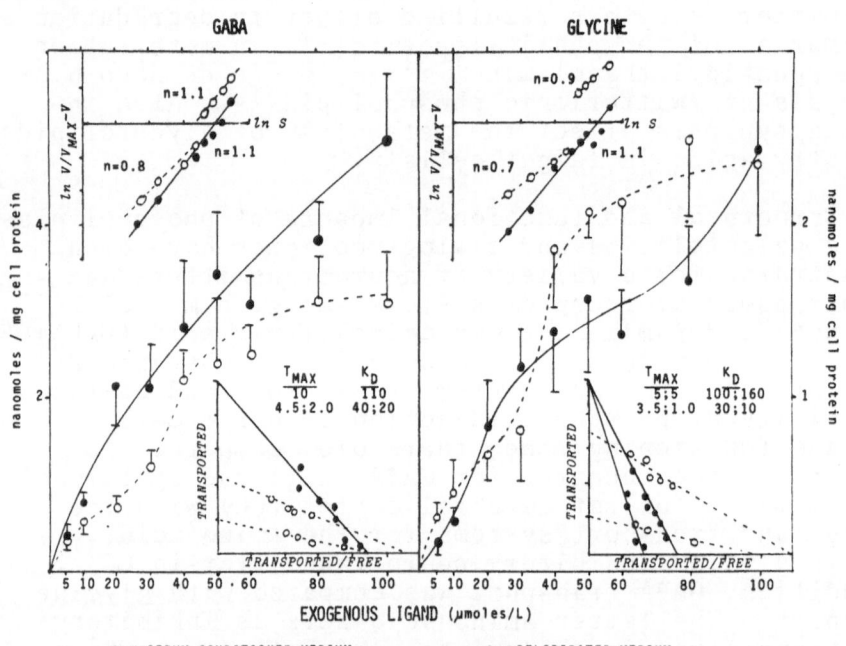

Fig. 1. Transport of GABA and glycine in CC explants at the neuronal stage after 8 days in vitro. The tissue was dissected from two-day-old donor rats and maintained in vitro in the presence[17] or absence[25] of serum lipids. The results from at least four individual determinations+SD are given. The data are expressed as Hill-coefficients (n) and in the form of Scatchard plots yielding the maximum transport site density (T_{MAX} = nanomole/mg protein) and dissociation constant (K_D = umole/L).

Table 1. Transport of GABA and Glycine in Cultures Labeled with Fluorescent Phospholipids[a].

Exogenous Phospholipid	GABA Transport		Glycine Transport	
	T_{MAX}	K_D	T_{MAX}	K_D
Phosphatidyl-choline				
di-(trans)18:4	5.0	70	1.0	10
-1-16:0-2-				
(trans)18:4	16.0;9.0	130;115	8.5;2.0	85;10
di-(cis)18:4	17.0;2.5	190;20	9.0;1.8	130;20
Phosphatidyl-ethanolamine N-dansyl-PE-1-16:0-2-18:2	4.0;4.5	75;30	6.0	90

[a] T_{MAX} is given in nanomole/mg protein, K_D in µmole/L. The values are averages from triplicates. Lipids were given as dispersions in a buffer solution (50 µg/ml) for 30 min. prior to the transport assay. Amino acids were given to the cultures in the absence of excess exogenous phospholipid.

Since lipid-free nutritioned cultures did not allow any differentiation between the transport systems, fluorescent labeled phospholipids were applied to cultures in order to alter the composition of the plasma membrane. The lipids were labeled either in the acyl chain in the form of parinaroyl derivatives or in the polar head that was previously dansylated. Introduction of the lipids was recently visualized in viable nerve cells [18]. Metabolism of the lipids did not occur during the incubation procedure as determined by mass spectrometry[15]. Table 1 shows the biological activity of the phospholipids.

Table 2. Transport of GABA and Glycine in Phospholipid-Enriched Cerebral Cortex Explants[a].

Exogenous Phospholipid	GABA Transport		Glycine Transport	
	T_{MAX}	K_D	T_{MAX}	K_D
Saturated PC Species				
di-12:0	10.0	140	4.5	90
di-14:0	7.6	150	14.0;1.0	420;20
di-16:0	1.9	30	2.0	80
di-18:0	6.0;1.2	70;20	8.0;0.7	250;20
Desaturated PC Species				
di-18:1	6.5	60	4.0	35
di-18:2	5.5	120	3.0	85
di-20:4	8.0	175	7.1	215
Methyl-PE Species				
PE	3.5	40	1.0	20
PME	2.5	30	1.5	20
PDE	3.5	50	2.0	30
PC	1.9	30	2.0	80
PE Species				
di-12:0	5.5	90	1.5	15
di-16:0	3.5	40	1.0	20
di:18:0	1.6	25	1:7	20
di:20:4	1.5	10	1.8	15

[a]. Incubation and assay conditions are given in the legend of Table 1. The data are average values from three independent determinations. Abbreviations: PC = phosphatidylcholine; PME = N-monomethylphosphatidyl-ethanolamine; PDE = N-dimethyl-phosphatidylethanolamine; PE = phosphatidylethanolamine. Comparison of the efficacy of polar heads in affecting amino acid transport was achieved with di-palmitoyl species.

In a more elaborate manner the cultures were incubated with a variety of phospholipid species to
evaluate the role of polar heads and fatty acids in determining the transport kinetics. Results are set out
in Table 2. GABA transport was affected in a fatty
acid specific manner in a way that increasing chain
length of acyl groups reduced T_{MAX} and K_D regardless of
the polar head. Phosphatidylcholines (PC) were more
potent. Desaturated PC species had the opposite effect
on the GABA system. In contrast to that the transport
of glycine was influenced in a fatty acid specific manner
only after introduction of PC species. This effect was
absent after phosphatidylethanolamine (PE) incorporation
and also after introduction of N-monomethyl-PE (PME) and
N-dimethyl-PE (PDE) into the cells. With respect to the
asymmetry of PL, i.e. PE and PME facing the cytoplasmic
side and PDE and PC being in the external leaflet[10],
the results strongly suggest that the two carriers differ
with respect to their transversal distribution.

Table 3. Binding of GABA and Glycine by Phospholipids[a].

Amino Acid Binding Cellular Phospho- lipid	GABA Carrier/n-hexane	Glycine Carrier/n-hexane
Total Phospholipids	2 x 10^3	0.4 x 10^2
Phosphatidylethanol- amine	57 x 10^3	51 x 10^2
Phosphatidylinositol	51 x 10^3	11 x 10^2
Phosphatidylcholine	0.7 x 10^3	0.5 x 10^2

[a]Phospholipids were isolated from the cultures as
indicated and solubilized in n-hexane (1 mg/ml). Amino
acids (10 µmole/L) in a buffer solution were layered
over the organic phase yielding a final volume of
1 ml n-hexane and 0.2 ml buffer. After vigorous shaking
for 30 min. the amino acid content was determined in
either phase. Postnuclear 100 000 x g pellets were incubated for 30 min. with the amino acids (10 µmole/L).
Specifically bound amino acid was determined by
filtration assays. An aliquot of the membrane preparation was analysed for PL distribution. The binding capacity of the particulate fraction was compared with
the PL-containing n-hexane layer on the grounds of
equimolar content of PL as indicated. Results are expressed as carrier - n-hexane ratio.

Evaluation of the capacity of lipids to bind either amino acid was achieved by comparing the distribution of the ligands in a partitioning system composed of n-hexane and buffer with the cellular activity of carrier binding (Table 3). The results show that the binding capacity of the PL-modulated carrier of GABA was more potent by at least one order of magnitude than the glycine carrier. It can be concluded that the results for GABA listed in Tables 1 and 2 are based on PL-carrier interaction, whereas those for glycine rather represent two components, namely ligand-PL and carrier-PL interaction.

The specificity of PL action on either carrier became even more obvious by studying the effect of temperature on transport activity in viable cells. Fig.2 shows that the GABA carrier had two discontinuities in non-treated cultures. Little temperature sensitivity was observed for the transport of glycine (Fig. 3). Transport of the latter amino acid also required lower activation energy (E_A). Introduction of lipids into the cells shifted the GABA curve essentially within the discontinuities, i.e. towards lower temperatures in dioleyl-PC treated cultures, towards higher temperatures in cholesterol-enriched cells. An intermediate situation was obtained after dipalmitoyl-PE incubation. According to the work of Thilo et al. (1977)[36] on ß-galactoside transport in E. coli the conclusion can be drawn that the GABA transport protein undergoes a partitioning between the fluid and ordered lipid domains in the plasma membrane that is dependent on the lipid composition. Partitioning will govern the activity of the protein either through changes in the rate constant or in the effective carrier concentration or in both. As shown in Tables 1 and 2, the PL action on GABA transport rather affects the effective carrier concentration, i.e. effective carrier proteins have a tendency for preferential partitioning in the fluid domains after the membranes have accumulated dioleyl-PC. Dipalmitoyl-PE and even more cholesterol treatment induces a certain part of the carriers to spend on the average more time in the ordered domain where they do not function any longer. As shown in Fig. 3, the transport of glycine is unlike that of GABA under the same experimental conditions indicative of the incapability of the protein to undergo preferential partitioning in the fluid or ordered lipid domains of the plasma membrane.

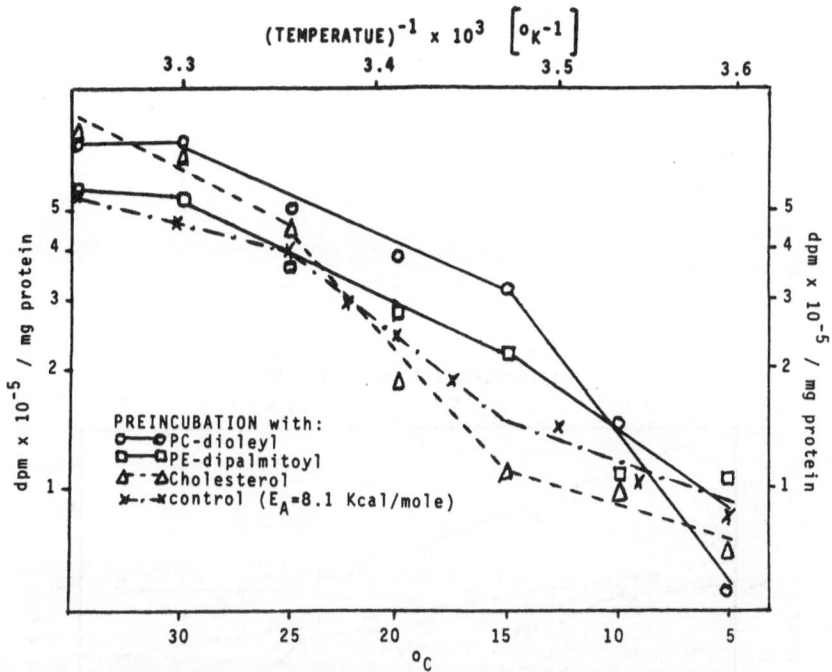

Fig. 2.   Effect of temperature on GABA transport on
lipid-enriched CC explants. The transmitter
was given at a concentration of 25 nanomole/L.
T_H = discontinuity at higher temperature; T_L =
discontinuity at lower temperature; E_A =
activation energy. Averages of quadruplicate
determinations are given.

    The two carrier systems were also to be distin-
guished after hydrolysis of membrane components. The
hydrolytic effects were different with respect to
application of the enzymes to living cells in comparison to poastnuclear homogenates (Figs. 4 and 5).
Proteolysis in homogenates potentiated glycine-carrier
binding in comparison to viable cells whereas the effect
on GABA was not even additive.

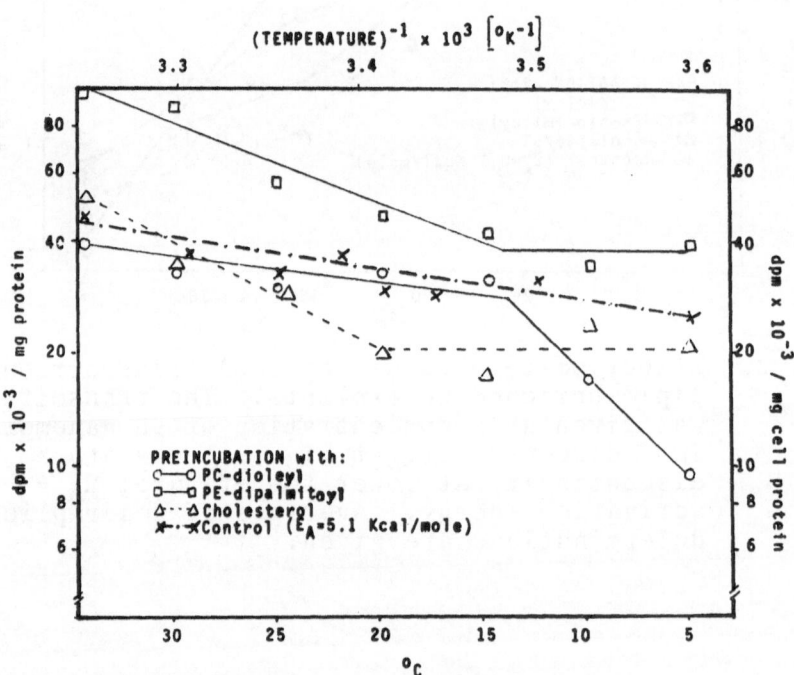

Fig. 3. Effect of temperature on glycine transport. Experimental conditions were the same as for GABA.

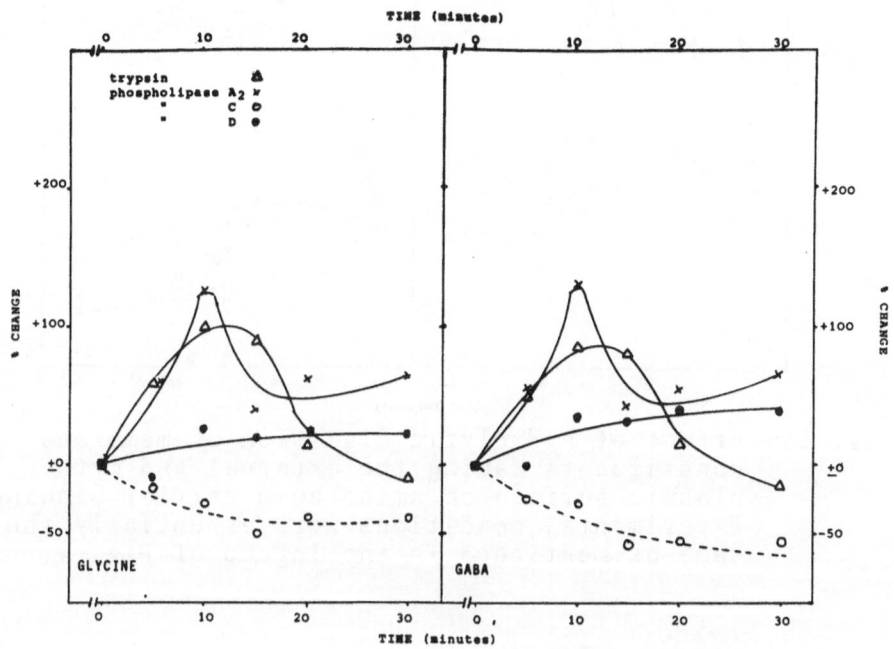

Fig. 4. Effect of hydrolytic digestion of external
surface components on carrier binding.
Enzymes were given to the cultures in a
buffer solution (Medium M-199): Trypsin at
1 mg/ml; phospholipase A_2 at 105 U/ml;
phospholipase C at 10 U/ml; phospholipase D
at 10 U/ml. Protease activity was terminated
through the addition of equivalent amounts of
trypsin inhibitor 0.8 mg/ml. After removal of
the enzymes amino acid carrier binding was
examined (2 µmoles/L for 30 min.) in post-
nuclear 100 000 x g pellets.

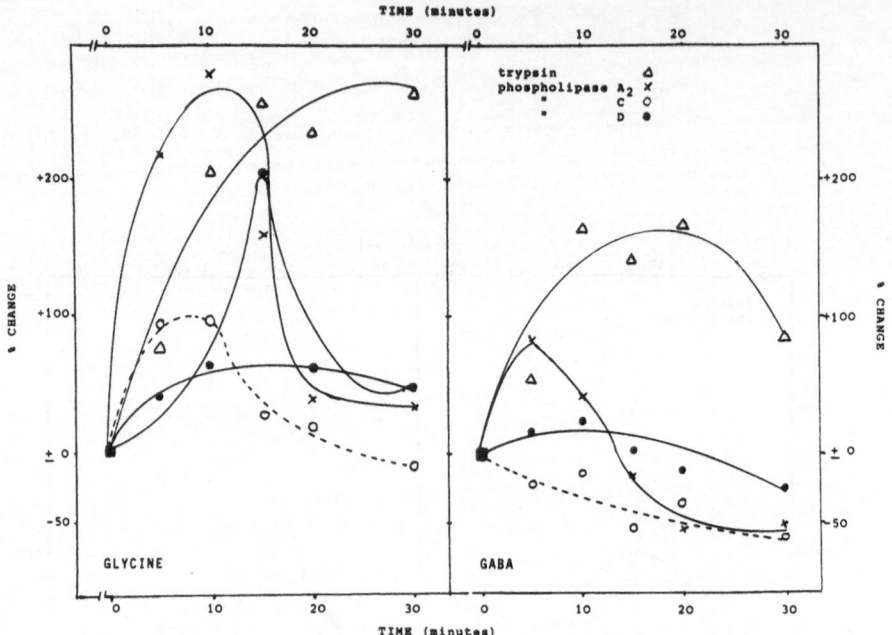

Fig. 5. Effect of hydrolytic digestion of membrane
constituents facing the external and cytoplasmic surface on amino acid carrier binding.
Experimental conditions were essentially the
same as mentioned in the legend of Fig. 4.

The glycine carrier could also be activated
through phospholipase A_2. The cytoplasmic and external surface contributed equally. Stimulation of
the GABA carrier, however, depended clearly upon removal of fatty acids on the external surface.
Hydrolysis of the polar head through phospholipase D
activity had a slight stimulatory effect on the two
carrier systems, apparently without exhibiting any
topological specificity. Liberation of the polar
heads from PL facing the external surface led to inhibition of both carriers. This was counteracted
through PL on the cytoplasmic side as far as glycine
is concerned. Removal of polar heads on the cytoplasmic surface virtually did not affect GABA-carrier

Table 4. Effect of Hydrolysis of External Membrane Constituents on GABA Binding Proteins[a].

| Enzymatic Digestion | High Affinity GABA Binding | | | |
| | Carrier | | Receptor | |
	B_{MAX}	K_D	B_{MAX}	K_D
None	250	1000	2.8	45
Phospholipase A_2	780	16500	42	450
Phospholipase C	550	4000	30	160
Trypsin	400	2500	3.2	20

[a]Enzyme concentration is given in Fig. 4. Phospholipase A_2 was given to viable cells for 10 min., phospholipase C for 15 min. and trypsin for 15 min.(carrier binding), and 30 min. (receptor binding). B_{MAX} is given in picomole/mg protein, K_D is expressed as nanomole/L.

Table 5. Effect of Successive Enzymatic Treatment of Viable Neurons on High Affinity GABA Receptor Binding.

| Incubation Conditions | B_{MAX} | K_D |
	% Change	
Phospholipase A_2	+1400	+900
Trypsin, then Phospholipase A_2	+ 185	+ 95
Phospholipase C	+ 970	+260
Trypsin, then Phospholipase C	-30	+ 65

binding. These results, however, do not tell which one
of the hydrolysis products accounts for the effects observed, but they provide further evidence that the
activity of the two carrier or transport systems is
specifically, i.e. in different ways, regulated through
other membrane constituents, preferably by PL.

GABA Receptor Binding

Several authors have shown that lipids affect GABA-
binding proteins [5,8,11-13,28, 30,37]. Most of the studies
have been carried out in membrane preparations stemming
from brain tissue. Interestingly, the membranes are
devoid of a high affinity system unless PL are removed
through organic solvents or detergents such as Triton
X-100 [37]. In contrast to this, preparations of the
receptor membranes derived from CC tissue specimens in
explant culture clearly show up with two receptor
systems, namely a high affinity system (B_{MAX} = 2.8
picomole/mg protein; K_D = 45 nanomole/L) and a low
affinity system (B_{MAX} = 24 picomole/mg protein; K_D = 570
nanomole/L)[15]. The discrepancy can be explained through
the artificial PL composition of membranes stemming from
brain tissue. Removal of the tissue from the donor
implies a period of hypoxia and ischemia in which liberation of fatty acids, mainly arachidonate, occurs
rapidly and to a substantial amount[6,29]. In contrast
to the animal model, phospholipase A_2 has to be applied
to cultures in order to abolish high affinity GABA receptor binding. As shown in Table 4, hydrolysis of PL
facing the external side of the plasma membrane in
living neurons greatly enhanced receptor activity.
Phospholipase C was relatively less potent than phospholipase A_2.

Receptor and carrier binding of the amino acid
reacted differently upon hydrolytic manipulation of the
membrane structure. Proteolysis increased the affinity
of the receptor whereas the carrier showed the
opposite effect. Liberation of PL fatty acids or
polar heads rather decreased the carrier affinity
whereas it increased considerably more the density
of effective receptor proteins. Two questions arose
from these experiments: first whether the receptor
is regulated through a trypsin-sensitive modulator
protein - a GABA receptor binding inhibitor protein
has been described previously [7, 20, 32] -, and second
whether the PL moiety acts directly upon the receptor
or on the modulator thereby governing receptor activity.

The latter question was investigated through successive hydrolysis of external surface components (Table 5).

The results clearly show that stimulation of receptor activity upon PL hydrolysis was almost abolished after preceding application of the protease. This indicates that the receptor per se is not a lipoprotein in nature. PL evidently affect a kind of receptor-modulator. The lipoprotein-like modulator prevents that the majority of available receptor proteins comes into operation and might therefore be functionally termed an inhibitor. However, the modulator can also increase the number of available receptor sites as will be shown. With respect to phospholipase A_2 application to viable cells, only about 7% of the receptor sites are operative in intact membranes. This "reserve" phenomenon, which obviously possesses an enormous capacity, has also been found in a variety of other systems [22,23]. The 7% range, however, is much lower than observed with brain membrane preparations. The reason for this discrepancy is obviously based on the asymmetric distribution of individual PL which can only be determined in the culture system. As shown in Table 6, phospholipase addition to post-

Table 6. Topological Specificity of the Effects of Membrane Hydrolysis on GABA Receptor Activity[a]

Enzyme Treatment Topology		High Affinity GABA Receptor Binding			
		B_{MAX}	% Change	K_D	% Change
Phospholipase A_2	Ext.	42	+1400	450	+900
	Ext.+Cyt.	9.0	+ 200	240	+120
Phospholipase C	Ext.	30	+ 970	160	+260
	Ext.+Cyt.	15	+ 400	330	+240
Trypsin	Ext.	3.2	+ 10	25	- 60
	Ext.+Cyt.	2.0	- 30	20	- 75

[a]Experimental conditions are given in Table 4. Ext. = external leaflet: Ext. + Cyt. = external and cytoplasmic leaflet. B_{MAX} is expressed in picomole/mg protein, K_D in nanomole/L.

nuclear homogenates, thereby hydrolysing PL on either surface of the membrane, gives results that are similar to those observed in brain membranes [37]. Stimulation of GABA receptor binding upon PL hydrolysis was then markedly reduced.

Assuming that no transbilayer movement occurred during the experiment[10], some elucidative conclusions can be drawn: intact PL facing the external side reduce the number of active binding sites, i.e. PL induced inhibitory activity of the modulator upon receptor activity; intact PL facing the cytoplasmic side of the plasma membrane increase the density of available sites, i.e. PL induced stimulatory activity of the modulator upon receptor activity. The affinity of the receptor for the ligand is increased by external PL whereas it is decreased by cytoplasmic PL. On the basis of the different types of phospholipases applied to viable cells and post-nuclear homogenates it seems that polar heads in external PL are more potent than external acyl residues with respect to B_{MAX} and K_D. Polar heads in PL facing the cytoplasmic side have little effect on the receptor. The stimulatory effect on the receptor binding is preferably caused by the acyl chains in these PL. The trypsin-sensitive modulator protein obviously is a transversal protein. The major part of it, however, should be in the cytoplasmic leaflet since trypsinization in post-nuclear homogenates was more potent in reducing receptor activity than application of the enzyme to viable cells.

Table 7. High Affinity GABA Receptor Binding in PL-Enriched Neurons.

Cellular Phospholipid	B_{MAX} (picomole/mg protein)	K_D (nanomole/L.)
Control	2.5	45
PE-di-16:0	7.2	85
PME-di-16:0	6.9	85
PDE-di-16:0	3.2	15
PC-di-16:0	5.2	30
PC-di-18:0	3.6	20
PC-di-18:2	2.8	15

At least some of these speculative conclusions can
be confirmed in PL-enriched cells (Table 7). PC reduced
receptor activity in a fatty acid specific manner. With
respect to the asymmetric distribution of PL a variety
of dipalmitoyl-species was given under routine incubation
conditions to viable cells. Similar to what could be predicted from the results presented in Table 6, it is
apparent that introduction of PL into viable cells that
are considered in a native membrane to face the cytoplasmic side, namely PE and PME, resulted in increased
binding of the amino acid to the receptor. Those PL
localized in the external leaflet under native conditions,
when introduced into viable nerve cells induced a reduction of receptor activity or at least a decrease in
K_D values. It is worthy of note that with respect to
the predictions made upon the results obtained from the
topological specificity of phospholipase actions,
PDE-di-16:0 turned out to be the most effective PL in
terms of reduction of receptor functioning. All these
results depicting so far only structural aspects of the
neuronal membrane should give a more functional impact
if changes in the metabolism could be shown. As described in the following chapter, we succeeded so far to
alter the metabolism of two PL facing the external
side, namely PDE and PC.

Regulation of GABA Receptor Binding by Cholinergically Induced Phospholipid Biosynthesis

Although not understood in detail it is certain
that the asymmetry of membrane PL is maintained
through more than a single mechanism[10]. For instance,
PC synthesized de novo in the rough endoplasmic reticulum might , after entering the plasma membrane,
be translocated from the inner to the outer leaflet
due to enzymatically introduced differences in the
physical properties of PL such as charge and size of
polar heads. Phosphatidic acid derived from external
PC through phospholipase D action was found in unilamellar vesicles to be translocated to the inner
monolayer, and an appropriately equal number of PC
molecules moved from the inner to the outer layer.
PC originating from methylation of PE appears in
the outer leaflet of a plasma membrane through
enzyme-mediated transbilayer movement of enzyme
substrates and products (for review see ref. 10).

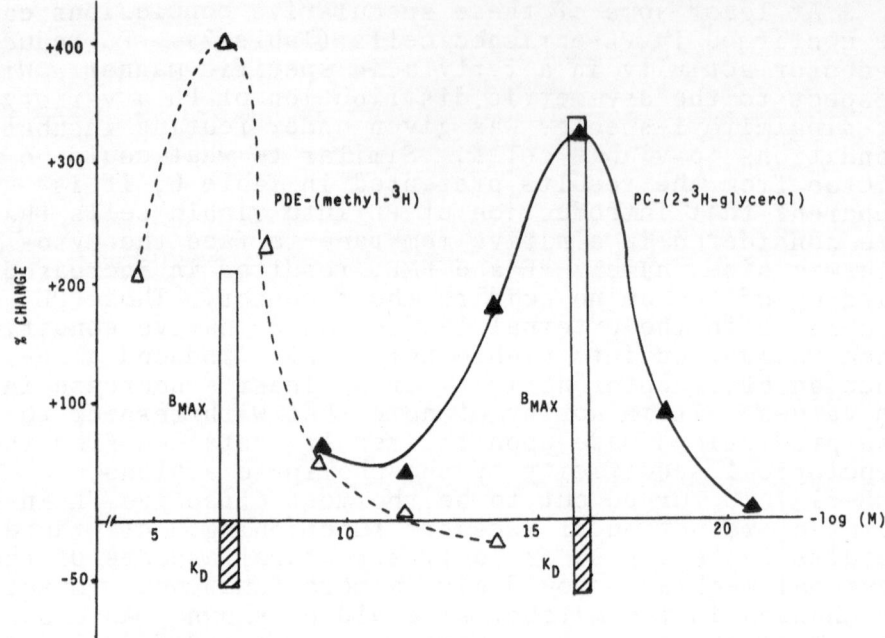

Fig. 6. Carbachol-induced stimulation of PL
methylation and de novo biosynthesis of PC.
The stimulant was given for 30 min. as
indicated. Methyl groups stemming from exogenous S-adenosylmethionine or methionine
appeared in PDE within 10 min. ("short-term
condition"). Introduction of glycerol into
cellular PC was observed 3 to 5 hours after
carbachol treatment ("long-term condition").
Percentage changes of high affinity GABA
receptor kinetics are given at maximum
activity of the cholinergic effects.

      We have recently reported that stimulation of PC
de novo biosynthesis can be achieved by previous bath
application of carbachol[16]. Deacylation of cellular
PC was observed simultaneously. The resulting lyso-
lipid will cause considerable perturbation of the
membrane properties if it is not removed, acting then,
for instance, similar to phosphatidic acid.

Methylation of PE was also greatly enhanced after
cholinergic stimulation of viable cells albeit at higher
carbachol concentration and within a shorter period of
time as compared to glycerol-labeled PC (Fig. 6).
If the two PL were capable of affecting GABA receptor
binding and if the predictions made in the previous
chapter were correct (see Table 7), activation of the
cholinergic receptor should cause a PL-mediated increase
in receptor affinity for GABA and, in addition,
fatty acid-specific changes in receptor density.
Fig. 6 clearly shows that the maximum effect on PL
formation via either pathway was paralleled by a
decrease in the values for K_D as predicted and also by
enhanced receptor site density.

However, looking into more details it turned out
that only the methylation pathway definitely affected
the GABA receptor as evaluated on the basis of carbamyl-
choline concentration (Fig. 7) and phospholipase activity
(Table 8). Treatment of viable nerve cells with
carbachol enhanced the values for B_{MAX} of both the
high and the low affinity GABA receptor in a surprisingly
similar fashion as PL methylation was stimulated.
Values for K_D in the two systems were reduced with
increasing concentration of the cholinergic stimulant.
This finding could be helpful in explaining desensitization[26,38] of the GABA receptor in lowering K_D
upon excess stimulation of the acetylcholine receptor
on the basis of a lipid theory: PDE activates a
trypsin-sensitive modulator protein that acts directly
upon the receptor leading to a stronger binding between
ligand and protein.

This fits also into the concept of PDE as a PL
facing the external side of the plasma membrane[10]. Not
only that exogenous PDE being introduced into the
membrane matrix exerted the same effects as carbachol
and that the opposite observation was made after hydrolysis of the outer leaflet-PL in unstimulated neurons,
the PDE-mediated activity of the modulator protein
became even more obvious by application of phospholipase C to post-nuclear homogenates stemming from
cholinergically stimulated nerve cells. As listed in
Table 8, the enzyme essentially failed to abolish
carbachol-induced high affinity GABA receptor binding as
it did in untreated neurons. This observation is
similar to trypsin application (Table 5). Both
findings might indicate that the modulator has reached
a state in which it is no longer governed by PL of the
outer leaflet nor by the hydrolysis products of these PL.

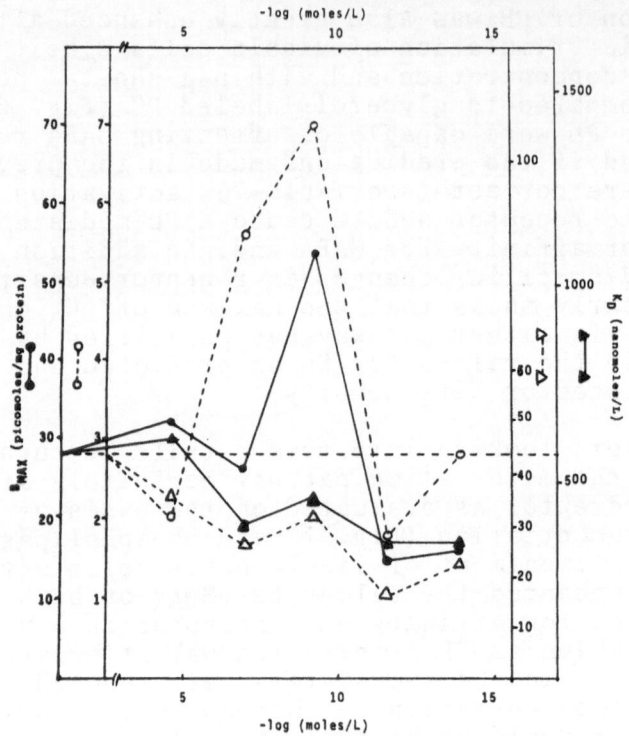

Fig. 7. Carbachol-induced effects on GABA receptor
        binding in viable nerve cells under
        "short-term conditions".

It seems that PL facing the external side do not
have access to the modulator protein. It is
interesting to note that the carbamylcholine-induced
increase of B_{MAX} could not be observed in the
100 000 x g pellet preparation as it appeared in
post-nuclear homogenates (Fig. 7). The decrease in
K_D, however, was persistent in either case. Activation of the modulator through external PDE then
acting as an inhibitor on the high affinity GABA
receptor and thereby increasing the strength of the
binding between ligand and receptor protein, must be
terminated through another mechanism unless the state
of desensitization is irreversible.

388

Table 8. Disappearance of Phospholipid-Mediated Regulation of the GABA Receptor in Excess Stimulated Neurons[a]

Incubation Conditions		$-Na^+$-GABA Binding in Post-Nuclear 100 000 x g Pellets	
Viable Cells	Post-nuclear Homogenates	B_{MAX}	K_D
None	None	3.0	75
		% Change	
Carbachol	None	-55	-80
None	Phospholipase C	+400	+340
Carbachol	Phospholipase C	-10	-65

[a]Cholinergic stimulation was achieved with carbachol (10^{-3} mole/L) under "short-term conditions". The enzyme was incubated with post-nuclear homogenates for 15 min. yielding a final concentration of 10 units/ml. PL-hydrolysis was terminated by the addition of excess icecold buffer and preparation of the 100 000 x g pellet. B_{MAX} is given in picomole/mg protein, K_D in nanomole/L. The average values from triplicate determinations are given.

On the basis of our PL concept such a lipid has to face the cytoplasmic side of the membrane (see Tables 6 and 7). This may also interact with the modulator and with the inhibitory action of the protein to a stimulatory activity. The tendency of this action can be termed supersensitivity[40] (See Table 7). The most potent PL in our experiments was PE. We have, however, not yet found the endogenous mechanism inducing PE formation. Nevertheless, it can be noted that the de novo formation of PE was greatly enhanced 3 to 5 hours after a single cholinergic stimulation for a 30 min. period.

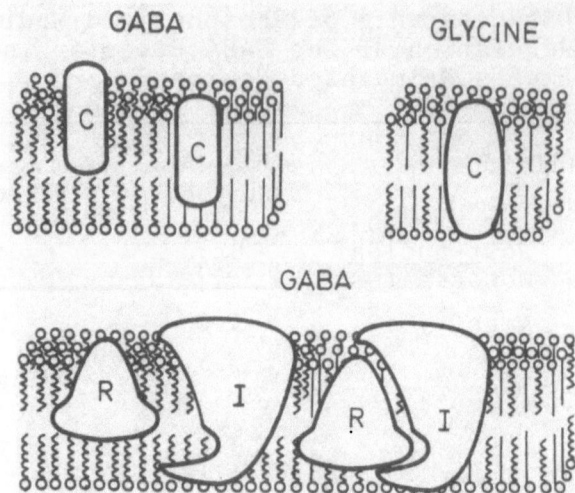

GABA

GLYCINE

GABA

Fig. 8. Hypothetic model of the distribution of carrier and receptor proteins for glycine and GABA in cerebral cortex.

CONCLUSIONS

Introduction of exogenous PL into viable cells as well as hydrolysis of membrane components facing the external or cytoplasmic side give rise for depicting hypothetic models of the distribution of the different carrier and receptor proteins for the two amino acids (Fig. 8). The transport of glycine which is not a relevant inhibitory neurotransmitter in cerebral cortex shows some sensitivity to PC species. The fatty acid pattern of PE does not play any role. The carrier is a transversal protein that does not undergo lateral partitioning between the fluid and ordered lipid domains of the plasma membrane. It seems that PL neither affect the rate constant nor the density of effective proteins. Individual PL have a considerable capacity to bind the amino acid. Therefore, the extent to which PL affect the carrier protein, the ligand or both, cannot be quantified. The GABA carrier exerts greater motility. The protein might be orientated rather

on the external leaflet. The fluidity of the membrane
determines through the asymmetric distribution of individual PL the lateral distribution of the protein
between ordered and fluid domains of the lipid matrix.
PL obviously affect the rate constant and/or the density
of effective proteins. The GABA receptor is not a
lipoprotein in nature, although PL on either side of
the membrane affect binding of the ligand in a fatty
acid-specific manner. External PC and PDE reduce
binding, cytoplasmic PE and PME oppose the effect.
PL act upon the receptor through a trypsin-sensitive
modulator protein(s) that is a lipoprotein in nature.
The modulator has inhibitory activity on the receptor
through external PL desensitization and stimulatory
capacity through cytoplasmic PL supersensitivity.
Excess cholinergic stimulation of the heterogenous
network of nerve cells in vitro leads to a PDE-in-
duced inhibition (I) of receptor binding having the
characteristics of a state of desensitization. According to a lipid theory of receptor modulation only PE
or PME should be capable of terminating the state of
desensitization by switching the inhibitory modulation
to a stimulatory effect, thus creating a state of
supersensitivity. Modulation is achieved either
through a single transversal protein that has both
inhibitory and stimulatory properties - as shown in
the figure - or through two asymmetric modulatory
proteins. The one on the outher leaflet inhibits GABA
receptor binding, the one on the inner leaflet
stimulates receptor activity. Both modulators are lipoproteins in nature.

REFERENCES

1. M.B. Anand-Srivastava and R.A. Johnson, Role of phospholipids in coupling of adenosine and dopamine
   receptors to striatal adenylate cyclase, J. Neurochem.
   36:1819 (1981).
2. T.J. Andreasen, D.R. Doerge, and M.G. McNamee, Effect
   of phospholipase A_2 on the binding and ion permeability
   control properties of the acetylcholine receptor, Arch.
   Biochem. Biophys. 194:468 (1979).

3. A.H. Aprison and E.C. Daly, Biochemical aspects of transmission at inhibitory synapses: The role of glycine, in: "Advances in Neurochemistry", Vol.3, B.W. Agranoff and M.H. Aprison, eds., Plenum Press, New York, pp. 203 (1978).

4. V.J.Balcar, J. Berg, J. Robert, and P. Mandel, Uptake of L-glutamate and taurine in neuroblastoma cells with altered fatty acid composition of membrane phospholipids, J. Neurochem. 34:1678 (198C

5. V.J.Balcar and P. Mandel, Inhibition of high affinity uptake of GABA by branched fatty acids, Experientia, 32:904 (1976).

6. N.G. Bazan, Effects of ischemia and electrocon-vulsive shock on free fatty acid pool in the brain, Biochim. Biophys. Acta, 218:1 (1970).

7. T.H. Chiu and H.C. Rosenberg, Endogenous modulator of benzodiazepine binding in rat cortex, J. Neurochem. 36:336 (1981).

8. A.Y. Chwelch and S.W. Uslie, Phosphatidylserine enhancement of [^{3}H]γ-aminobutyric acid uptake by rat whole brain synaptosomes, J. Neurochem. 38:691 (1982).

9. M. Criado, H. Eibl, and F.J. Barrantes, Effects of lipids on acetylcholine receptor. Essential need of cholesterol for maintenance of agonist-induced state transition in lipid vesicles, Biochemistry 21:3622 (1982).

10. L.L.M. van Deenen, Topology and dynamics of phos-pholipids in membranes, FEBS Lett. 123:3 (1981).

11. F.V. DeFeudis, A.N.K. Yusufi, L. Ossola, M. Maitre, P. Wolfe, G. Rebel, and P. Mandel, Antiserum to gangliosides inhibits (^{3}H) GABA binding to a synaptosome-enriched fraction of rat cerebral cortex, Gen. Pharmacol. 11:251 (1980).

12. G.E. De Medio, G. Trovarelli, A. Hamberger, and G. Porcellati, Synaptosomal phospholipid pool in rabbit brain and its effect on GABA uptake, Neuro-chem. Res., 5:171 (1980).

13. S. Fiszer de Plazas and E. de Robertis, Isolation of hydrophobic proteins binding amino acids. GABA binding in the rat cerebral cortex, J. Neurochem. 25:547 (1975).

14. M. Giesing, "Explantatkulturen des Nervensystems: Ein neues Modell für die Neurochemie. Bericht von

der Regulation einiger Lipidbausteine", Habili-
tationsschrift, Universität Bonn (1978).

15. M. Giesing and U. Gerken, The role of asymmetrically distributed phospholipids in the binding
of gamma aminobutyric acid, in: "Basic and
Clinical Aspects of Molecular Neurobiology",
A.M. Giuffrida Stella, G. Gombos, G. Benzi,
and H.S. Bachelard, eds., Fondazione Inter-
nazionale Menarini, Milano, pp. 135 (1982).

16. M. Giesing and U. Gerken, The effects of carba-
mylcholine on extrasynaptic phosphatidylcholine
biosynthesis in grey matter of cerebral cortex,
in: "Phospholipid Metabolism in the Nervous
System", L.A. Horrocks, G.B. Ansell, and G.
Porcellati, eds., Raven Press, New York, in
press (1982).

17. M. Giesing, G. Neumann, H. Egge, and
F. Zilliken, Lipid metabolism of developing
central nervous tissues in organotypic cultures.
I. Lipid distribution and fatty acid profiles of
the medium for rat brain cortex in vitro, Nutr.
Metabol. 19:242 (1975).

18. M. Giesing, B. Schmitz, B. Kempfle, H. Egge, and
F. Zilliken, Effect of phosphoglycerolipids
related to nutrition on GABA transport in cultured
neurons, in: "New trends in nutrition, lipid
research and cardiovascular diseases", N.G.Bazan,
J.M. Iacono, and R. Paoletti, eds., Alan R. Liss
Inc., New York, pp. 45 (1981).

19, M. Giesing and F. Zilliken, Lipid metabolism of
developing central nervous tissues in organotypic
cultures. III. Ganglionic control of glycerolipids and fatty acids in cortex grey matter,
Neurochem. Res. 5:257 (1980).

20. A Guidotti, G. Toffano, and E. Costa, An endogenous
protein modulates the affinity of GABA and benzodiazepine receptors in rat brain, Nature 275:553(1978).

21. J.N. Hawthorne and M.R. Pickard, Phospholipids in
synaptic function, J. Neurochem. 32:5 (1979).

22. D.S. Heron, M. Israeli, M. Hershkovitz, D. Samuel, and
M. Shinitzki, Lipid-induced modulation of opiate
receptors in mouse brain membranes. Eur. J.
Pharmacol. 72:361 (1981).

23. D.S. Heron, M. Shinitzki, M. Hershkovitz, and
    D. Samuel, Lipid fluidity markedly modulates
    the binding of serotonin to mouse brain
    membranes, Proc. Natl. Acad. Sci. U.S.A.
    77:7463 (1980).
24. F. Hirata, J.F. Tallmann, R.C. Henneberry,
    P. Mallorga, W.J. Strittmatter, and J. Axelrod,
    Phospholipid methylation: A possible mechanism
    of signal transduction across biomembranes,
    in: "Membrane transport and neuroreceptors",
    Alan R. Liss Inc., New York, pp. 383 (1981).
25. A.F. Horwitz, M.E. Hatten, and M.M. Burger,
    Membrane fatty acid replacement and their
    effect on growth and lectin-induced agglutinability,
    Proc. Natl. Acad. Sci. U.S.A., 71:3115 (1974).
26. R.J. Levkovitz, D. Mullikin, C. Wood, T. Goore,
    and C. Mukherjee, Regulation of Prostaglandin
    Receptors by Prostaglandins and Guanine Nucleo-
    tides in Frog Erythrocytes, J. Biol. Chem.252:
    5295 (1977).
27. F. Lembeck, A. Saria, and N. Mayer, Substance P:
    Model studies of its binding to phospholipids,
    Naunyn-Schmiedeberg's Arch. Pharmacol. 306:189
    (1979).
28. K.G. Lloyd and K. Beaumont, Possible role of
    phospholipids in GABA receptor function in
    human and rat brain, Brain Res. Bull., 5:285
    (1980).
29. M.D. Majewska, R. Manning, and G.Y. Sun, Effects
    of postdecapitative ischemia on arachidonate
    release from brain synaptosomes, Neurochem. Res.
    6:567 (1981).
30. R.H. Ng and B.D. Howard, Inhibition of neurotoxic
    phospholipase A_2 on synaptosomal uptake of
    γ-aminobutyric acid, J. Neurochem. 36:310(1981).
31. A. Nistri and A. Constanti, Pharmacological
    characterization of different types of GABA
    and glutamate receptors in vertebrates and in-
    vertebrates, Progr. Neurobiol. 13:117 (1979).
32. R.W. Olsen, J.D. Bayless, and M. Ban, Potency of
    inhibitors for γ-aminobutyric acid uptake by
    mouse brain subcellular particles at 0°,Mol.
    Pharmacol. 11:558 (1975).

33. J. Robert, P. Mandel, and G. Rebel, Neutral lipids and phospholipids from cultured astroblasts, J. Neurochem. 26:771 (1976).
34. E.J. Simon, Studies on membrane-bound opiate receptors, in: "Membrane mechanism of drugs and abuse", Alan R. Liss Inc., New York, pp. 51 (1979).
35. H. Tamir, W. Brunner, D. Casper and M.R. Rapport, Enhancement by gangliosides of the binding of serotonin to serotonin binding proteins, J. Neurochem. 34:1719 (1980).
36. L. Thilo, H. Träuble, and P. Overath, Mechanistic interpretation of the influence of lipid phase transitions on transport functions, Biochemistry 16:1283 (1977).
37. G. Toffano, C. Aldinio, M. Balzano, A. Leon, and G. Savoini, Regulation of GABA receptor binding to synaptic plasma membrane of rat cerebral cortex: the role of endogenous phospholipids, Brain Res. 222:95 (1981).
38. L.T. Williams and R.J. Lefkovitz, Slowly reversible binding of catecholamine to a nucleotide-sensitive state of the ß-adrenergic receptor, J. Biol. Chem. 252:7207 (1977).
39. J.R. Yandrasitz, R.M. Cohn, B. Masley, and D. Rowe, Evaluation of the binding of serotonin by isolated CNS acidic lipids, Neurochem.Res. 5:465 (1980).
40. Y. Yoneda and K. Kuriayama, Presence of a low molecular weight endogenous inhibitor on ^{3}H-muscimol binding in synaptic membranes, Nature 285:670 (1980).

PARTITION OF PROTEINS BETWEEN WATER AND NON-POLAR PHASES

Gregorio Weber

Department of Biochemistry
School of Chemical Sciences
University of Illinois
Urbana, IL   61801, USA

The partition of small molecules between water and a non-polar
phase follows simple and well defined rules. The most important
macroscopic characteristic that determines partition between these
phases is their difference in dielectric constant. The high dielectric constant of water favors the separation of charges with formation of independent ions of opposite charge but in media of much
lower dielectric constant, separation of charges cannot take place
and in consequence the ions present in the water cannot be transferred to the second phase. The high dielectric constant of water
results from the permanent dipole moment of the water molecules and
the high dipole density. These properties determine the very different molecular interactions prevalent in water and in a typical
non-polar medium. On the water side we have the strong attractive
forces between permanent dipoles and in a lipid phase the much weaker dispersion forces that result from the attractive effects of
mutually induced dipoles. Partition is determined by the work necessary to transfer a given molecule from one of the phases to the
other. We can consider it as the result of two successive operations: transfer from one phase to the vacuum, followed by transfer
from the vacuum to the second phase. In the first operation it will
be necessary to break the bonds between the solute and the molecules
of the solvent phase. For this purpose work is required because of
the attractive forces always present between molecules. This requirement for work will be partly conterbalanced by the spontaneous
decrease in free energy that takes place when additional interactions between the solvent molecules appear as the solute molecule is
removed. Solution of the transferred molecule in the second medium
requires the two converse operations: Bonds between the solvent

molecules must be broken up and the energy required for this process
is to be partially or totally paid for by the interactions of the
solute with the molecules of the second phase[1]. The difference
between the energies required to carry out these two independent
processes will determine the partition coefficient of the molecule. The molecules of an ideal non-polar solvent are attracted to
each other by the weakest forces (mutually induced dipoles). The
forces between the solute molecules and this solvent will therefore
be at least as strong as those that they replace when a solute molecule is introduced and some solvent-solvent bonds are replaced by
solvent-solute bonds. One could say that an ideal non-polar solvent
is unable to reject a molecule that attempts to penetrate it. Quite
a different state of affairs exists when the solvent molecules are
permanent dipoles which interact strongly with each other. In these
cases the solvent will ´resist´ penetration by the solute molecule
unless the bonds to be formed between solute and solvent are at
least of strength comparable to the solvent-solvent bonds that have
to be broken to insure penetration. It is easy to see that because
of the much larger energies involved in the case of the polar solvent the partition to or away from it will be determined largely by
the strength of the interactions within it and that those in the
non-polar medium will be much less important, and can be taken as
small and constant, at least in a grossly qualitative sense. The
rejection of a non-polar solute from a polar solvent has been seen
as a ´solvophobic´ (in the case of water a ´hydrophobic´) effect.
As water is one of the media with the highest dipole density the effects are particularly strong in its case, but they need no special
explanation which relies on the existence of specific water structures. This has been directly demonstrated by application of the
methods of molecular dynamics (Rahman & Stillinger, 1971) to the
determination of the equilibrium properties of liquid water (for
review see Stillinger, 1980). Many complex structural regularities
of liquid water, including the hydrogen bonds, appear naturally when
the time-dependent interactions among the water dipoles are computed
(Rahman & Stillinger, 1973). We can then be sure that in a discussion of the interactions of water with solute molecules we need not
consider any particular water structures, but only their initial
cause, the elementary molecular properties. From this brief qualitative introduction we can see that in order to transfer a molecule
from water to a lipid phase it will be necessary to neutralize to a
large extent the sources of electrical interaction that provide a
link between the molecule and water. Although it will not be

[1]Solubility of a molecular species in either phase is limited by the
excess of the strength of solute-solute interactions over those
between solute and solvent. As we assume that solubility of the
transferred species in either phase does not limit partition,
solvent-solvent and solute-solvent interactions are the only ones
that need be considered here.

possible to reach appreciable partition of an ion into the lipid phase the transfer of an ion pair is possible as in this case the charges have been neutralized. The free energy released as a result of interaction between two ions with opposite charge will increase with the strangth of the ionization of each of the two interacting groups. As a consequence we expect the stability of the ion pair formed by guanidinium (pK app. 14) and a carboxyle group to be considerably greater than the ion pair that this latter group could form with an aliphatic amine (pK app. 11). Also the strength of these two interactions would be respectively increased if the carboxyle group (pK=4-5) is replaced by a sulphonic group (pK-1.5).

PARTITION OF PROTEINS

It is a common observation that the globular proteins cannot be dissolved, either in a lipid phase or in a moderately non-polar phase like that provided by butanol or pentanol. Reciprocally the intrinsic membrane proteins cannot be dissolved in water without the help of molecules that interact with both polar and non-polar groups. In all prbability the introduction of such amphipathic detergents is accompanied by structural changes which modify the original properties of the system to an unknown extent. Consequently it would seem that great interest ought to be attached to the study of the converse process: the modification of typical globular proteins to make them able to penetrate a non-polar phase, as in this case we are starting with a protein species of which the structure and properties in water are known in great detail. It is therefore surprising to notice that such studies have been carried out to only a limited extent (e.g. Flanagan & Barondes, 1975). In what follows I shall therefore refer almost exclusively to our own publications on this subject. If we wish to transfer a globular protein to a non-polar medium immiscible with water we find from the start some important experimental restrictions: The necessity of neutralizing both the acidic and basic groups of the protein would demand the introduction in the solvent of both anionic and cationic ligands and thus deprive us of the possibility of carrying out experiments in which the effect of one kind of charge upon the transfer properties may be separately explored. We can obviate this difficulty by operating at a pH value at which virtually only one kind of charge is present. To repress the ionization of cationic groups the pH of the solvent must be increased to at least 13 and most globular proteins placed under these conditions undergo irreversible covalent bond reactions. On the other hand at pH 2.5 to 3 the ionization of the anionic groups is virtually supressed and the protein behaviour as regards charge effects is that of a cationic polyelectrolyte. A good number of globular proteins may be kept at this pH for hours without detectable covalent chemistry and the properties of the native protein are recovered completely on reversion to a neutral pH. We chose therefore to study the transfer of two characteristic globular proteins, bovine serum album and lysozyme, from water at pH 2.5 to 1-butanol. The solvent choice is a logical one if one wishes to study the transfer to a medium with the least apolar character, yet capable of forming a two phase system in contact with pure water. Transfer to such a medium may be done with much less difficulty than transfer to a true lipidic phase, and may be expected to entail less drastic changes in the characteristic properties of the protein. As might be expected the partition of lysozyme and bovine serum albumin from a water solution of low ionic strength buffered at pH 2.5 into 1-butanol was undetectable. To increase the partition one needs to introduce an anion capable of forming stable ion pairs with the cationic groups, mostly amino groups of lysine and guanidinium groups of arginine, and for

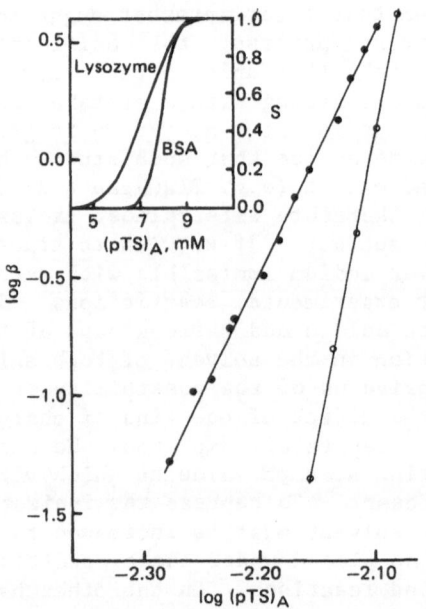

Fig. 1. Plots of log β against log[PTS] (concentration of p-toluenesulfonate in the water phase) for dansyl-albumin (○) and dansyl-lysozyme (•) conjugates at pH 2.5. [PTS] for equipartition was 7.1 mM for lysozyme, 7.9 mM for albumin. Inset: Saturation S of the butanol layer as function of [PTS]. (From Mustacich & Weber, 1978.)

this purpose and for reasons already discussed we chose the p-toluensulfonate anion (PTS). A plot of the partition coefficient β of the protein between the two phases as a function of the PTS concentration is shown in figure 1. One is immediately struck by the very steep dependence of the partition upon the concentration of added p-toluensulfonate. A plot of log β against log [PTS] has a slope of 25 for serum albumin and 10 for lysozyme. It may be tempting to assume that the high order of the PTS-induced transfer from one phase to the other is a strongly cooperative phenomenon. However, as discussed below, theory shows that partition of a molecule having many independent identical groups capable of binding a neutralizing anion will almost certainly show the high order of the partition reaction observed in these cases.

PARTITION OF A POLYELECTROLYTE UNDER THE INFLUENCE OF A COUNTERION LIGAND. PHENOMENOLOGICAL DESCRIPTION

The theory is given by Mustacich and Weber (1978). The protein at pH 2.5 is virtually a polyamine, which at vanishingly small ionic strength has a fixed partition coefficient into the non-polar phase. The partition-inducing ligand X (PTS in our case) must be appreciably soluble in either phase, with a partition coefficient of the order of unity. For PTS this ratio is app. 2. It is evident that to obtain such partition of the ligand this must have a rather large non-polar part to greatly reduce its affinity for water. A typical inorganic ˊnakedˊ ion pair would have negligible partition into the non-polar phase. The partition coefficient of the species PX(J), that is the protein with J liganded counterions is determined by the free energy of transfer of the unliganded protein -- a constant value that need not be explicitely known -- and the difference in the free energy of binding of J ligands to the protein in the two phases. Obviously the energy of the electrostatic bond between protein and counterion will be greater in the non-polar, low dielectric phase. As a result the partition into this phase increases steadily with the number of bound ligands. If all binding groups are assumed to have equal affinity the slope in the plot of log β against log free ligand, at the equipartition point, is given by the simple expression:

$$H(1/2) = N[(\gamma-1)/(\gamma+1)] \qquad (1)$$

where N is the number of anion binding sites in the protein and $\gamma = \exp-[(\Delta F(X) + \Delta F(S))/RT$, with $\Delta F(X) =$ free energy of transfer of ligand between the phases; $\Delta F(S) =$ difference in the free energy of binding of one mole of ligand in the two phases.

N, the number of protein cationic groups that can transport an anion to the non-polar phase was determined by direct chemical analysis. It was found to be 80 for albumin and 12 for lysozyme. H(1/2) is determined from /lots of log β against log [PTS] as in figure 1, and since $\Delta F(X)$ is independently known through the partition coefficient of the free ligand, the free energy of stabilization of ligand binding in the non-polar phase $\Delta F(S)$ is then calculable from eq.1: It is 0.47kcal/mol of bound PTS for albumin and 0.85 kcal/mol for lysozyme. This difference in the average stabilization of the protein-ligand complexes in the non-polar phase arises most probably from the larger ratio of arginine to lysine in lysozyme as compared to albumin. The superiority of arginine over lysine in the formation of electrostatic bonds with anions commonly found in organisms (carboxyle, phosphate) is of common observation e.g. in enzymes that can bind anions at the active center (Riordan et al., 1977) or in the detailed interactions present in the crystal of Phosphoryl-choline with a specific antibody (Padlan et al., 1976). It is easy to see from application of the theory to our cases how the high order of the partition reaction induced by PTS (25 for albumin, 10 for lysozyme) arises: Each additional anionic ligation changes the free energy of transfer by the same amount. When the free energy of transfer approaches zero ligation of one or two additional anions causes the partition coefficient to increase from near zero to a large value. Thus transfer appears to take place when a critical number of ligands is bound. The very high orders of reaction for the transfer between phases under the influence of ligands deserve consideration as one possible origin of the all-or-none character of many physiologic and pharmacologic responses.

Under these simple conditions: pH 2.5, addition of PTS we could obtain partition of the proteins into butanol but not into pentanol or hexanol. It was reasonable to expect partition into a less polar medium at the same pH, or into 1-butanol at a higher pH, by increasing the non-polar groups in the protein through chemical modification, or by altering in similar fashion the pK of the cationic groups and thereby stabilizing the ion pairing, or by using a ligand with a lower pK or a partition more favorable to the polar medium than that of PTS (e.g. octylsulphonate). We examined to some extent each of these possibilities. Modification of the protein groups through some ´mild´ chemical treatment results in preparations which are grossly heterogeneous in the sense that they are a mixture of molecular species with different numbers of modified groups. This heterogeneity considerably reduces the steepness of the transfer process as measured by the order of the transfer reaction. Thus on partial ethylation of the carboxyle groups with triethyloxonium fluoroborate a preparation of albumin was obtained that could be transferred into butanol at a lower concentration of PTS but which had a much less steep dependence of β upon [PTS]. The order of the transfer reaction was app. 7 instead of the original 25. To obtain ion pairs of greater stability we converted, also in albumin, the ε-amino groups of lysine into homoarginine by treatment with O-methyl isourea. In these modified proteins (ethylated and/or guanidated) partition was not achieved for the totality of the preparation: A fraction of the protein separated at the interphase, a behaviour that must be attributed attributed to the heterogeneous populations formed. This was inferred because the fraction that passed into the polar medium could be transferred reversibly between water and the polar medium without further loss. The protein that in these cases, and to a very much smaller extent in intact albumin, remained at the interphase correspond to species which, though uncharacterized, have considerable interest: They can undergo partial transfer into the polar phase. Their characteristics are such that some parts can interact strongly with the polar medium while others do it with water, neither interaction being sufficiently strong to completely overcome the other. Here we have a prototype of what may be the behaviour of a peripheral membrane protein: One which is unable to penetrate completely into the membrane but which can associate to it modifying the membrane functions in the process. Table I gives the results of transfer experiments with the modified proteins. It will be noticed that the partition into alcohols higher than butanol at

Table I: Partitioning of Modified Proteins into 1-butanol by Binding of p-toluenesulfonate (From Mustacich & Weber, 1980).[a]

	BuOH		
	pH 2.4	pH 4.8	pH 7.1
dansyl-BSA	C_o = 7.5 mM H = 25	-	-
ethylated dansyl-BSA	+	C_o = 3.8 mM H = 5.0	-
dansyl-lysozyme	C_o = 7.1 mM H = 10	-	-
ethylated dansyl-lysozyme	C_o = 5.5 mM H = 7.6	+	C_o > 10 mM
guanidated dansyl-BSA	C_o = 4.8 mM H = 6.2	-	-
guanidated, ethylated dansyl-BSA	+	+	+

[a]Critical partition concentrations are given where available; otherwise, "+" and "-" indicate partitioning and the absence of partitioning, respectively. C_o, ligand concentration in aqueous phase required for partition coefficient $\beta=0.5$. H, slope in the plot of log (ligand concentration) vs. log β.

pH values higher than 2.5 was possible in the modified proteins and an ethylated and guanidated species could be transferred at low pH into 1-octanol. One important question to be answered is that of th physical state of the proteins transferred to the polar solvents. Observations in the ultracentrifuge are difficult to carry out in phases with a considerable content of a second solvent as is the cas here. Some information about the state of the proteins in these media can be obtained by measurements of the rotational diffusion through a study of the fluorescence polarization of their dansyl conjugates. The fluorescence polarization was found to be independent of protein concentration thus indicating the absence of aggregation of the proteins in the apolar medium. This is not surprising because of the considerable residual charge of these proteins and the low dielectric constant of the alcohols, particularly the higher ones. The polarization observations can be given straightforward interpretation as the linear dependence of 1/pol. upon the ratio temperature/viscosity was well followed in all cases over a range of temperature of 2 to 38 C. Uniformly we found rotational relaxation times several times shorter than those observe in the equilibrium water phase. This can be rationalized as arising from the association of the alcohols with non-polar residues in the protein. This association breaks internal protein bonds and facilitates the independent rotations of some regions of the molecule wit respect to others. The penetration would be expected to be optimal for the alcohols of intermediate length and it is in fact found that the lowest rotational relaxation times are observed in these cases. The S-100 protein can be transferred to solvents like pentanol at neutral pH by addition of salts of divalent and even monovalent cations (Morero & Weber, 1982a) and in this case a noticeable reduction of the rotational relaxation time can also be observed. S-100 is known to associate with lipid vesicles inducing in them an increased cation leakage (Calisano et al. 1974) but it is noteworthy that this association is accompanied by an increase in rotational relaxation time, indicating an interaction of the protein with the surface of the vesicles rather than penetration into the lipid phase (Morero & Weber, 1982b). One would conclude from all these observations that the independent motions of parts of the protein ma be increased to a variable extent when the protein interacts with a membrane, the general rule being that the greater the contact with the lipid phase the larger the independent motions. Although the protein may show a compact well defined structure in other circumstances (e.g. in a crystal structure or in water solution) one can expect that association with a membrane will result in appreciable polymorphism. This must set definite limits to our ability to establish correlations between the protein structure, as it may become known to us through observations in the crystalline state, an its functions resulting from interactions with a membrane or a lipid vesicle. These limitations are discussed by Morero and Weber for th case of the S-100 protein (1982b); they may be expected to arise in many similar cases. It is a general difficulty in all branches of

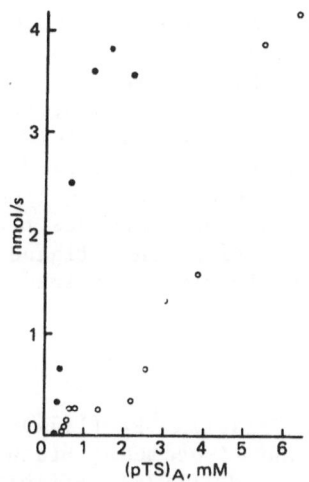

Fig. 2 Ligand concentration dependence of the rate of electro-
phoretic migration, under a constant electric field, of dansyl-
albumin (○) and dansyl-lysozyme (•) across a water-butanol
interphase. (From Mustacich & Weber, 1978.)

chemistry that the molecular states directly responsible for the
chemical reactions are not accessible to direct observation because
of their rarity in comparison with the inert ´ground´ state. A very
small free energy gap (2-3 kcal) between these states is all that is
required to make the active species undetectable. We may find that
in membrane proteins such difficulty is compounded by the polymorphism of the protein and its ability to fluctuate among many possible
ground states. An aspect of some interest, that derives from the
stability of the ion pairs upon the acid and basic strangth of the
ligands is that sulphonic groups are superior in their ability to
form stable ion pairs in comparison with carboxyle or phosphate, and
therefore in favoring partition into a membrane. The presence of
sulfonic groups in peptides like gastrone or fibrinopeptide may perhaps be related to this advantage.

PARTITION OF PROTEINS BETWEEN PHASES UNDER AN ELECTRIC FIELD

Employing dansyl-labelled proteins it is possible to follow
their migration when an electric field is applied across the interface (Mustacich & Weber, 1978). It is then found that a modest field
of a few volts/cm can produce the penetration into the second medium

when the concentration of PTS is somewhat below that required to achieve appreciable partition. In this concentration range the rate of interfacial migration to the apolar phase of lysozyme and serum albumin under a fixed field was found to be proportional to the PTS concentration with slopes that were inversely proportional to the number of bound ligands, N as shown in Fig. 2. This cooperation between the field and a partition-inducing ligand will in all probability furnish the explanation of membranes processes that are gated, in the sense that a small increase in a ligand concentration or a small change in a field previously maintained constant can produce ´critical´ effects.

REFERENCES

Calisano, P., Alema, S. & Fassella, P. (1974)   Interaction of S-100 protein with cations and liposomes.   Biochemistry 13, 4553-4560.
Flanagan, S.D. & Barondes, S.H. (1975)   Affinity partitioning.   J. Biol. Chem. 250, 1484-1489.
Morero, R.D. & Weber, G. (1982)   Properties of S-100 protein studied by fluorescence methods.   Biochim. Biophys. Acta, 703, 231-240.
Morero, R.D. & Weber, G. (1982)   Interaction of bovine S-100 protein with detergents and phosphatidylcholine vesicles.   Biochim. Biophys. Acta, 703, 241-246.
Mustacich, R.V. & Weber, G. (1978) Ligand promoted transfer between phases: spontaneous and electrically helped.   Proc. Natl. Acad. Sci. USA 75, 779-783.
Mustacich, R.V. & Weber, G. (1980)   Ligand-induced transfer between phases.   Dependence upon the strength of ion pair interactions. Biochemistry 19, 990-985.
Padlan, E.A., Davies, D., Rudikoft, S. & Potter, E.M. (1976) Structural basis for the specificity of binding of phosphorylcholine to immunoglobulins.   Immunochemistry 13, 945-949.
Rahman, A. & Stillinger, F.H. (1971)   Molecular dynamics study of liquid water. J. Chem. Phys. 55, 3336-3359.
Rahman, A. & Stillinger, F.H. (1973)   Hydrogen bonded patterns in liquid water.   J. Amer. Chem. Soc. 95, 7943-7948.
Riordan, J.F., McElvany, K.D. & Borders, C.L. Jr. (1976) Arginyl residues: anion recognition sites in enzymes.   Science, 195, 884-886.
Stillinger, F.H. (1980)   Water Revisited.   Science, 209, 451-457.

PHOTOMODIFICATION OF BIOMEMBRANES

Angelo A. Lamola

Bell Laboratories
Murray Hill, New Jersey 07974, USA

INTRODUCTION

Scope

This paper concerns two aspects of light-induced modifications of biomembranes: the degradative action of light on biomembrane function and structure, and the use of photochemical reactions to probe membrane structure.

The cooperative action of light and sensitizing agents to effect biomembrane damage is of considerable interest because consequential cellular events can be devastating (Pooler 81). Indeed, much human photosensitivity disease, especially that associated with light of wavelengths greater than that of the "sunburn" range, appears to be associated with photochemical processes that affect membranes (Blum 64; Spikes 82). Investigations of molecular mechanisms involved membrane photodegradation have been spurred by the recent interest in porphyrin-sensitized tumor photochemotherapy (Doughterty 82; Spikes 82).

Much membrane damage caused by the combination of visible light and colored sensitizers requires molecular oxygen and is mediated by activated oxygen species. These reactive intermediates are promiscuous leaving no membrane component immune to attack. As a consequence it has proven difficult to connect eventual cellular responses with particular primary photochemical events.

The first and major portion of this paper is a review of sensitized photooxidation of biomembranes with an emphasis on molecular mechanisms. Examples of photosensitized oxygen-independent membrane destructive phenomena, for which there is some knowledge of associated molecular events, are also presented.

The second part of this paper is a discussion of "photochemical probes" of structure in biomembranes. Primarily focussed on "photoaffinity" reagents that contain the azide, diazirine or diazo groupings, the discussion contains a treatment of general principles of the approach and several examples from the recent literature.

Some Photochemical Preliminaries

While the chemical reactions of electronically excited molecules are no more predictable than those of ground state molecules, there are simple rules that govern whether or not photochemistry may occur. These rules are often overlooked by investigators engaged in photobiochemistry only as an incidental part of their work and this has led to misinterpretation of results. Concise treatments of the fundamentals of molecular spectroscopy, of molecular photochemistry and of photosensitization, all oriented towards biomolecules, have been published (see Lamola 77b; Turro 77). Such treatments cannot, of course, be repeated here. However, the "basic laws" of photochemistry as well as some definitions are worth a brief review before discussion of membrane photochemistry.

A photochemical reaction is defined as a chemical reaction that starts in one of the electronically excited states of a reactant and ends with the appearance of ground-state product(s). Since, for our purposes, promotion of molecules to electronically excited states requires the absorption of light, we arrive at the first law of photochemistry (Grotthus - Draper law): only absorbed light causes photochemistry.

The energy difference between the ground electronic state and lowest energy optically accessible excited state varies tremendously depending upon molecular structure, and so compounds differ widely with regards to the wavelength regions in which they absorb light. Absorption spectra of a variety of biochemicals and "photosensitizers" are shown in Fig. 1.

Most biochemicals are colorless; they absorb light of wavelengths lower than 400 nm. In fact peptides and nucleic acids do not absorb light of wavelengths greater than about 320 nm, the long wavelength edge of what the dermatologists call the erythema or "sunburn" range. Unsaturated fatty acids and cholesterol show only end absorption at wavelengths higher than ~200 nm. However, mild oxidation of unsaturated lipids produces of chromophores that absorb at longer wavelengths. "Aging" or oxidation of peptides and lipids eventually leads to compounds that absorb visible light.

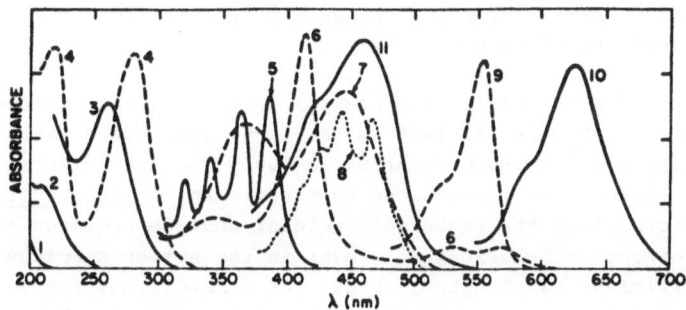

Fig. 1. Absorption spectra of some biomolecules and some photosensitizing dyes. The spectra are not scaled to indicated relative absorption strengths (extinction coefficients). In most cases only the longest wavelength absorption band is shown. Code: 1 unsaturated fatty acids; 2 mildly oxidized fatty acids; 3 deoxyribonucleic acid; 4 polypeptide containing assorted aromatic amino acids; 5 anthracene; 6 oxyhemoglobin; 7 riboflavin; 8 β-carotene; 9 rose bengal; 10 methylene blue; 11 bilirubin.

In biosystems it is the prosthetic groups and pigments that lend color (absorb visible light). Flavins and bile pigments, such as bilirubin, are yellow or orange. Porphyrins and chlorins are purple or green, etc., and may absorb far into the red region. Thus, visible light-induced alteration of a biosystem is usually initiated by excitation of prosthetic groups or pigments.

A photosensitizer may be defined as a component of a system that causes another component to react (or exhibit enhanced reaction) to the light. Generally the mechanism involves absorption of the light by the sensitizer, which light is not absorbed by the system without sensitizer. The long wavelength portions of the absorption spectra of some commonly employed photosensitizers are shown in Fig. 1. Polycyclic aromatic hydrocarbons, exemplified by anthracene, generally absorb in the near ultraviolet. Dyestuffs, such as rose bengal and methylene blue, absorb visible light.

The lessons should be obvious. For example, purified mixtures of lipids and proteins are not reactive towards visible light. If one wants to discern, for example, the photosensitizing effect of a flavin in a biosystem one should irradiate with blue light and avoid ultraviolet wavelengths that would directly excite protein, etc. It is imperative to know the spectrum of the light source employed as well as the absorption spectra of key components of the system. Such experimental parameters are often overlooked.

The Grotthus-Draper law can be used to determine which component(s) of a complex system is involved in the light absorption step leading to the photoresponse. Using monochromatic (very narrow wavelength band) light of various wavelengths one constructs an <u>action</u> <u>spectrum</u>, a plot of the reciprocal of the number of incident photons required to produce a given effect vs. wavelength. Peaks in the action spectrum represent the most efficient wavelengths. Under ideal conditions the action spectrum matches the absorption spectrum of the component(s) that absorb the light leading to the effect of interest (see Jagger 77).

The second law of photochemistry, known as the Einstein law, derives from the quantum nature of light absorption. It states that the quantum yield (number of events/number of absorbed photons) of a primary photochemical process cannot exceed 1. A primary photochemical process is one that leads to deactivation of the excited molecule. Therefore an observed quantum yield for a particular event, e.g., fatty acid side chain destruction, that greatly exceeds 1, indicates the operation of an amplification step "in the dark" sequential to the primary photochemical event.

At this point we note that light (photons) is a particularly convenient reagent (cofactor) in two ways. Firstly, specificity can be achieved through wavelength selection. Secondly, one can control delivery. The light can be introduced at various times with respect to the introduction of other components, the fluence (dose) and fluence rate (intensity) can be varied widely, and the delivery may be continuous or pulsed. The ability to control delivery, generally made little use of in biomembrane studies, can facilitate control of or study of kinetics.

The absorption spectra of Fig. 1 represent transitions from the ground electronic states which are singlet spin states (total electron spin = 0) to excited singlet states. For each excited singlet state there exists an electronically excited state of total electron spin = 1, a triplet state. Optical transitions between singlet and triplet state are "forbidden", that is, occur with only very low probability. However, molecules of many classes are excited indirectly to triplet states via light absorption. In these cases excitation to the lowest excited singlet state is followed by radiationless relaxation (intersystem crossing) to a lower lying triplet state. The variation in intersystem crossing efficiency among different classes of molecules is enormous; some, such as bilirubin exhibiting virtually no intersystem crossing, while others, such as protoporphyrin can achieve triplet yields greater than 0.5.

Triplet state excited molecules are very often key intermediates in photobiochemical reactions simply because they live longer than do excited singlet states and consequently have a greater probability to engage in bimolecular reactions. In addition, triplet excited molecules can effect the formation of singlet oxygen (see below).

An alternate way to promote molecules to excited triplet states is by excitation energy transfer from an appropriate donor molecule. In excitation energy transfer a donor molecule in an excited state interacts with an acceptor molecule leading to the simultaneous demotion of the donor and excitation of the acceptor. Excitation energy transfer processes are well understood and rules predicting efficiencies are well developed (Lamola 69; Lamola 77). For example, the triplet excitation energy transfer process,

$$D(T_1) + A(S_O) \rightarrow D(S_O) + A(T_1)$$

in which an excited triplet (T_1) donor (D) reacts with a singlet ground state (S_O) acceptor (A) to give the triplet acceptor and ground state donor, requires that the excitation energy $[D(S_O) \rightarrow D(T_1)]$ of the donor be equal to or greater than that of the acceptor (energy conservation), and that the donor and acceptor molecules approach within collisional distances. Such rules must, of course, be appreciated in designing or choosing an appropriate donor molecule with the aim of exciting a particular grouping (chromophore) in a complex biochemical system to a triplet state.

PHOTOOXIDATION

Definition and Classification

The term photooxidation is used here to mean the light induced chemical reaction of some substrate with molecular oxygen or an intermediate derived from molecular oxygen. When the photooxidation is mediated through the action of a sensitizing compound the process is termed "sensitized photooxidation" or "photodynamic action", an older term that refers specifically to sensitized photooxidation-based degradation of biosystems. Because until recently photodegradation of biomembranes was synonomous with photooxidation and the bulk of ongoing work on biomembrane photodegradation is concerned with sensitized photooxidation, a brief review of fundamental aspects of sensitized photooxidation is in order. Several excellent reviews of the field have been published and some of these refer specifically to the photooxidation of compounds of biological interest (Foote 76; Foote 80; Jori 80; Jori 81; Pooler 81; Spikes 82).

Schenck and coworkers described two general classes of sensitized photooxidation reactions (Gollnick 68): Type 1, in which the oxidized products are similar to those obtained by thermally initiated free-radical oxidation pathways, in which the sensitizer is generally consumed, and which may continue to proceed in the dark after an initial irradiation period; and Type 2, in which quite specific product types with selective stereochemistry are formed, in which the sensitizer is generally not consumed, and in which require continued irradiation for continued product formation.

It is now clear that in Type 1 reactions the sensitizer (S) participates in a primary photochemical process to produce an intermediate(s) that can enter into reactions leading to the oxidized products. Primary photochemistry may involve hydrogen abstraction from the substrate (RH) by excited triplet or singlet sensitizer (S^*) or electron transfer between excited sensitizer and substrate or oxygen.

Type 1 Photooxidation:

$$^3S^* \text{ or } ^1S^* + RH \rightarrow \cdot SH + R\cdot \xrightarrow{O_2} \text{ product(s)}$$

$$^3S^* \text{ or } ^1S^* + RH \rightarrow S^{\pm} + RH^{\pm} \xrightarrow{O_2} \text{ product(s)}$$

$$^3S^* \text{ or } ^1S^* + O_2 \rightarrow S^{\pm} + O_2^{\pm} \xrightarrow{RH} \text{ product(s)}$$

Free radical intermediates formed either as a result of primary photochemistry or in a subsequent step may lead to a radical chain oxidation process that does not require light after initiation. Unsaturated fatty acids, with their very labile allylic hydrogens, are good substrates for radical chain oxidation.

Radical Chain Oxidation:

Initiation:
$$(S^*, \cdot SH \text{ or } R'\cdot) + RH \rightarrow (\cdot SH, SH_2 \text{ or } R'H) + R\cdot$$

Propagation:
$$R\cdot + O_2 \rightarrow RO_2\cdot$$
$$RO_2\cdot + RH \rightarrow ROOH + R\cdot$$

Termination:
$$2R\cdot \rightarrow R_2$$
$$RO_2\cdot + R\cdot \rightarrow ROOR$$

A variety of photoinitiated processes can lead to a free radical intermediate capable of efficient abstraction of a fatty acid allylic hydrogen. The key to the chain process is that the hydroperoxy radical ($RO_2\cdot$), formed efficiently via the combination of the allylic radical $R\cdot$

with molecular oxygen, is, itself, capable of allylic hydrogen abstraction that leads both to product hydroperoxide and regeneration of $R\cdot$. The chain will continue until some process removes chain-carrying radicals. Termination processes include radical combination such as those shown above. Reactions of $R\cdot$ or $RO_2\cdot$ with substances called inhibitors (IH) may lead to "termination" of a chain because the radical produced in the reaction is incapable of abstracting a hydrogen atom from a fatty acid chain.

Inhibition:
$$IH + RO_2\cdot \rightarrow RO_2H + I\cdot$$
$$I\cdot + RH \rightarrow \text{no reaction}$$

The melding of molecular spectroscopy with the methods of mechanistic organic photochemistry developed in the early 60's led to the association of the Schenck Type 2 photooxidation with the intermediacy of "singlet oxygen" ($^1O_2$) molecular oxygen in an excited electronic state. Singlet oxygen is formed by energy transfer from an appropriate triplet state donor.

Type 2 Photooxidation:

$$^3S\cdot + O_2 \rightarrow S + {}^1O_2$$
$$^1O_2 + RH \rightarrow \text{products}$$

Singlet Oxygen

The ground electronic state of molecular oxygen is a triplet spin state. There are two low lying excited singlet states, $^1\Sigma^+_g$ and $^1\Delta_g$, at energies corresponding to transitions in the red and infrared regions of the spectrum, respectively. However, these transitions are highly forbidden. Gaseous oxygen is colorless for all intents and purposes and direct excitation of oxygen to these low-lying singlet states is exceedingly inefficient.

Singlet oxygen can be formed efficiently via energy transfer from a triplet state donor. The excitation energy of the donor must, of course be equal to or greater than 22 kcal/mole, the excitation energy of $^1\Delta_g$ oxygen. Because the transfer process involves the "exchange" of electrons, the donor and ground state oxygen molecules must collide to effect the process. In solvents of viscosities like that of water at $20°$ the diffusion-controlled (bimolecular) rate constant for molecular oxygen and a low molecular weight donor is of the order of 10^{10} $M^{-1}s^{-1}$. For a water oxygen concentration of $2.5 \cdot 10^{-4}M$ (equilibrium with 1 atmosphere of air), this gives a pseudounimolecular transfer rate constant of $2.5 \cdot 10^6$ s^{-1}. Thus, any appropriate triplet state donor with a lifetime of 0.4 μs or greater (in the absence of oxygen) will transfer excitation to oxygen with an efficiency of 0.5 or greater, respectively. Should the excitation energy of the donor approach or exceed 37 -kcal/mole $^1\Sigma g^+$ oxygen will be produced. However, $^1\Sigma^+ g$ oxygen is extremely short-lived in solution due to rapid conversion to $^1\Delta g$ oxygen. The latter undergoes radiationless decay to ground state oxygen with a lifetime of 2µs in water (20µs in D_2O). The term singlet oxygen, $^1O_2$, in solution refers, then, to $^1\Delta g$ oxygen.

Singlet oxygen is extremely reactive, capable of undergoing several kinds of chemical reactions in solution within its short lifetime (Foote 76). $^1O_2$ can undergo reactions in which it simultaneously forms two bonds, a reaction mode not allowed ground state (triplet) oxygen. Thus, $^1O_2$ undergoes addition reactions of the 1,2-type, to form four-membered ring products, and of the 1,4-type, to form six-membered ring products. For example, $^1O_2$ adds across the center ring of anthracene, at the 9 and 10 positions, to form the polycyclic anthracene peroxide.

Perhaps, the most characteristic reaction of $^1O_2$ is the so called ene-reaction in which $^1O_2$ reacts with olefins that possess allylic hydrogens to give the allylic hydrogen peroxide product with a double bond shift.

The reaction involves a six-membered transition state and proceeds in a concerted fashion so that both regioselectivity and stereospecificity are high.

For example, the reaction of $^1O_2$ with cholesterol gives the 5α-hydroperoxide (Δ^6) practically exclusively. The free radical initiated hydroperoxidation of cholesterol gives a mixture of the more thermodynamically stable 7α and 7β hydroperoxides with the double bond in the "normal" position. This sort of product specificity provides perhaps the best test for the occurrence of $^1O_2$ in a reaction system.

414

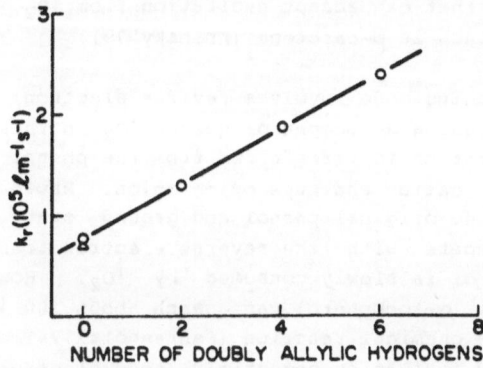

A substantial dictionary of reaction rate constants for reactions of $^1O_2$ in homogeneous solutions exists (Wilkinson 82), from which informative relationships can be derived. For example, the rate constant'for reaction of $^1O_2$ with unsaturated fatty acids increases linearly with the number of allylic double bonds (number of allylic hydrogens) as is shown in Fig. 2 (Doleiden 74). Cholesterol reacts with $^1O_2$ at a rate nearly equal to that of a fatty acid with one double bond.

Singlet oxygen is promiscuous in its reactions (Foote 76). In addition to reaction with unsaturated lipids, $^1O_2$ reacts rapidly with a number of amino acid side chains in membrane proteins, namely tyrosyl, tryptophanyl, histidyl and thyl. Polysaccharides are less reactive and

Fig. 2. Bimolecular rate constants for the reaction of singlet oxygen with the methyl esters of fatty acids with 0, 2, 4 and 6 doubly allylic hydrogens :   oleate, linoleate, linolenate and arachidonate, respectively. The triangle represents cholesterol.

give products of lower molecular weight that contain keto and carboxyl groups, but are otherwise uncharacterized. Reactions with minor membrane components are also known, for example, $^1O_2$ reacts in at least two ways with tocopherol (see below).

The structure of many of the products of reaction between $^1O_2$ and free amino acids have been determined (Foote 76). Some are distinctly different from those that arise from free radical initiated oxidation but many are similar. Very few of the isolated products are primary products, most represent stable products of multistep pathways. In the cases of histidine and tryptophan the isolated products have lost their imidazoyl and pyrryl rings, respectively. These products probably arise from primary products of addition of $^1O_2$ to the rings. The primary step of reaction with the tyrosyl side chain is undoubtedly electron transfer from the phenolic ring to $^1O_2$. Complex products arise as a consequence. Little work has been done to characterize products of $^1O_2$ and polypeptides. Certainly the side chains mentioned above are much more reactive towards $^1O_2$ than other side chains and the peptide bond itself.

Not all interactions of $^1O_2$ with organic compounds lead to product formation. There are two main modes which lead to deactivation of $^1O_2$. The first quenching mode discovered involved excitation energy transfer (Foote 76). Any compound possessing a triplet state excitation energy approaching 22 Kcal/mole or lower can accept excitation energy from $^1O_2$ in reverse of the process by which the $^1O_2$ is usually produced. Of the very few compounds that can accept excitation from $^1O_2$, the best known are the polyenes, such as β-carotene (Krinsky 79).

A second quenching mode involves reverse electron transfer (Foote 76). Phenols including α-tocopherol quench $^1O_2$ in this manner. In the initial step an electron is transferred from the phenol to $^1O_2$ to produce the phenolic cation and superoxide anion. Rapid reverse electron transfer leads to the original phenol and ground state oxygen. Other reactions may compete with the reverse electron transfer and so, for example, α-tocopherol is slowly consumed by $^1O_2$. However, depending upon the solvent, α-tocopherol can quench about 100 $^1O_2$ molecules for each non-reversible chemical reaction (Fahrenholtz 74). Other compounds with relatively low ionization potentials (good electron donors) undergo facile and reversible electron transfer reactions with $^1O_2$ and act as effective quenchers. Among these are tertiary amines, diazabicyclooctane (DABCO) being most notable, and azide ion, notable because of its water solubility. Generally, the electron transfer reactions of $^1O_2$ proceed at rates significantly lower than the diffusion-controlled rates.

416

An excellent review of $^1O_2$ processes especially as related to bio-logical systems is given by Foote (Foote 76). Look there for references to the original literature which, for expediency, were not included in the discussion above.

Differentiation of Reaction Type

It is not difficult to determine whether a product arises via a simple Type 2 mechanism or via a more complicated pathway when the system under examination consists of relatively low molecular weight components in homogeneous solution. The structure of the product, the quantum yield of its formation (quantum yield >1 is inconsistent with Type 2), and determination of whether or not continued irradiation is obligatory for continued product formation provide suggestive answers. The effect of a change in sensitizing compound reveals additional information. The products of a Type 2 photooxidation are independent of the sensitizer structure. Of course, the quantum yield of Type 2 product formation will depend upon the quantum yield of formation of the triplet state sensitizer and the efficiency of excitation transfer to oxygen (the latter is dependent upon the lifetime of the triplet sensitizer).

Typical Type 1 photosensitizers include ketones, such as benzo-phenone and anthraquinone derivatives, and flavins, such as riboflavin. These compounds in their excited triplet states are good hydrogen abstracting reagents or good electron acceptors. In either case, reactive radicals, are formed from the excited sensitizer. To avoid Type 1 reactions and achieve high yield Type 2 photooxidation, the sensitizers of choice should form triplet states with high quantum efficiency and the triplet state sensitizer should have relatively low chemical reactivity. Typical Type 2 sensitizers include porphyrins, chlorophyll, halogenated fluorescein dyes such as eosin, and acridine-type dyes such as methylene blue.

The use of inhibitors (quenchers) is, perhaps, the most common approach to differentiate radical mediated oxidation and the singlet oxygen pathway. While most compounds that act as inhibitors of photoox-idation can react with (quench) both singlet oxygen and radical species, many of them exhibit large differences in rates of reaction with these two kinds of intermediates. Therefore some attention needs to be paid to inhibition kinetics (e.g. run the inhibition experiment over a range of inhibitor concentration) in the design of an inhibition study. Exci-tation transfer quenchers, such as β-carotene, quench $^1O_2$ at the diffusion-controlled rate. Such compounds are also reactive towards most carbon-centered radicals albeit with rates much slower than diffusion-controlled. Classical free radical inhibitors include the highly-substituted phenols, e.g., tocopherol. These compounds are also excellent quenchers of singlet oxygen, operating by the reversible electron transfer mechanism.

Another approach is to add a compound to the system that is known to react with singlet oxygen to give a characteristic product. Already mentioned is the reaction of cholesterol with singlet oxygen to give the 5α-hydroperoxide. The oxidation product of diphenylisobenzofuran, o-dibenzoylbenzene, is commonly taken to be indicative of singlet oxygen. However, it is likely that other oxidation modes could give this product.

Definitive elucidation of the photooxidation pathway may require a detailed kinetic study. Alteration of substrate and inhibitor concentrations in rate studies, especially with the use of an inhibitor with known reactivity towards a suspected intermediate such as singlet oxygen, can yield key information. For example, singlet oxygen possesses a characteristic decay time that is solvent dependent. Determination that a quenchable intermediate has the same lifetime is proof that it is singlet oxygen.

Photooxidation in Biological Systems

Much activity in studies of oxygen mediated photoactivated degradative phenomena in biosystems is aimed at the relating cellular or system response to primary photochemical events. Because many individual chemical events often intervene between primary photochemistry and the observed response, elucidation of such relations is often difficult. The complexity and non-homogeneous nature of the systems under study also contribute to the difficulty. For example, the locations of sensitizers, key substrates and inhibitors and whether they are bound or undergo free diffusion are factors that must be understood before mechanistic conclusions can be drawn. Many of the approaches to mechanism that apply to simple homogeneous solution systems many not be applicable to biosystems.

An often overlooked problem in the determination of mechanism is the possibility of confluence of pathways (see Fig. 3). For example consider photooxidation in a biomembrane. Hydroperoxides, primary products of singlet oxygen attack, may not be stable under the experimental conditions. Decomposition of primary hydroperoxides gives rise to radical species and initiate radical chain oxidation that swamps the primary photochemical damage. Depending on the system it might be quite difficult then to demonstrate that the primary photochemistry involved singlet oxygen and not a Type 1 pathway.

418

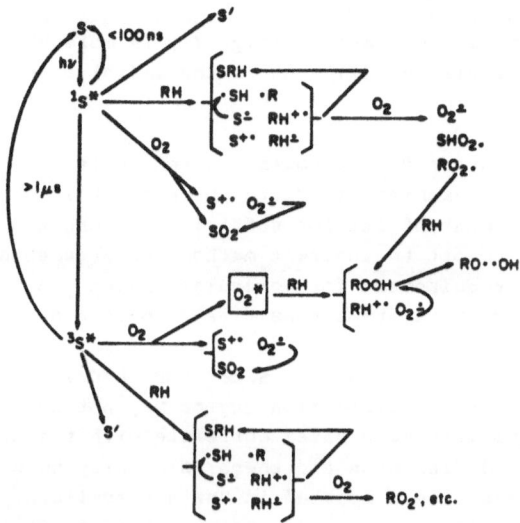

Fig. 3. This scheme shows a number of ways that oxygen, an excited sensitizer (S^*) in a singlet or triplet state, and some substrate with a labile hydrogen (RH) can interact.

PHOTODEGRADATION OF BIOMEMBRANES

Sensitized Photoinactivation

The inactivating effect on cells caused by the combination of light and a sensitizing dye was first described in those terms by Oscar Raab at the turn of the century (Blum 64). It was realized subsequently that the majority of the effects required the presence of molecular oxygen so that the general term employed for sensitized photoinactivation of bio-logical (or biochemical) activity, "photodynamic inactivation," is now reserved for those cases in which molecular oxygen plays a role.

Cellular photoinactivation, ranging from obvious destruction, to inability to reproduce or to carry out specified function, has been associated with modifications of various cellular components (Pooler 81). In sensitized photoinactivation the pertinent modified component is frequently the surface membrane with the modification often being a change in permeability or transport properties. Such modification of the membranes of intracellular organelles, such as lysosomes or mito-chondria, can lead to subsequent adverse effects for the whole cell (Allison 66; Sandberg 81). DNA is always the prime suspected target for inactivation caused by ultraviolet light in the range between 320 and 250 nm but is also often a key target in sensitized photoinactivation processes.

The first questions investigated about some sensitized cellular photoinactivation process are usually: (1) is molecular oxygen required (photodynamic action)? and (2) what is the (main) site of modification leading to inactivation?

The first question is, of course, often quite easy to answer for systems that can be examined in vitro, because it is usually possible to remove oxygen (or most of it) for sufficiently long times without undue harm. One must resort to indirect methods or arguments from analogy to decide about the requirement for molecular oxygen in photoinactivation phenomena that involve systems sensitive to oxygen deprivation.

The identification of target site for sensitized cellular photoinactivation or photomodification may or may not be easy to establish. Generally, photomodification rates correlate with the sensitizer concentration at the modified site and there is usually no significant remote effect. Thus, sensitizers located in the extracellular medium are usually much less effective than sensitizers associated with the cells. When extracellular photosensitizers are effective it is usually because the medium is low in reactive substrates and the excited sensitizer or reactive intermediates live long enough to diffuse to cells. Intracellular sensitizers are usually bound to some site or have greatly reduced diffusivities and the concentrations of potential reaction sites in the vicinity of the sensitizer is usually high. Both reasons support the notion that modification is localized to the vicinity of the sensitizer. Thus, the solubility and binding properties of the sensitizer are major factors in the determination of target site(s) in the cell. To assure specific photomodification of cellular membranes one must choose sensitizers that partition such that little resides elsewhere. Targeting of internal membranes, e.g. lysosomal membranes, rather than the plasma membrane, would require some special binding properties and or the use of the time domain. Membrane soluble sensitizers introduced from without the cell will, of course, first reside in the plasma membrane and then with time equilibrate among binding sites.

Assay of Photodegradation of Biomembranes

There are a variety of specific indicators of biomembrane photodegradation ranging from assays of functionality to in situ spectroscopic indications of chemical alteration (Pooler 81). Indications of functional alteration include lysis, release of cytosolic component(s), increased permeability to small molecules and ions, modification of active transport rates decreased activity of membrane associated enzyme(s), and altered conductance of excitable membrane channels. Indications of membrane photodegradation based on chemical analysis include loss of amino acids, loss of unsaturated fatty acids, cross-linking of membrane proteins, and detection of lipid oxidation

420

product(s). A widely employed spectroscopic assay for biomembrane oxidation is the fluorometric detection of the so called lipofuscin pigment (Mead 76). With progressive oxidation of a biomembrane there is a concomitant increase in the intensity of blue fluorescence (430-480 nm) elicited by near ultraviolet excitation (370-390 nm). There is good evidence that the fluorophore is a mixture of various N,N'-disubstituted 1-amino-3-iminopropenes (-NH-CH=CH-CH=N-) derived from the reaction of malondialdehyde (CHO-CH$_2$-CHO) and protein amino groups). These blue emitting pigments derived from membrane oxidation have been called lipofuscin pigments and have been equated with the so-called lipofuscin granules (age pigments) of postmitotic cells thought to derive from accumulated damaged membrane components. However, the connection between the blue emitting pigments and the lipofuscin granules which emit yellow fluorescence appears to be an open question (Eldred 82).

Malondialdehyde, CH^2(CHO)$_2$, the purported precurser of the lipofuscin pigments, is itself detectable in a variety of ways including a facile reaction with this barbituric acid (Mead 76). It must be noted that malondialdehyde, and hence also the photoinduced lipofuscin pigments, represent advanced oxidative damage since this dialdehyde must derive from polyunsaturated fatty acids which have undergone two oxidative cleavage reactions at neighboring unsaturated centers.

Sensitized Photodegradation of Lipids and Liposomes

The relative reactivities of the common natural fatty acids and cholesterol towards $^1O_2$ are given in a previous section. Reactivity was found to correlate well with the number of ways the compound can undergo the "ene" reaction which is essentially equal to the number of allylic hydrogen centers (Doleiden 74).

An enormous amount of work has been done to understand fat oxida-

421

tion and its inhibition because, of course, of the importance of these processes in the preservation of foods. The free radical chain oxidations of the fatty acids have been investigated in great detail and the kinetic constants for various steps in the chain reaction (see above), propagation steps in particular, have been determined (Emanuel 67; Mead 76). It is inappropriate to review this work here. Suffice it to say that reactivities in radical chain oxidation parallel those for $^1O_2$ attack, that is, they increase with degree of unsaturation. It appears, therefore, not possible to differentiate between a $^1O_2$ mediated pathway (Type 2) and a photoinitiated radical oxidation (Type 1) pathway on the basis of relative rates of reaction of the various fatty acids in a mixture. It is also expected that under conditions in which the hydroperoxide products of $^1O_2$ attack are unstable, radical chain oxidation should also contribute to the overall oxidation. The contribution of $^1O_2$ may be assessed via analysis for specific $^1O_2$ products, e.g. cholesterol-5-α-hydroperoxide, or via the kinetic effects of various inhibitors (see above).

The effects of motional inhibition in phospholipid bilayers are on oxidation and radical reaction kinetics just now receiving attention. It was found that escape of geminate free radicals from the formation "cage" is depressed in lecithin bilayers compared with hydrocarbon solvents of the same chain length, but that escape is more efficient than expected from bilayer microviscosities determined by spin label or fluorescence techniques (Winterle 80). The kinetics of $^1O_2$ reactions in bilayer vesicles of egg lecithin and some synthetic surfactants were examined recently and found to be essentially unchanged from those in micellar media (Rodgers 82). The same investigators showed that $^1O_2$ produced in the bilayer region of relatively unreactive vesicles, can diffuse out of the vesicle and through the aqueous dispersing medium to react with substrates in another vesicle.

Liposomes of egg lecithin and of dipalmitoyl lecithin (DPL) have been used as model systems for the action of activated oxygen species on biomembranes (Anderson 73, 74; Suwa 78; Muller-Runkel 81). Destruction of fatty acid chains, detection of oxidation products and, "lysis" of the vesicles have been followed. In these studies lysis was operationally defined as either the escape of relatively small species (glucose, eosin) from the internal water compartment or loss of red light scattering. These two assays were found to correlate very well. Lysis due to the action of $^1O_2$ generated external to the liposomes by toluidine blue photosensitization of electrodeless radio frequency discharge, or generated in the bilayer region by porphyrin photosensization was demonstrated for egg lecithin liposomes. Loss of unsaturated fatty acids in the ratios expected for $^1O_2$ reactivity was observed, as well as the formation of cholesterol-5α-hydroperoxide. Liposomes of DPL and other "saturated" lipids did not lyse under similar conditions.

Type 1 photolysis of egg lecithin liposomes was effected via the "sensitizer-oxygen" pathway (Muller-Runkel 81). Superoxide anion (O_2^-) was generated photochemically in the external medium via the riboflavin/EDTA/O_2 system and led to effective lysis of egg lecithin liposomes at a pH dependent rate but no lysis of DPL liposomes. The pH dependence suggested that $HO_2\cdot$ may be a much more efficient radical chain initiator than is O_2^-. Parallel studies using photosensitized $^1O_2$ generation indicated that $^1O_2$ is about 200 times more effective than O_2^- for inducing liposome lysis when both are produced external to the liposomes.

Egg lecithin liposomes into which cholesterol-5-α-hydroperoxide was incorporated, also undergo radical chain oxidation and lysis when heated to 35° in the dark (Lamola unpublished). This opens the questions as to whether the hydroperoxides first formed upon $^1O_2$ attack are lytic themselves or act as free radical chain oxidation initiators or both. That "dark" events contribute to the lysis is evidenced by the fact that lysis proceeds in the dark after a sufficient period of irradiation in the presence of $^1O_2$ sensitizers.

Grossweiner and coworkers (Grossweiner 82) have recently observed an oxygen independent pathway for sensitized photolysis of liposomes made up of lecithins with unsaturated fatty acid side chains. When the liposomes contained 0.1% hematoporphyrin, photolysis required molecular oxygen and inhibitor studies as well as the enhancing effect of D_2O in the medium suggested a $^1O_2$ mediated pathway. However, when the hematoporphyrin was present at 1%, at which concentration there is formation of aggregates of the sensitizer, an oxygen independent pathway emerges. It was suggested that the sensitizer aggregates support photoionization and the ejected electrons react with unsaturated chains forming products that lead to lysis. DPL liposomes are unaffected by irradiation in the presence of 1% hematoporphyrin.

Differentiation between $^1O_2$ and free radical mediated lysis of liposomes in these studies depended primarily upon the use of inhibitors which have different reactivities towards the intermediates in question (Anderson 74). In addition, a specific test for superoxide anion in the aqueous medium was provided by the effect of the O_2^--destroying enzyme superoxide dismutase. Other useful water soluble enzymes for delineating oxidation mechanisms include catalases and peroxidases which use hydrogen peroxide and other peroxides, respectively, as substrates.

The product(s) of phospholipid oxidation responsible for the instability of the liposomes are not known. Neither is it known what fraction of the phospholipid molecules must be affected for lysis to occur. That unsaturated fatty acid side chains and not head groups are involved is certain since liposomes of saturated lipids were not affected by either $^1O_2$, $O_2^{\pm}$ or free radical sources. Introduction of hydroperoxy or hydroxy groups at the unsaturated sites in the fatty acid chains should cause a significant perturbation in the bilayer structure since this puts a polar group in a normally nonpolar region. However, as monitored by several techniques the effect of incorporation of cholesterol-5-α-hydroperoxide into DPL liposomes was found to be virtually identical to that of cholesterol (Lamola, unpublished); for example, the effect on the "melting" temperature is the same. Perhaps oxidative cleavage of the hydrocarbon chains leads to lysis because the products would resemble lysolecithins in having one short chain, but in addition would have a polar group at the chain terminus. However, there appears to be no data concerning the relation of lysis to the fraction of oxidized fatty acid side chains. Finally, Grossweiner's (Grossweiner 82) of an oxygen-independent liposome lysis mechanism that also involves the unsaturated fatty acid chains introduces new possibilities.

PHOTODEGRADATION OF ERYTHROCYTE MEMBRANES

The human erythrocyte membrane has been a favorite subject of membrane physiologists and biochemists for many reasons including accessibility of red cells, their relative simplicity, ease of preparing the membranes, accessibility of resealed membranes with preserved functions, ease of lysis assay and others. For these reasons and because of the existence of abnormally photosensitive erythrocytes in porphyria diseases, the sensitized photodegradation of red cells has been a very active field of study involving investigators from various disciplines (Pooler 81).

Photohemolysis

Irradiation of erythrocyte suspensions with ultraviolet light in the "sunburn" range or with visible light in the presence of a wide variety of sensitizing compounds causes the cells to lyse. Lysis is easily followed either by colorimetry of hemoglobin in the supernate after spinning down intact cells or by the decrease in light scattering by the suspension which under proper circumstances can be linear with the lysed fraction.

Lysis is a critical phenomenon and different kinds of membrane modification or degradation can cause it to occur. In photolysis of red cells a common observation is that the passive permeability of the membrane towards small ions is greatly increased and leads to loss of osmotic equilibrium, swelling and eventual bursting of the cell because of intake of water due intracellular impermeable solutes ("colloid osmotic lysis") (Cook 75).

424

Generally, in light-induced hemolysis, there is a lag period followed by cell lysis which occurs with kinetics that are often of order greater than one (velocity increases with time); see Fig. 4. Because upon alteration of an experimental parameter, e.g., wavelength, sensitizer, concentration of inhibitor, etc., it is possible to effect the lag time, the order, (shape of the curve), or the "velocity", individually or simultaneously, it is important to record the complete hemolysis curve. Very often the time required for 50% hemolysis is used as a measure of the hemolysis "rate". Depending upon the purpose of these "rate" data, such a procedure may or may not be valid. While such an extreme case has not been reported it is conceivable to arrive at the same time for 50% hemolysis yet have distinctly different hemolysis curves (exemplified in Fig. 4) indicative, perhaps, of the operation of a different lysis mechanism. In even more dangerous "short-cut" is to take the extent of hemolysis at some arbitrary time as a measure of hemolysis "rate". Unfortunately, much information in the literature on the effects of various potential inhibitors of photohemolysis is based upon such single time assays. It is then not possible to know whether the inhibitor affected the lag time, the kinetics of hemolysis once begun, or both.

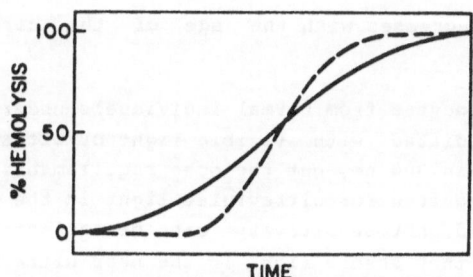

Fig. 4 Two hemolysis vs. time plots that have the same 50% hemolysis time.

Because of the obvious complexity of the hemolysis event and the further complication that red cells in any population are not homogeneous with respect to various potentially important parameters such as, for example, natural antioxidant content, no attempt to interpret the shape of hemolysis curves has been made. Hemolysis curves are, at this point, used in a semiquantitative sense. For example, Fig. 5 shows the photohemolysis curves for five fractions of red cells from a patient with protoporphyria. The cells were fractionated on a discontinuous density gradient with fraction 1 representing the least dense and youngest fraction and fraction 5 the most dense and oldest fraction. The monotonic decrease in hemolysis "rate" (increase in time required) with fraction number correlates with the free protoporphyrin content of the

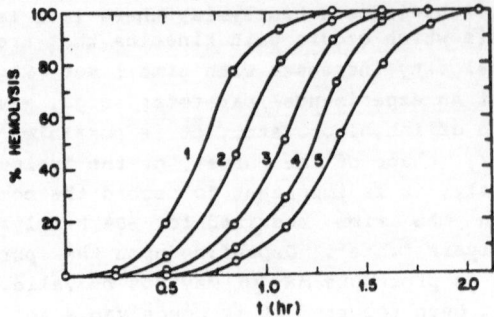

Fig. 5. Washed red cells from a patient with erythropoietic protoporphyria were fractionated in a discontinuous density gradient (Piomelli 75). Equal members of cells from each fraction were suspended in buffered saline and irradiated with blue light under identical conditions. The progress of hemolysis with irradiation time for the five fractions are plotted. These are unpublished results of Lamola.

cells which rapidly decreases with the age of the circulating cells (Piomelli 75).

Untreated erythrocytes from normal individuals undergo accelerated hemolysis when irradiated with visible light or ultraviolet light of wavelengths longer than 300 nm, but the dose requirements are very much higher than the required for ultraviolet light in the range below 300 nm. Greatly increased photosensitivity can be affected by a large variety of compounds that absorb light in the near ultraviolet and visible ranges. Among the effective sensitizers are fluorescein derivatives, acridine derivatives, anthracene derivatives, flavins, porphyrins, chlorophyll, bilirubin, sulfanilimide and chlorpromazine. Many of these sensitizers of red cell hemolysis are known phototoxic agents for cells in vitro and are associated with human cutaneous photosensitivity syndromes (Blum 64; Spikes 82; Pooler 81).

Photohemolysis with ultraviolet light and, with few exceptions, that with near ultraviolet or visible light in the presence of sensitizers either does not proceed in the absence of molecular oxygen or is greatly inhibited under deoxygenated conditions. Inhibition of photohemolysis can be effected by one or more classes of oxidation inhibitors depending on the sensitizer and conditions of incubation of the red cells with the sensitizer. Thus in the various cases there is evidence for both free radical mediated pathways and singlet oxygen mediated pathways for the membrane modification(s) that leads to hemolysis.

In most cases the hemolysis proceeds in the dark after some initial irradiation period with the "rate" of the dark hemolysis increasing with the irradiation dose. In some cases there is evidence that the bulk of the damage leading to subsequent dark hemolysis occurs during irradiation and the delay simply reflects the time required for the hemolytic osmotic processes to operate. In other cases this dark hemolysis is undoubtedly due to additional free radical chain oxidation damage initiated by unstable primary photoproducts, e.g., hydroperoxides.

The site(s) of the light-induced modifications responsible for hemolysis has been the subject of considerable study especially for the case of protoporphyrin sensitization (see below). Information about critical sites should derive from data concerning the site(s) occupied by various sensitizers in relation to the effectiveness of hemolysis and specific modifications associated with them. Unfortunately not much is known about sensitizer binding sites in the erythrocyte membrane. However, it is clear from studies to date that binding of sensitizers to specific sites in the membrane is not a requirement for photohemolysis. Compounds such as riboflavin sensitize hemolysis via the production of active oxygen species in the extracellular medium (Michelson 77). Chlorpromazine and protriptyline may be irradiated independently of the cells to give products that cause lysis (Kochevar 79). Of the fluorescein dyes, those that associate with the erythrocyte membrane are most effective at sensitizing photohemolysis (Valenzeno 82). Similar observations have been reported for porphyrin derivatives (see Pooler 81). All this indicates that multiple pathways for sensitized photohemolysis exist and that mechanistic conclusions for one sensitizer may not hold for another.

Perhaps the most interesting series of sensitizers that have been studied are the porphyrins. The large quantity of work with the porphyrins has been prompted by their connection, on the one hand, with human photosensitivity diseases, and, on the other hand, with their use as potential tumor photochemotherapeutic agents (Doughterty 82).

Under most conditions heme is photochemically inactive, excitation energy being rapidly degraded via relaxation pathways that involve the iron. In contrast, metal free porphyrins and metalloporphyrins containing diamagnetic metal ions such as Zn^{+2} and Mg^{+2} can be potent sensitizers since they yield triplet states with high efficiency upon photoexcitation. There is a family of genetic disorders of heme biosynthesis called porphyrias in which one or more metal free porphyrins accumulate and render the patients photosensitive (Harber 81). The cutaneous phototoxic effects differ greatly in nature and severity with the particular porphyrins present. It was also observed that the red cells form patients with porphyrias exhibit abnormal photosensitivity to various degrees in vitro due to the metal free porphyrins associated with the cells (Harber 64). Although there is no evidence for significant in vivo photohemolysis, the in vitro photohemolysis has received much attention as a model for the general cellular photosensitization due to porphyrins. The in vitro work has included both studies of red cells from patients with porphyria and studies of red cells incubated with the various porphyrins (Schothorst 71). The observations are instructive with respect to the question of the relationship between sensitizer site and photohemolysis efficiency.

Erythrocytes irradiated in the presence of protoporphyrin are much more photosensitive than those irradiated in the presence of coproporphyrin or uroporphyrin. The latter two are much more water soluble than protoporphyrin which is immediately taken up by the cells (See Pooler 81). Membrane association appears to be an important factor for sensitizing ability since all three compounds have similar high potential for singlet oxygen production.

Erythrocytes from patients with erythropoietic protoporphyria, "EPP-cells", contain free protoporphyrin and are exquisitely photosensitive (Harber 64). On the surface it would appear that the EPP-cells and red cells incubated with protoporphyrin in vitro (Schothorst 71), "pseudo EPP cells", are the same with respect to free porphyrin distribution. However they differ both with respect to distribution of the porphyrin among the cells and within the cells. In EPP cells more than 99% of the protoporphyrin is bound to hemoglobin in the cytosol, characterized by a fluorescence emission maximum of 625 nm (Fig. 6), and only about 0.5% is associated with the membrane, characterized by a fluorescence maximum at 634 nm (Lamola 75). In vivo there is a rapid diffusion of the protoporphyrin out of the cells to the plasma which is rapidly cleared by the liver (Piomelli 75). The half life of the cellular protoporphyrin is only a few days so that only the younger EPP cells contain significant quantities of the porphyrin and the oldest cells have close to normal concentrations. When normal cells are incubated with protoporphyrin to produce pseudo EPP cells there is a rapid uptake of the porphyrin by the cell membranes as demonstrated by the 634 nm fluorescence maximum (Fig. 7). Most of this porphyrin eventually (1/e time ~1 day) makes its way to the cytosol hemoglobin, but for the first few hours most of the porphyrin resides in the membranes and is homogeneously distributed among the cells.

Now, for the EPP cells what are the relative contributions of the membrane protoporphyrin and hemoglobin-associated protoporphyrin to the photohemolytic effect? Experiments have shown that the hemoglobin-associated protoporphyrin makes the major contributions (see Fig. 8).

These observations also explain why all red cells in a specimen from a patient with EPP undergo photohemolysis although only a fraction

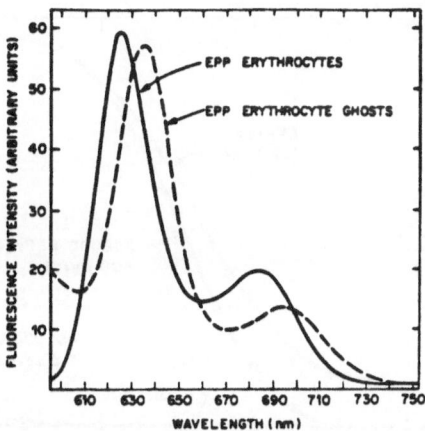

Fig. 6. Fluorescence spectra of intact red cells from a patient with erythroproetic protoporphyria and from the washed membranes prepared from the cells. The wavelength of excitation was 405 nm.

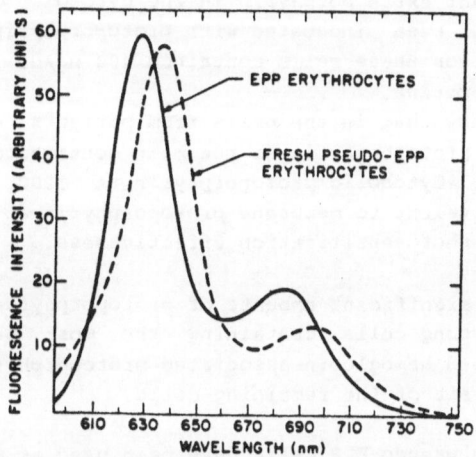

Fig. 7. Fluorescence spectra of red cells that were incubated with an aqueous solution of protoporphyrin and then washed (pseudo-EPP cells) and of washed red cells from a patient with erythropoietic protoporphyria (EPP cells). The wavelength of excitation was 405 nm.

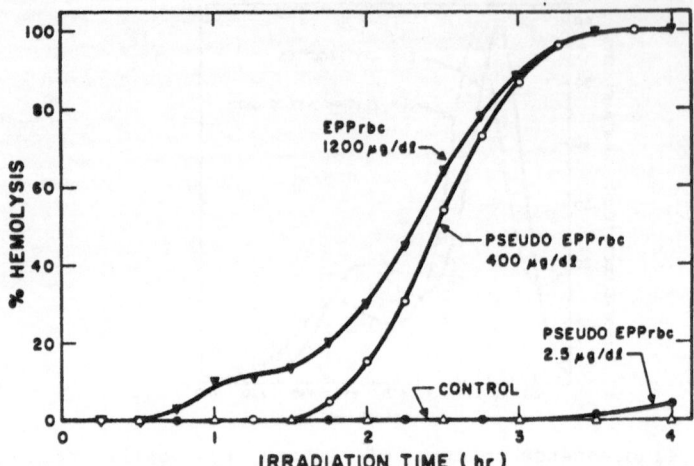

Fig. 8. Comparison of photohemolysis rates for protoporphyrin-sensitized red cells irradiated with blue light under identical conditions. △:Control cells were washed cells from a normal donor. ▼: Cells from a patient with EPP. These cells contained 1200 µg/dl protoporphyrin of which 2.5 µg/dl was located in the membrane. ●: Normal cells that had been incubated with protoporphyrin. At the time of irradiation these cells contained 2.5 µg/dl protoporphyrin in the membrane and no significant extra porphyrin in the cytosol. O: Normal cells that had been incubated with protoporphyrin. At the time of irradiation these cells contained 400 µg/dl protoporphyrin in the membrane.

These results show that in the cells from patients with EPP, the cytosolic protoporphyrin is the main contributor to photosensitization. Cytosolic protoporphyrin at 1200 µg/dl is apparently equivalent to membrane protoporphyrin at 400 µg/dl with respect to photosensitization effectiveness.

of the cells contain significant amounts of protoporphyrin (See Pooler 81). Evidently the young cells containing the most sensitizer lyse first but the released hemoglobin-associated protoporphyrin then sensitizes the photohemolysis of the remaining cells.

Freshly prepared pseudo-EPP cells have been used as models for EPP cells (Schothorst 70, 71, 72). While certain characteristics of photohemolysis appear to be independent of the protoporphyrin distribution there are inhibition data in the literature that indicates that at least the early events in the hemolysis pathway may be different (Schothorst 70). The results of the inhibitor study given in Table 1, showed that various potential photohemolysis inhibitors gave different results for

TABLE 1

Inhibition of photohemolysis by various agents

Expressed in % increase of the irradiation time, required to obtain 50% hemolysis. Negative figures indicate augmentation of the process.

Agent	Concentration (mM)	Inhibition EPP cells	Pseudo-EPP Cells
2,3-dimercapto-I-propanol	10	>+500	+15
Carotene	5	+45	+54
Cysteine	5	+77	-60
Ergothioneine	5	0	-30
Glutathione	5	+20	-7
Hydroquinone	10	+47	+35
2-Mercaptoethanol	25	+150	-34
Ascorbic acid	10	+33	+30
Cysteamine-HCL	15	>+500	-13

(from Schothorst 70)

the two kinds of cells. These data are consistent with the difference in sensitizer localization in EPP cells and in fresh pseudo-EPP cells in that the water soluble inhibitors were, in general, more effective in protecting the EPP cells compared with the pseudo-EPP cells.

Photohemolysis with Ultraviolet Light

The photohemolysis of human erythrocytes in the absence of added sensitizers has been investigated in some detail (Cooke 75; Roshchupkin 75; Michelson 77). All of the work employed light in the ultraviolet region below about 320 nm.

The early and careful work of Cook and coworkers set a high standard (see Cook 75). They showed that the hemolysis induced is of the colloid-osmotic type, due to an increase in the passive permeability of the membrane to small ion rather than the destruction of activated ion pumps.

In the work of Cook and coworkers the UV-dose was given in a short time and the resultant hemolysis was allowed to proceed in the dark. Kinetics were characterized by a significant lag time followed by lysis with some high kinetic orders. The time required for 50% hemolysis, which for the most part was the lag time, was found to be inversely

431

proportional to the square of the light dose. Removal of molecular oxygen greatly inhibits the UV-induced hemolysis (4 to 10 times slower) but does not halt it; there appears to be an oxygen independent mechanism which is much less efficient than an oxygen dependent one. Both water soluble and lipid soluble thiol reagents inhibit photohemolysis in the presence of oxygen as does α-tocopherol. Roshchupkin suggested that the lag time may be due to inhibition by naturally occurring α-tocopherol and other reducing agents (Roshchupkin 75).

What is the important primary photochemistry remains an open question. The action spectrum shows a maximum near 280 nm and a minimum near 265 nm indicative of aromatic amino acids as the primary light absorbers for the events that lead to hemolysis. From work with erythrocyte ghosts and liposome systems, Kochevar (unpublished) suggests that tryptophan moieties in membrane proteins serve as the natural sensitizers. She has collected data that suggests that tryptophan sensitization of $^1O_2$ formation in addition to other pathways, perhaps the well-known tryptophan photoionization, are operative. However, one cannot dismiss photochemistry in the cytosol as an important initiating event. For example, Salhaney and coworkers (unpublished) have demonstrated that oxyhemoglobin can yield methemoglobin and superoxide when irradiated with UV-light. Michelson and Durosay provide evidence that initiating events leading to hemolysis do arise in the cytosol (Michelson 77). However, they dismiss superoxide as contributing directly to hemolytic events and argue for $^1O_2$ and hydroxyl radicals as the key activated oxygen species responsible.

Little data has been reported concerning the alterations that occur in the membranes of intact erythrocytes irradiated with UV-light. However, Kochevar (unpublished) has investigated isolated red cell membranes irradiated with light of wavelengths from 250 to 380 nm. She found covalent crosslinking of membrane proteins, oxidation of polyunsaturated fatty acid side chains and inactivation of several membrane enzymes. Protein crosslinks were formed by both oxygen dependent and oxygen-independent pathways and crosslinks were detected at lower dose than were required to achieve observable loss of polyunsaturated fatty acids or enzyme inactivation.

Protoporphyrin-Sensitized Hemolysis

The protoporphyrin sensitized photohemolysis of human red cells (Harber 64) is to date the most carefully studied example of membrane photodegradation. The major contributors have been the groups of van Steveninck and Girotti. In the majority of these studies normal cells were incubated with exogenous protoporphyrin and experiments performed within times such that most of the sensitizer was located in the plasma membranes. The results reviewed below refer specifically to such prepared cells and not to the abnormally photosensitive red cells from patients with EPP.

Two observations have been confirmed in several laboratories: the photohemolysis is greatly inhibited (<0.1 of control) when oxygen is removed from the system; hemolysis is of the osmotic type, due to increased permeability of the membrane to small ions. Many functional and chemical alterations in the erythrocyte membrane can be observed as a result of irradiation before hemolysis occurs. There are large decreases in the activities of the enzymes acetylcholinesterase, glyceraldehyde-3-phosphate dehydrogenase, Mg ATPase, and Na-K ATPase. Associated with the latter is, of course, reduction of active Na^+ and K^+ transport. Carrier mediated transport of glucose, L-leucine, glycerol and sulfate is degraded considerably (Dubbelman 80b). The nonspecific (passive) permeabilities of glycerol, thiourea and, most notably, Na^+ and K^+ increase (Dubbelman 80b). Associated changes in the osmotic properties are observed. Finally, there is significant loss of cell deformability.

Chemical alterations that can be observed before hemolysis occurs include loss of unsaturated lipid (fatty acid side chains and cholesterol) in ratios indicative of oxidation, formation of malondialdehyde, loss of reduced glutathione, loss in quantities of all five photomodifiable amino acids, and the formation of covalently cross linked membrane proteins, most notably spectrin and glyceraldehyde-3-phosphate dehydrogenase. Glycoproteins appear relatively unaltered.

One aim of these investigations was to determine the relative contributions of the observable structural (chemical) alterations to lysis. While important details remain to be filled in, the general question of the relative importance of lipid alteration vs. protein alteration seems to have been settled in favor of the latter.

It is clear that the increased transmembrane fluxes, of Na^+ and K^+ in response to irradiation are too large to be ascribed to the shut down of the activated ion pumps and that increased passive permeabilities obtain. Extracellular solutes as small as sucrose inhibit the lysis of photodamaged cells so that the photoinduced "channels" have effective "diameters" that are very small (<1 nm). Distortions in the dynamical bilayers due to the introduction by oxidation, of more polar groups into normally nonpolar regions might supply a basis for such changes in permeability characteristics.

Lipid oxidation products were suggested as candidates responsible for the permeability changes. Lamola and coworkers isolated 5α-cholesterol hydroperoxide, the product of singlet oxygen attack upon cholesterol, from irradiated membranes (Lamola 73). Incorporation of this compound into membranes of erythrocytes at quantities up to 5% of the cholesterol present did, in itself, not appear to modify the osmotic properties of the cells. However, incubation at 37°C of cells with 1% of the cholesterol replaced by cholesterol hydroperoxide leads to efficient lysis of the colloid osmotic type (cellular K^+ leaks out before swelling and lysis). Under these conditions the hydroperoxide decomposes and initiates free radical chain oxidation of unsaturated lipid. Little or no protein cross linking is induced (Lamola 80). Enzyme activities are decreased suggesting protein damage also results. These results demonstrate a mechanism for hemolysis that begins with a photoproduct that is known to be produced in red cell membranes containing protoporphyrin. The lysis induced by cholesterol hydroperoxide occurs in the absence of protein cross linking and appears to be due to lipid oxidation. However, van Steveninck and coworkers performed careful studies of the kinetics of K^+ release and hemolysis with respect to the levels of lipid and protein damage (amino acid loss) and were able to conclude that lipid oxidation plays a minor role in hemolysis of irradiated porphyrin-doped erythrocytes (Schothorst 71; de Goeij 76).

The van Steveninck and Girotti groups suggested that protein cross linking, a very effective process, may be the direct cause of changes that lead to hemolysis (Girotti 76; de Goeij 76; Dubbelman 78a). Protein cross linking has been shown not to be due to disulfide formation nor to malondialdehyde, a known cross linking agent (Girotti 79). The chemistry involved in the cross linking is still a matter of controversy (Dubbelman 78b; Girotti 80) but it is clear that oxidized lipids are not involved (Girotti 78, 79; Lamola 80; Dubbelman 78b). There appears to be at least two pathways, an oxygen dependent one, and an oxygen independent one that accounts for about one-third of the links (Lamola 80). Since photohemolysis is nearly totally oxygen dependent it follows that the oxygen independent cross linking of membrane protein does not lead to hemolysis. Photohemolysis and protein cross linking were also shown to be uncoupled on the basis of the temperature dependences of the processes (Dubbelman 80a).

The present conclusion is that the cation permeability changes responsible for hemolysis of irradiated red cells containing membrane localized protoporphyrin are due mainly to the effects of photooxidized amino acid residues on membrane proteins (Dubbelman 80a). Protein cross linking is another consequence of amino acid photooxidation but does not lead to hemolysis. Severe oxidation of lipids can also lead to hemolysis.

Consistent with the above conclusions, Girotti and coworkers have shown (Deziel 80) that in resealed erythrocyte membranes Na$^+$ permeability is continuously increased, with no lag, when the ghosts are irradiated in the presence of bilirubin which appears to sensitize many of the same oxidation reactions as does protoporphyrin. In contrast, glucose-6-phosphate is not released from the ghosts until a significant time has passed and is then released abruptly. This abrupt "lysis" of the ghosts is correlated with advanced lipid oxidation.

The primary photochemistry that is most important for the lytic event following irradiation of protoporphyrin sensitized erythrocytes has not been defined. Singlet oxygen is formed in the system because a specific $^1O_2$ product, 5α-cholesterol hydroperoxide, is produced (Lamola 73). However, Type 1 pathways cannot be excluded. Other sensitizers such as methylene blue and bilirubin sensitize photohemolysis with characteristic similar to that when protoporphyrin is used. While methylene blue is a "good" $^1O_2$ sensitizer, bilirubin, because of its extremely low triplet state yield, is not expected to cause effective $^1O_2$ production. Results of inhibitor studies are inconclusive. β-Carotene, α-tocopherol, other phenol antioxidants and thiols all inhibit photohemolysis but not impressively (Schothorst 71; Lamola unpublished). The phenol antioxidants inhibit lipid oxidation much more effectively than they inhibit protein cross linking (Girotti 79; Deziel 80). It would appear that free radical mediated oxidation pathways operate subsequent to electron transfer between excited protoporphyrin molecules and membrane proteins (Lamola 81). There is little information concerning the localization of protoporphyrin in the erythrocyte membrane. The fluorescence maximum of membrane associated protoporphyrin 634 nm, is characteristic of the porphyrin both when it is in a lipid environment and when it is bound to hydrophobic proteins.

Nonoxidative Photosensitized Hemolysis

In 76 this author stated that the sensitized photodegradation of biomembranes is synonomous with light-induced oxidation (uptake of molecular oxygen) of membrane components (Lamola 77a). Within the next year it was demonstrated that two compounds, chlorpromazine and protriptyline, were effective sensitizers for the photolysis of red cells in the absence of oxygen (Kochevar 79). Both chlorpromazine and protriptyline are used as tranquilizing drugs and both are reported to cause cutaneous photosensitivity to ultraviolet light in the wavelength region where they absorb (Magnus 76).

Structures of Chlorpromazine (left) and Protriptyline (right)

The two drugs appear to sensitize the photolysis of red cells in the same way. The drugs themselves are lytic at high concentrations since they are detergent-like. Irradiation of either drug in aqueous solution or in organic solvents produces several products among which are dimers and higher multimers (Kochevar 81, 83). These dimers and multimers are several orders of magnitude more lytic towards red cells than are the parent compounds. In the case of protriptyline one lytic photoproduct has been shown to be a cyclobutane-type photodimer (Kochevar 80).

Structure of Protriplyline Photodimer

The lysis induced by the chlorpromazine and protriptyline photoproducts appears to be of the "colloid osmotic" type whereby membrane alterations occur that allow rapid passive transport of potassium and sodium ions (Kochevar 79). At drug concentrations (0.05 mM) that cause cell lysis in the dark, some drug molecules may be near enough to each other to cause local disruption of the bilayer structure and allow ion transport. Photodimers may cause similar bilayer disruption at much lower concentrations because the ring structures are permanently attached together (Kochevar 80). Whether or not formation of such photodimers and multimers is responsible for the cutaneous photosensitivity of the patients receiving these drugs is still an open questions.

Recently, another drug, benoxaprofen, 2-(4-chlorophenyl)-α-methyl-S-benzoxazole acetic acid, under investigation as a non-steroidal anti-inflammatory agent, was shown to cause immediate erythemal and urticarial responses in patients exposed to near ultraviolet light (Kligman 82). Benoxaprofen also sensitizes the photolysis of red cells. Kochevar (unpublished) has shown that the sensitized lysis of red cells does

436

not require molecular oxygen. One rapid photochemical reaction of benoxaprofen is decarboxylation. The lytic ability of the decarboxylated product is under investigation.

Photodegradation of Excitable Membranes

Electrophysiological functions in excitable cells can be blocked or modified through the action of light and sensitizing dyes. Most of the investigations carried out in this interesting area have employed photodynamic sensitization chiefly with fluorescein dyes. An excellent review of the subject has been prepared by Pooler and Valenzeno (Pooler 81).

Excitable systems studied include lobster main axon; giant axons of squid; and cardiac, smooth and skeletal muscles. Generally, irradiation of photodynamic dye-treated cells with large doses, of light blocks electrical impulse propagation via sodium channel inhibition and depolarization of the membranes. Potassium channels appear to be very much less sensitive to photodynamic alteration. At low light doses at which only a fraction of the sodium channels are affected, impulses are elongated because the remaining open sodium channels exhibited prolonged closing times. Stimulation of impulses due to the creation of unspecified ion leakage can also be observed.

The requirement for membrane binding of the sensitizing dyes was demonstrated in exquisite fashion by Pooler and Valenzeno (Pooler 79) who were able to predict the relative sensitizing effectiveness of nine fluorescein dyes for blocking sodium channels in the lobster axon. The predictions were based on three properties of the dyes: absorption cross section at the excitation wavelength, the triplet state yield and the octanol/water partition ratio. The latter provided the major differences among the dyes. Predicts and observed efficiencies correlated over a 30,000-fold range of effectiveness.

PHOTOCHEMICAL PROBES OF BIOMEMBRANE STRUCTURE

Photochemically induced covalent bond formation between components molecules in biomembranes would indicate that those components come into contact. Photochemical cross-linking indigenous components can be induced by direct irradiation of the components or via the use of a photosensitizer. Cross-linking of some added photoactivible probe with membrane components or of indigenous components with involvement of an added photoactivible reagent as the cross-linking agent can also occur. Such photoactivible reagents fall into the category of photolabeling agents (photoaffinity probes) and have received much attention. But almost no use for structural information has been made of the direct or sensitized oxidative cross-linking of membrane proteins discussed in the sections on photodegradation. One of the very few examples of the use of photodynamic modification for structural information addressed the problem of the organization of the mitochondrial ATPase (Hackney 80).

Photochemical Labeling and Biomembrane Structure

The use of photochemically generated reagents to probe biomembrane structure derives from the older idea of photoaffinity labeling introduced by Singh, Thorton and Westheimer in 1962. The idea of photoaffinity labeling is that latent reactivity of a probe molecule bound at a specific site on a protein or other macromolecule is activated by irradiation and the activated probe rapidly forms a covalent bond with its immediate surrounding molecular environment to provide a marker during subsequent analysis. Derivations of this idea have expanded rapidly such that hundreds of papers now appear each year that describe photolabile reagents for such environmental probing. Almost all the reagents contain the azide, diazirine or the diazo grouping as the photolabile component. Photolysis of these compounds produce very reactive divalent nitrogen or carbon species or other reactive intermediates.

Because the literature is large and good reviews exist (Peters 77, Das 79, Tometsko 80, Bayley 83). the discussion here will center on the basic principles and some examples of probes of general membrane structure.

Westheimer has enumerated the characteristics of an ideal photolabeling reagent (Westheimer 80). The reagent should, of course, exhibit high specificity for the site or region being probed. It should be easy to synthesize and to tag with a radioactive (or fluorescent) marker and, except for its photolability, it should be stable. The reagent should be stable in visible light, at least towards red light, so that it can be conveniently manipulated. At the same time the reagent should be labile towards activation by light of wavelengths that would otherwise not perturb the system under study. The reagent should strongly absorb activating light (high extinction coefficient) and the photoactivation should proceed with reasonable quantum efficiency. The activated species should react rapidly and indiscriminately with any and every molecule in its environment (except water) but it should not undergo rearrangement that leads to loss of reactivity.

The photoactivable groups that provide the best compromises of the above mentioned specifications are the azide, diazo and diazirine groups, shown below in example phenyl-based reagents a,b and c, respectively. The azides are photolyzed to give the highly reactive nitrene intermediate. The diazo compounds and diazirines are photolyzed to the generally more reactive carbene intermediate.

438

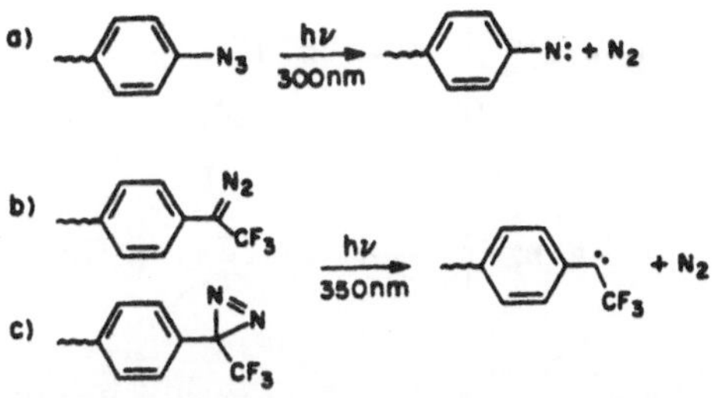

a)

b)

c)

a) Azido, b) Diazo, and c) Diazirine Photoaffinity Reagents

A detailed review of the chemistries of carbenes and nitrenes and of their precursor compounds is not appropriate here. Two comprehensive books on these organic intermediates, although a decade old, are still timely (Lwowski 70; Kirmse 71).

The deficit of two bonds provides the carbenes and the nitrenes with high reactivity. These so-called divalent intermediates can form two new bonds by reaction with a single substrate via the so-called insertion reactions. These insertion reactions are shown below using a carbene as an example. Besides insertion into -X-H bonds, with reactivity generally in the order shown, addition to carbon-carbon double bonds to give three-membered ring products is also a facile reaction type.

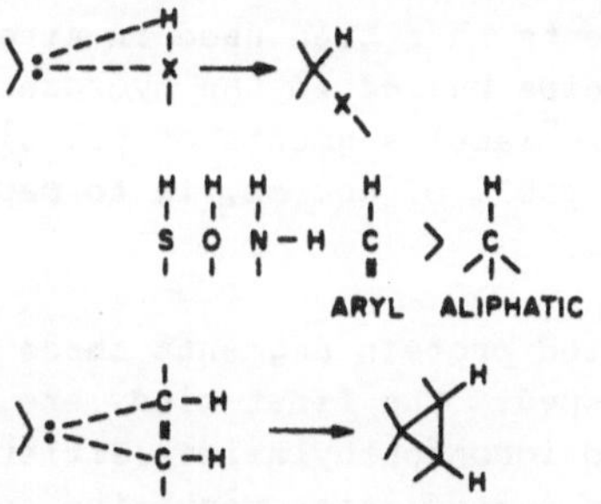

ARYL ALIPHATIC

When the carbene or nitrene has its two unsatisfied valence electron in a singlet spin state these insertion and addition reactions can proceed in a concerted fashion, both bonds formed at the same time or at least fast compared to bond rotation times. When the divalent species are in a triplet spin state, the bonds must form one at a time and it is possible for the two radical species formed in the case of hydrogen abstraction, exemplified below for a nitrene, to separate. The effect of spin state on reactivity is discussed by Lwowski (Lwowski 70) and Kirmse (Kirmse 71) as well as by Turro (Turro 80).

R—N:⟍⟍△C ⟶ R—N(H)—C

R—N: ⟍CH ⟶ RNH—C·

Generally carbenes react faster and are more indiscriminate than are nitrenes. However, depending upon the group into which the carbene or nitrene center is incorporated, reactivities vary many orders of magnitude, so that some nitrenes are more reactive than some carbenes. The wavelengths of light that can effectively transform diazo and diazirine precursors to carbenes and azido precursors nitrenes are also very much dependent upon the structures of the precursors. Generally, however, azido compounds to not absorb as far to the red as do diazo compounds and diazirines. Many interesting and useful reagents can be activated with light between 300 and 400 nm so that it is easy to avoid light of wavelengths that are absorbed by membrane proteins and lipids (Bayley 77).

Photolabeling of Membrane Proteins

Photoactivated reagents have been used in attempts to specifically label segments of proteins buried in the hydrocarbon region of biomembranes and to specifically label segments of proteins accessible to the aqueous surround. The goal, of course, is to map the protein in relation to the lipid bilayer.

For labeling of buried protein segments three different kinds of reagents have been developed. The first kind, exemplified by adamantane diazirine (Bayley 80) and iodonaphthylazide (Gitler 80) are hydrophobic reagents that can be activated after partition and distribution within the hydrocarbon care of the bilayer.

The second kind contains a photolabile group in the hydrophobic region of a amphipatic single-chained molecule. An example is steroylglucosamine substituted at carbon 12 with a nitroazidophenolate group (Wisnieski 79). The third kind are phospholipids with photolabile groups exemplified by the series, shown below, that was synthesized by the Khorana group (Radhakrishnan 80).

440

$$\begin{array}{l}
\text{O} \\
\| \\
H_2C - OCR \\
| \quad\quad O \\
R'CO - CH \quad \| \\
\| \quad\quad CH_2 - P - OCH_2CH_2 N^+(CH_3)_3 \\
O \quad\quad\quad | \\
\quad\quad\quad\quad O_-
\end{array}$$

$$R = -(CH_2)_{16} CH_3$$

$$R' = -(CH_2)_{11} \ OC - C - CF_3$$
with C=O and N_2 on the central carbon

$$-(CH_2)_{10} - O - \text{(benzene ring)} - \overset{+}{N} \equiv N$$

$$-(CH_2)_{10} - O - \text{(benzene ring)} - N_3$$

$$-(CH_2)_7 \cdot \underset{H}{C} = \underset{H}{C} - CH_2 CH (CH_2)_5 CH_3 \quad \text{with } N_3$$

$$-(CH_2)_8 - CH - (CH_2)_5 CH_3 \quad \text{with } N_3$$

Some Phospholipid-based Photoactivable Reagents (from Radhakrishnan 80)

Hydrophobic photolabeling reagents of the first kind have the advantage of being relatively simple to synthesize and to incorporate into biomembranes. Furthermore, they should, in the absence of specific strong binding to some membrane protein, diffuse reasonably freely within the membrane and widely explore the hydrophobic region. The phospholipid reagents have the advantage of greater specificity with respect to localization. Exploration of labeling at various depths within the membrane is limited only by the patience and skill of the synthetic chemist. It is obvious that with sufficient forbearance, much detailed structural information would result from this approach.

The hydrophobic buried segments of intrinsic membrane proteins appear to exhibit a preponderance of hydrocarbon side chains to the surrounding lipid. The photogenerated reagents must be able to label such unreactive hydrocarbon groups and this certainly points to carbenes and nitrenes. Reaction within the hydrophobic region must be rapid when the photogenerated probe can freely diffuse within the bilayer because reaction with more polar groups, such as the thiol moieties of extrinsic proteins, can be very rapid. Thus, for example, phenyl diazirine which yields the very reactive phenyl carbene (Bayley 78b), labels the hydrocarbon chains in phospholipid vesicles much more efficiently than does phenylazide which gives the much less reactive phenyl nitrene (Bayley 78a). Phenyl nitrene generated in the bilayer interior reacts much more efficiently with scavengers in the aqueous medium than it does with the lipid hydrocarbon chains. Such considerations are much less important for other nitrene reagents such as azido derivatized phospholipids, that

are more strongly confined to the hydrocarbon region. Tritiated adamantane diazirine was found to bind to red cell ghosts with a partition
ratio 1500 compared to the aqueous buffer. The label was efficiently
incorporated (38% could not be removed by washing) with 20% of the
incorporated label on protein and 80% in lipid molecules. While spectrin was chiefly labeled using phenylazide, the adamantane reagent most
heavily labeled the intrinsic proteins glycophorin A, PAS I and II, the
anion channel (Band 3), Band 4.5 and glycophorin B.

Radioiodine tagged 5-iodonaphthyl-azide appears to be even more
strongly confined to the hydrophobic region of the erythrocyte membrane
than is the adamantane diazirine. Despite the expected lower reactivity
of the nitrene intermediate the iodonaphthylazide reagent appears to be
at least as selective towards intrinsic proteins as is the adamantane
reagent. Analysis (Kahane 78) of the distribution of the iodonaphthyl
label along the polypeptide chain of glycophorin extracted from labeled
red cell ghosts showed that 80-90% of the label was located in the so-
called trypsin-insoluble segment, a hydrophobic domain thought to function as a membrane anchor for the protein by insertion into the lipid
bilayer. Iodonaphthylazide has been used to identify the intrinsic proteins of several other membranes including intestinal brush border membranes (Sigrist-Nelson 77), sarcoplasmic reticulum membranes (Bercovici
78), T-and B-lymphocytes, the ciliary, pellicle and endoplasmic reticulum membranes of Tetrahymene pyriformis, and acetylcholine receptor-rich
fragments (Gitler 80).

While there has been considerable success in devising photolabeling
reagents that operate from within the membrane and label protein segments in the hydrophobic region there has been less success in devising
reagents that are highly specific in mapping those protein segments that
face the external bulk aqueous phase. Early work (Staros 75) centered
on nitroazidophenyltaurine or NAP-taurine.

It was thought that the great water solubility conferred by the sulfonic
acid group would confine the reagent to the aqueous compartment and that
the photogenerated nitrene would react only with those protein segments
in contact with the bulk aqueous phase. However, it was found that the
reagent was not as selective as expected (Richards 80). In experiments
with intact erythrocyte membranes it was found the NAP-taurine did pho-
tolabel the Band 3 protein heavily but also labeled PAS staining regions
significantly. Later it was found that the heavy Band 3 labeling was
due not so much to the fact that this protein does have an external face
that does contact the aqueous compartment, but that there is specific
binding of the reagent to the anion transport protein. In fact the use

of this reagent for specific photoaffinity labeling of the anion transport channel has yielded much mechanistic information about this prototype transport protein (Knauf 80).

Hydrophobic Heterobifunctional Proximity Probes

A number of reagents have been synthesized for the purpose of probing the spatial relations between proteins in membranes. The general idea centers around reagents with at least two reactive linking groups, one which operates in the dark and one which requires photogeneration, hence the name heterobifunctional probes. The dark reaction is allowed to proceed and this labels certain proteins. Specificity depends upon structural details. Next the system is irradiated thereby activating the second group if the latter reacts to form a covalent bond with a second protein, the two membrane proteins are cross-linked. If the reaction lifetime of the photogenerated group is short compared to times for rearrangement of the protein-protein contacts, the reagent probes the static average protein distribution. If the reaction times are long compared to protein diffusion, the reagent probes a dynamical average.

Two examples of such reagents are 5-isothiocyanato-1-naphthalene azide (Sigrist 82) (INA) and the unsymmetrical di(nitrophenylazido)cystamine dioxide (Richards 80) (DNCO).

INA undergoes dark binding and reaction at specific sites on bacteriorhodopsin and the anion-exchange protein (Band 3) of human red cells. So far experiments have been performed in systems with high concentrations of either of these proteins with the result that irradiation subsequent to dark binding gives homopolymers of the proteins. The DNCO reagent reacts with protein-SH groups by displacement of the sulfinic acid analog of cysteine. Photoactivation may lead to cross-linking with a second protein via the generated nitrene. The -S-S-link with the first protein can be reductively cleaved after separation steps to allow for convenient identification of the protein.

The main problem with arylazido-based reagents is that the arylnitrene group is relatively slow to react (1 msec) compared to protein diffusion. Thus it is difficult to distinguish between long-lived and collisional protein complexes. Reagents based upon carbene intermediates might reduce the probing time to a few microseconds during which there is not much membrane protein movement.

Phospholipid Photolabling Reagents and Phase Separation

The Khorana group have performed studies aimed at testing whether photogenerated phospholipid-based reagents that undergo cross-linking with phospholipids in bilayers can be used to sense phospholipid distributions within the bilayer (Radhabrishnan 80). In one study they used a diazirine phospholipid with a phosphatidylcholine head group, one C_{16}-side chain and one C_{11}-side chain terminating with a phenyldiazirine. They first showed that at temperatures above 30°C the cross-linking of the reagent, denoted PC** with DPL and DOL (diolyl lecithin) in vesicles containing mixtures of the three components was statistical (linear with mole fraction of one or the other lecithin) except for a preference for cross-linking to DOL. The temperature dependence of cross-linking in vesicles containing DPL/DOL/PC** in the ratio 1:1:0.2 is shown in Fig. 9. Above about 30°C, the distribution of cross-linked products was invariant with temperature. At temperatures below about 30°, the PC** reacts more preferentially with the DOL. From other work it is known that there is phase separation and so it appears that the PC** preferentially partitions into the more fluid DOC phase. Thus, the cross-linking method using phospholipid photolabeling reagents is able to detect lateral phase separation and can give information about which particular molecular species undergo phase separation.

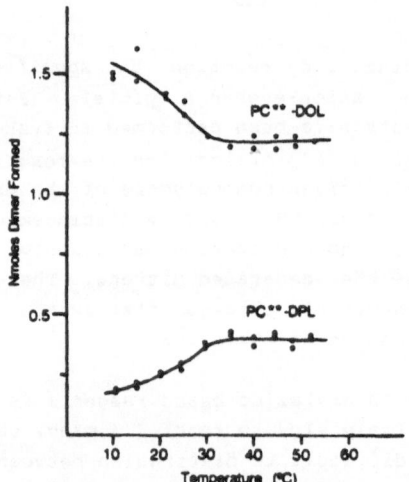

Fig. 9. The distribution of cross-linked products generated by photolysis of DPL/DOL/PC** (1:1:0.2) as a function of temperature (from Radhakrishnan 80).

ACKNOWLEDGEMENTS

The author thanks Dr. Irene E. Kochevar for permission to quote unpublished results from her laboratory and for her help in the preparation of these lectures. Valuable suggestions were also made by Dr. William E. Blumberg, who also greatly aided in the preparation of this manuscipt.

REFERENCES

Allison, A. C., Magnus, I. A., and Young, M. R. (1966) Role of lysosomes and cell membranes in photosensitization. Nature 209:874-878.

Anderson, S. M., and Krinsky, N. I. (1973) Protective action of carotenoid pigments against photodynamic damage to liposomes. Photochem. Photobiol. 18:403-408.

Anderson, S. M., Krinsky, N. I., Stone, M. J., and Clagett, D. C. (1974) Effect of singlet oxygen quenchers on oxidative damage to lipsomes initiated by photosensitization or by radiofrequency discharge. Photochem. Photobiol. 20:65-69.

Bayley, H., and Knowles, J. R. (1977) Photoaffinity labelling. Meth. Enzym. 46:120-172.

Bayley, H., and Knowles, J. R. (1978a) Photogenerated reagents for membrane labeling. Phenylnitrene formed within the lipid bilayer. Biochem. 17:2414-2419.

Bayley, H., and Knowles, J. R. (1978b) Photogenerated reagents for membrane labeling. Phenylcarbene and adamantylidene formed within the lipid bilayer. Biochem. 17:2420-2423.

Bayley, H., and Knowles, J. R. (1980) Photogenerated, hydrophobic reagents for intrinsic membrane proteins. Ann. N.Y. Acad. Sci. 346:45-54.

Bayley, H. (1983) Photogenerated reagents in biochemistry and molecular biology. In "Laboratory Techniques in Biochemistry and Molecular Biology," Work and Burdon, (eds) North Holland Press.

Bercovici, T., and Gitler, C. (1978) 5-Iodonaphthyl azide, a reagent to determine the penetration of proteins into the lipid bilayer of biological membranes. Biochem. 17:1484-1489.

Blum, H. F. (1964) Photodynamic Action and Diseases Caused by Light. Hafner Pub. Co., N.Y.

Cook, J. S. (1975) Photopathology of the erythrocyte membrane, in Pathobiology of Cell Membranes. Trump, B. F., and Arstella, A. (eds) Academic press, NY.

Das, M., and Fox, C. F. (1979) Chemical cross-linking in biology. Ann. Rev. Biophys. Bioeng. 8:165-194.

de Goeij, A. F. P. M., Ververgaert, P. H. J. T., and van Steveninck, J. (1975) Photodynamic effects of protoporphyrin on the architecture of erythrocyte membranes in protoporphyria and in normal red cells. Clin. Chem. Acta. 62:287-292.

de Goeij, A. F. P.M., and van Steveninck, J. (1976) Photodynamic effects of protoporphyrin on cholesterol and unsaturated fatty acids in erythrocyte membranes in protoporphyria and in normal red cells. Clin. Chem. Acta 68:115-122.

Deziel, M. R., and Girotti, A. W. (1980) Bilirubin-photosensitized lysis of resealed erythrocyte membranes. Photochem. Photobiol. 31:593-596.

Doleiden, F. H., Fahrenholtz, S. R., Lamola, A. A., and Trozzolo, A. M. (1974) Reactivity of cholesterol and some fatty acids towards singlet oxygen. Photochem. Photobiol. 20:519-521.

Doughterty, T. J., Boyle, D. G., and Weishaupt, K. R. (1982) Photoradiation therapy of human tumors, in The Science of Photomedicine Regan, J. D., and Parrish, J. A. (eds) Plenum Press, NY, 625-638.

Dubbelman, T. M. A. R., de Goeij, A. F. P. M., and van Steveninck, J. (1978a) Protoporphyrin-sensitized photodynamic modification of proteins in isolated human red blood cell membranes. Photochem. Photobiol. 28:197-204.

Dubbelman, T. M. A. R., de Goiej, A. F. P. M., and van Steveninck, J. (1978b) Photodynamic effects of protoporphyrin on human erythrocytes. Nature of the cross-linking of membrane proteins. Biochim. Biophys. Acta 511:144-141.

Dubbelman, T. M. A. R., Haasmost, C., and van Steveninck, J. (1980a) Temperature dependence of photodynamic red cell membrane damage. Biochim. Biophys. Acta 601:220-227.

Dubbelman, T. M. A. R., de Goeij, A. F. P. M., and van Steveninck, J. (1980b) Protoporphyrin-induced photodynamic effects on transport processes across the membrane of human erythrocytes. Biochem. Biophys. Acta 595:133-139.

Eldred, G. E., Miller, G. V., Stark, W. S., and Feeney-Burns, L. (1982) Lipofuscin:resolution of discrepant fluorescence data. Science 216:757-758.

Emanuel, N. M., and Lyaskovskaya, Y. N. (1967) The Inhibition of Fat Oxidation Processes. Pergamon Press, Oxford.

Fahrenholtz, S. R, Doleiden, F. H., Trozzolo, A. M., and Lamola, A. A. (1974) On the quenching of singlet oxygen by α-tocopherol. Photochem. Photobiol. 20:505-509.

Foote, C. S. (1976) Photosensitized oxidation and singlet oxygen: Consequences in biological systems, in Free Radicals in Biology, Pryor, W. A. (ed) Academic Press, NY Vol. 2, 85-133.

Foote, C. S. (1980) Photooxidation and toxicity, in Molecular Basis of Environmental Toxicity Bhatnager, R. S. (ed) Ann Arbor Science Pub., Ann Arbor.

Girotti, A. W. (1976) Photodynamic action of protoporphyrin on human erythrocytes:cross-linking of membrane proteins. Biochem. Biophys. Res. Commun. 72:1367-1374.

Girotti, A. W. (1978) Bilirubin-photosensitized cross-linking of polypeptides in the isolated membrane of the human erythrocyte. J. Biol. Chem. 253:7186-7193.

Girotti, A. (1979) Protoporphyrin-sensitized photodamage in isolated
    membranes of human erythrocytes. Biochem. 18:4403-4411.

Girotti, A. W. (1980) Photosensitized cross-linking of erythrocyte mem-
    brane proteins. Evidence against participation of amino groups in
    the reaction. Biochem. Biophys. Acta 602:45-56.

Gitler, C., and Bercovici, T. (1980) Use of lipophilic photoactivable
    reagents to identify the lipid-embedded domains of membrane pro-
    teins. Am. N.Y. Acad. Sci. 346:199-211.

Gollnick, K. (1968) Type II photooxygenation reactions in solution.
    Advances Photochem 6:1-122.

Grossweiner, L. I., Patel, A. S., and Grossweiner, J. B. (1982) Type 1
    and type 2 mechanisms in the photosensitized lysis of phosphatidyl-
    choline liposomes by hematoporphyrin. Photochem. Photobiol.
    36:159-167.

Hackney, D. D. (1980) Photodynamic action of bilirubin on the inner
    mitochondrial membrane. Implications for the organization of the
    mitochondrial ATPase. Biochem. Biophys. Res. Commun. 94:875-880.

Harber, L. C., Fleischer, A. S., Baer, R. L. (1964) Erythropoietic pro-
    toporphyria and photohemolysis. J. Am. Med. Assoc. 189:191-194.

Harber, L. C., and Bickers, D. R. (1981) Photosensitivity Diseases. W.
    B. Saunders, Philadelphia. 189-223.

Jagger, J. (1977) Phototechnology and biological experimentation, in The
    Science of Photobiology Smith, K. C. (ed) Plenum Press, NY.

Jori, G. (1980) The molecular biology of photodynamic action. Springer
    Series in Optical Sciences 23:58-66.

Jori, G., and Spikes, J. D. (1981) Photosensitized oxidations in complex
    biological structures, in Oxygen and Oxy-Radicals in Chemistry and
    Biology, Rodgers, M. A. J., and Powers, E. L. (eds) Academic Press,
    NY, 441-457.

Kahane, I., and Gitler, C. (1978) Red cell membrane glycophorin labeling
    from within the lipid bilayer. Science 201:351-352.

Kirmse, W. (1971) Carbene Chemistry Academic Press, New York, 2nd ed.

Kligman, A. M., and Keidbey, R. H. (1982) Phototoxicity to benoxaprofen.
    Eur. J. Rheum. Inflam. 5:124-137.

Knauf, P. A., and Rothstein, A. (1980) Use of NAP-taurine as a photoaf-
    finity probe for the human erythrocyte anion exchange system. Ann.
    N.Y. Acad. Sci. 346:199-211.

Kochevar, I. E., and Lamola, A. A. (1979) Chlorpromazine and protripty-
    line phototoxicity: photosensitized, oxygen independent red cell
    hemolysis. Photochem. Photobiol. 29:791-796.

Kochevar, I. E. (1981) Phototoxicity mechanisms: chlorpromazine pho-
    tosensitized damage to DNA and cell membranes. J. Invest. Derma-
    tol. 10:59-64.

Kochevar, I., and Hom, J. (1983) Photoproducts of chlorpromazine which
    cause red blood cell lysis. Photochem. Photobiol. 37:163-168.

447

Kochevar, I. E. (1980) Possible mechanisms of toxicity due to photochemical products of protriptyline. Toxicol. Applied Pharmacol.
54:258-264.

Krinsky, N. I. (1979) Carotenoid protection against oxidation. Pure
Appl. Chem. 51:649-660.

Lamola, A. A. (1969) Electronic energy transferin solution:theory and
applications, in Leermakers, PA and Weissberger, A. (eds) Energy
Transfer and Organic Photochemistry, Interscience, NY, 17-132.

Lamola, A. A., Yamane, T., and Trozzolo, A. M. (1973) Cholesterol hydroperoxide formation in red cell membranes and photohemolysis in
erythropioetic protoporphyria. Science 149:1131-1133.

Lamola, A. A., Piomelli, S., Poh-Fitzpatrick, M. D., Yamane, T., and
Harber, L. C. (1975) Erythropoietic protoporphyria and lead
intoxication:the molecular basis for difference in cuaneous photosensitivity II. J. Clin. Invest. 56:1528-1535.

Lamola, A. A. Electronic energy transfer in solûtion:Theory and applications, in Techniques of Organic Chemistry, Leermakers, P. A., and
Weissberger, A. (eds), Wiley-Interscience, NY 14:17-126.

Lamola, A. A. (1977a) Photodegradation of biomembranes in Research in
Photobiology Castellani, A. (ed) Plenum, NY, 53-63.

Lamola, A. A., and Turro, N. J. (1977b) Spectroscopy, in The Science of
Photobiology Smith, K. C. (ed) Plenum Press, NY, 27-62.

Lamola, A. A., and Doleiden, F. H. (1980) Cross-linking of membrane proteins and protoporphyrin-sensitized photohemolysis. Photochem.
Photobiol. 31:597-601.

Lamola, A. A. (1981) Fluorescence methods in the diagnosis and manage-
ment of diseases of tetrapyrrole metabolism. J. Invest. Derm.
77:114-121.

Lwowski, W., ed. (1970) Nitrenes Wiley-Interscience, New York.

Magnus, I. A. (1976) Dermatological Photobiology Blackwell, Oxford.

Mead, J. F. (1976) Free radical mechanisms of lipid damage and consequences for cellular membranes in Free Radicals in Biology. Pryor,
W. A. (ed) Academic Press, NY 1:51-68.

Michelson, A. M., and Durosay, P. (1977) Hemolysis of human erythrocytes
by activated oxygen species. Photochem. Photobiol. 25:55-63.

Muller-Runkel, R., Blais, J., and Grossweiner, L. I. (1981) Photodynamic
damage to egg lecithin liposomes. Photochem. Photobiol. 33:683-
687.

Peters, K., and Richards, E. M. (1977) Chemical cross-linking:Reagents
and problems of membrane structures. Ann. Rev. Biochem. 46:523-51.

Piomelli, S., Lamola, A. A., Poh-Fitzpatrick, M. D., Seaman, C., and
Harber, L. L. (1975) erythropoietic protoporphyria and lead
intoxication:the molecular basis for difference in cutaneous photosensitivity I. J. Clin. Invest. 56:1519-1527.

Pooler, J. P., and Valenzeno, D. P. (1979) Physicochemical determinants
of the sensitizing effectiveness of nerve membranes by fluorescein
derivatives. Photochem. Photobiol. 30:491-498.

Pooler, J. P., and Valenzeno, D. P. (1981) Dye-sensitized photodynamic inactivation of cells. Med. Phys. 8:614-628.

Radhakrishnan, R., Gupta, C. M., Erni, B., Robson, R. J., Curatolo, W., Majumdar, A., Ross, A. H., Takagaki, Y., and Khorana, H. G. (1980) Phospholipids containing photoactivable groups in studies of biological membranes. Ann. N.Y. Acad. Sci. 346:165-197.

Richards, F. M., and Brunner, J. (1980) General labeling of membrane proteins. Ann. N.Y. Acad. Sci. 346:144-163.

Rodgers, M. A., and Bates, A. L. (1982) A laser flash kinetic spectrophotometric examination of the dynamics of singlet oxygen in unilammellar vesicles. Photochem. Photobiol. 35:473-477.

Roshchupkin, D. I., Pelenitsyn, A. B., Potapenko, A. Y., Talitsky, V. V., and Vladimirov, Y. A. (1975) Study of the effects of ultraviolet light on biomembranes. The effect of oxygen on uv-induced hemolysis and lipid photoperoxidation in rat erythrocytes and liposomes. Photochem. Photobiol. 21:63-70.

Sandberg, S. (1981) Protoporphyrin-induced photodamage to mitochondria and lysosomes from rat liver. Clinica Chim. Acta 111:55-60.

Schothorst, A. A., van Steveninck, J., Went, L. N., and Surrmond, D. (1970) Protoporphyrin-induced photohemolysis in protoporphyria and in normal red blood cells. Clin. Chim. Acta 28:41-49.

Schothorst, A. A., van Steveninck, J., Went, L. N., and Summond, D. (1972) Photodynamic damage of the erythrocyte membrane caused by protoporphyrin in protoporphyria and in normal red blood cells. Clin. Chim. Acta. 39:161-170.

Schothorst, A. A., van Steveninck, Went, L. N., and Surrmond, D. (1971) Methods aspects of the photodynamic effects of protoporphyrin in protoporphyria and in normal red blood cells. Clin. Chim. Acta. 33:207-213.

Sigrist-Nelson, K., Sigrist, H., Bercovici, T., and Gitler, C. (1977) Biochem. Biophys. Acta. 468:163-76.

Sigrist, H., Allegrini, P. R., Kempf, C., Schnippering, C., and Zahler, P. (1982) 5-Isothiocyanato-1-naphthalene azide and p-azidophenylisothiocyanate. Synthesis and application in heterobifunctional photoactive cross-linking of membrane proteins. Eur. J. Biochem. 125:197-201.

Spikes, J. D. (1982) Photodynamic reactions in photomedicine, in The Science of Photomedicine Regan, J. D., and Parrish, J. A. (eds) Plenum Press, NY, 113-144.

Staros, J. V., and Richards, F. M. (1974) Photochemical labeling of the surface proteins of human erythrocytes. Biochem. 13:2720-2726.

Suwa, K., Kimura, T., and Schaap, A. P. (1978) Reaction of singlet oxygen with cholesterol in liposomal membranes. Photochem. Photobiol. 28:469-473.

Tometsko, A. M., and Richards, F. M. (eds) (1980) Applications of Photochemistry in Probing Biological Targets Ann. N.Y. Acad. Sci. 346.

Turro, N. J., and Lamola, A. A. (1977) Photochemistry, in The Science of Photobiology Smith, K. C. (ed) Plenum Press, NY, 63-86.

Turro, N. J. (1980) Structure and dynamics of important reactive intermediates involved in photobiological systems. Ann. N.Y. Acad. Sci. 346:1-15.

Valenzeno, D. P., and Pooler, J. P. (1982) Cell membrane photomodification:relative effectiveness of halogenated fluoresceins for photohemolysis. Photochem. Photobiol. 35:343-350.

Westheimer, F. H. (1980) Photoaffinity labeling-retrospect and prospect. Ann. N.Y. Acad. Sci. 346:134-143.

Wilkinson, F., and Brummer, J. G. (1981) Rate constants for the decay and reactions of the lowest electronically singlet state of molecular oxygen in solution. J. Phys. Chem. Ref. Data 10:809-999.

Winterle, J. J., and Mill, T. (1980) Free-radical dynamics in organized lipid bilayers. J. Am. Chem. Soc. 102:6336-6338.

Wisnieski, B. J., and Bramhall, J. S. (1979) Labeling of the active subunit of cholera toxin from within the membrane. Biochem. Biophys. Res. Commun. 60:308-312.